Krzysztof Jajuga · Jacek Batóg · Marek Walesiak
Editors

Classification and Data Analysis

Theory and Applications

Springer

Editors
Krzysztof Jajuga
Department of Financial Investments
and Risk Management
Wroclaw University of Economics
and Business
Wroclaw, Poland

Jacek Batóg
Institute of Econometrics
and Statistics
University of Szczecin
Szczecin, Poland

Marek Walesiak
Department of Econometrics
and Computer Science
Wroclaw University of Economics
and Business
Wroclaw, Poland

ISSN 1431-8814 ISSN 2198-3321 (electronic)
Studies in Classification, Data Analysis, and Knowledge Organization
ISBN 978-3-030-52347-3 ISBN 978-3-030-52348-0 (eBook)
https://doi.org/10.1007/978-3-030-52348-0

Mathematics Subject Classification: 62Hxx, 62H25, 62H30, 62H86, 62-07, 62-09, 68Uxx, 68U20, 62Pxx, 62P12, 62P20, 62P25

This Springer imprint is published by the registered company Springer Nature Switzerland AG
The registered company address is: Gewerbestrasse 11, 6330 Cham, Switzerland

Studies in Classification, Data Analysis, and Knowledge Organization

More information about this series at http://www.springer.com/series/1564

Preface

This volume presents the papers from the 28th Conference of Section of Classification and Data Analysis of Polish Statistical Society held at University of Szczecin on September 18–20, 2019. The papers presented referred to a set of studies addressing a wide range of recent methodological aspects and applications of classification and data analysis tools in micro and macroeconomic problems. In the final selection, we accepted 20 of the papers that were presented at the conference. Each of the submissions has been reviewed by two anonymous referees and the Authors have subsequently revised their original manuscripts and incorporated the comments and suggestions of the referees. The selection criteria were based on the contribution of the papers to the theory and applications of modern classification and data analysis.

The chapters have been organized along the major fields and themes in classification and data analysis: Methodology, Application in Finance, Application in Economics and Application in Social Issues.

The part on Methodology contains five papers. The paper by Batóg and Wawrzyniak focuses on modifications of selected formulas which allow to receive a transformation of nominant into stimulant that ensures that the order of objects before and after the transformation is consistent with the real values of the nominant. Dudek in his paper presents the recommendations on the principles of correct application of the Silhouette index, indicating that the "mechanical" use of the index leads to results that do not correspond to the actual structure of the classes. The paper by Grzenda discusses how discretization of continuous variables can improve the classification accuracy of machine learning models, with an application of supervised discretization of continuous variables based on the entropy criterion and the Gini criterion in demography. Jefmański in his paper proposes an intuitionistic fuzzy synthetic measure for ordinal data based on the Hellwig's linear ordering method that allows for a comparative analysis of objects due to the complex phenomenon described by ordinal measurement scales as well as to take into account the uncertainty in comparing objects expressed in the form of neutral points on the ordinal measurement scales. Pełka in his paper conducts research on the usefulness and prediction power of extracting variables from neural networks (multilayer perceptron for symbolic data) as the method of variable selection for the purposes of ensemble learning for symbolic data.

The paper on Application in Finance contains also five papers. The paper by Doszyń using the so-called Szczecin algorithm of real estate mass appraisal is aimed to analyze if an econometric model with restrictions may support the process of real estate mass appraisal, providing a more precise determination of the impact of real property attributes on the prices than an analogous model without restrictions. Krężołek in his paper address the issue of estimation of tail index of probability distribution using Hill estimator and its modification, comparing selected non-parametric and parametric models. The paper by Mikulec and Misztal using prediction error curves based on the bootstrap cross-validation estimates of the prediction error estimation provides the evidence that survival function made in each of the obtained subsets of objects with the use of Kaplan-Meier method enables more precise estimate of firm's duration than the use of Kaplan-Meier function for the total data. Pawełek and Pociecha in their paper compare the effectiveness prediction of the logit leaf model as a hybrid classification algorithm that enhances logistic regression and decision tree with the use of individual classifiers. The paper by Trzpiot examines the relation between economic, financial and demographic variables and longevity in terms of long-term investment portfolios that are sensitive to risk factors according to the APT portfolio factor model, using the Principal Component Regression.

The part on Application in Economics contains four papers. The paper by Markowicz and Baran investigates the issue of mirror data concerning intra-Community supplies of goods, with the use of their original indicators of data asymmetry and an empirical example based on data from the Eurostat COMEXT database. Cheba and Bąk in their paper explore the relationships between sustainable development and green economy and assess the results obtained for the EU countries in four particular areas using a taxonomic development measure based on the Weber median. Misztal and Kupis-Fijałkowska in their paper analyze the ICT development level in Poland against other European Union countries in the individual users and households perspective, using the exploratory data analysis methods and the Hellwig's method of linear ordering. Sagan and Grabowski in their paper identify cause-effect relationships as the impact of unknown disturbing variables affecting both the mediation and focal dependent variables by applying a simulations of correlated disturbances effect of dependent variables in the technology acceptance models on the degree of average causal mediation effect bias.

The part on Application in Social Issues contains six papers. Bieszk-Stolorz in her paper verifies whether risk of subsequent registrations in the labour office depends on the characteristics of the unemployed persons using Prentice-Williams-Peterson's conditional models, which consider the time until the event occurs from the beginning of observation, and the time from the previous event. The paper by Głowicka-Wołoszyn and Wysocki answers the question whether correction of ideal values occurring in Hellwig's and TOPSIS methods by the quartile criterion, contributed to the improvement of consistency between the identified levels of the Polish communes financial autonomy and the synthetic measure values assigned to them. Konarzewska in her paper conducts research on the problem of statistical independence of chosen properties of objects and especially the choice of adequate weights in multi-criteria rankings, applying the values of Variance Inflation Factors, Principal Component Analysis and Multi-Criteria Principal Components. Landmesser in her paper presents the comparison of personal income distributions taking into account the gender income gap for 28 European countries and using the Oaxaca-Blinder decomposition procedure, the decomposition procedure to different quantile points along the whole income distribution, and finally the counterfactual distribution based on the Recentered Influence Function—Regression approach. The paper by Majewska and Trzpiot evaluates different approaches to identification of the existence of the common mortality trends and derives the mortality time-varying indicator from the Lee-Carter model to obtain the similarities of different countries via a semi-parametric comparison approach to prove that multi-population mortality models are superior to individual mortality forecasting models. Matuszewska-Janica in her paper verifies whether selected attributes of employees affect the level of their wages, considering the impact of outliers on changes in relative importance of analysed features.

We wish to thank the Authors for making their studies available for our volume. Their scholarly efforts and research inquiries made this volume possible. We are also indebted to the anonymous referees for providing insightful reviews with many useful comments and suggestions.

In spite of our intention to address a wide range of problems pertaining to classification and data analysis theory there are issues that still need to be researched. We hope that the studies included in our volume will encourage further research and analyses in modern data science.

Wroclaw, Poland Krzysztof Jajuga
Szczecin, Poland Jacek Batóg
Wroclaw, Poland Marek Walesiak
January, 2020

Contents

About the Editors

Krzysztof Jajuga is a professor of finance at Wrocław University of Economics and Business, Poland. He holds master, doctoral and habilitation degree from Wrocław University of Economics and Business, Poland, title of professor given by President of Poland, honorary doctorate from Cracow University of Economics and honorary professorship from Warsaw University of Technology. He carries out research within financial markets, risk management, household finance and multivariate statistics.

Jacek Batóg is professor of economics and a Director of the Institute of Economics and Finance at University of Szczecin. He earned his Ph.D. and habilitation degrees at University of Szczecin. He was granted by the Foundation for Polish Science and the assistantship at University of Massachusetts. His research interests include econometric as well as classification and data analysis methods and its applications. He was also employed as a credit risk expert at Bank Pekao S.A. for several years. Currently he is a member of Advisory Committee of West Pomeranian Found of Funds JEREMIE 2, editorial team of Folia Oeconomica Stetinensia journal, International Federation of Classification Societies as well as the Polish Academy of Science (Econometric and Statistics Committee). He has authored books, chapters in edited volumes and over 100 articles in scholarly journals and did research work and expertise for many companies and local governments.

Marek Walesiak is a professor of economics at Wroclaw University of Economics and Business in Department of Econometrics and Computer Science. He holds master, doctoral and habilitation degree from Wrocław University of Economics and Business, Poland, title of professor given by President of Poland. He is a member of the Methodological Commission and Scientific Statistical Council in Statistics Poland (GUS) and an active member of many scientific professional bodies (i.e. Section of Classification and Data Analysis SKAD). His main areas of interest include: classification and data analysis, multivariate statistical analysis, marketing research, computational techniques in R.

Methods

Comparison of Proposals of Transformation of Nominants into Stimulants on the Example of Financial Ratios of Companies Listed on the Warsaw Stock Exchange

Barbara Batóg and **Katarzyna Wawrzyniak**

Abstract In case of linear ordering, it is important to determine the character of the variables describing the examined objects. When the set of variables contains nominants next to stimulants and destimulants, there is need to transform nominants into stimulants in order to have comparable variables. The paper focuses on the formulas of transformation of nominants with a recommended range of values into stimulants with the range of [0; 1]. Because during the linear ordering of some companies due to their financial condition, it turned out that after the transformation of nominant indicators with a recommended range of values, the obtained standardized stimulants for companies with values outside the range of recommended values did not always order companies in accordance with the expectations resulting from the original values of these indicators. Therefore, the aim of the study was to compare selected formulas of transformation of nominants into stimulants and to indicate those transformations on the basis of which the order of companies before and after the transformation was characterized by greater consistency. The Authors also proposed modifications for selected formulas which allow to maintain this consistency. The data on two financial indicators: current ratio and debt ratio which are considered in the literature as nominants with the recommended range of values were used. The data on the ratios come from Notoria Serwis and concern companies from the *Machinery industry* sector listed on the Warsaw Stock Exchange in 2016.

Keywords Nominants with the recommended range of values · Stimulants · Financial ratios

B. Batóg (✉)
University of Szczecin, Szczecin, Poland
e-mail: barbara.batog@usz.edu.pl

K. Wawrzyniak
West Pomeranian University of Technology Szczecin, Szczecin, Poland
e-mail: katarzyna.wawrzyniak@zut.edu.pl

K. Jajuga et al. (eds.), *Classification and Data Analysis*, Studies in Classification, Data Analysis, and Knowledge Organization,
https://doi.org/10.1007/978-3-030-52348-0_1

1 Introduction

Linear ordering of multi-variable objects can be done by using formula with or without pattern. In the case of formulas without pattern, usually, the basis for ordering is averaged standardized values of variables, and in the case of formulas with pattern—different types of distances of individual objects from the pattern (Gatnar and Walesiak 2004, pp. 352–355). Regardless of the applied ordering formula, it is important to determine the character of the variables describing the examined objects and then unify them and make them comparable by means of an appropriate normalizing transformation. Since the examined objects—from the point of view of the analysed phenomenon—are usually ordered from the best to the worst, then if the set of variables contains only stimulants, there is no need to unify them. However, when a set of variables contains not only stimulant but also destimulants and nominants, they are usually transformed into stimulants and normalized. It is worth mentioning that the terms "stimulant" and "destimulant" were introduced in Polish literature by Hellwig (1968), and the term "nominant" (unimodal and multimodal) was introduced by Borys (1978, 1984). In the paper of Kukuła (2000, p. 54), the term "nominant" was extended, and he used not only nominants with a single nominal value but also nominants with a recommended range of values.

In the case of destimulants, the literature offers some ways of their use without prior transformation into stimulants, e.g.: the use of zero unitarization as a standardization formula (Kukuła 2000, Sokołowski and Markowska 2017), when the basis for ordering of objects is the averaged values of standardized variables or the application of GDM distance to ordering and the adoption of a pattern at the level of the minimum value (Walesiak 2011, pp. 73–78). Also, there are a lot of proposals in the literature for transformation nominants into stimulants—they concern both nominants with a single nominal value (Walesiak 2011, p. 18) and nominants with a recommended range of values (Strahl and Walesiak 1997; Kukuła 2000, pp. 182–188; Kowalewski 2002, 2006; Wójciak 2003).

The paper focuses on the formulas of transformation of nominants with a recommended range of values into stimulants with the range of [0; 1]. The reason for that was the following phenomenon: during the linear ordering of some listed companies on the base of their financial condition with the mentioned formulas, it turned out that after the transformation of nominant indicators with a recommended range of values, the obtained standardized stimulants for companies with values outside the range of recommended values did not always order companies in accordance with the expectations resulting from the original values of these indicators. Therefore, the main aim of the study presented in the paper was to identify the reason of this phenomenon, to compare selected formulas of transformation of nominants into stimulants and to propose the modification of these transformations so that the order of companies before and after the transformation was characterized by greater consistency.

The achievement of the objective was based on artificial data and real data concerning two selected financial indicators—the current ratio and the debt margin (debt ratio) which are considered in the literature of the economic analysis as nominants with the recommended range of values. Annual data on these ratios concern the companies from the *Machinery industry* sector listed on the Warsaw Stock Exchange in 2016.

2 Methodology

2.1 Justification of the Choice of Indicators-Nominants with the Recommended Range of Values

Because the research concerns financial ratios of listed companies, hereafter the Authors decided to use term indicator instead of variable and company instead of object. Table 1 presents financial indicators, for which in the literature, one can find proposals of theoretical recommended ranges of values. Most Authors give the same ranges of values for the current ratio and for the debt margin. However, in the case of other indicators, it is not so unambiguous, which means that some Authors provide proposals for ranges, while others do not. For example, for a quick ratio, the value 1 is often given as a recommended value, i.e. the indicator is treated as a unimodal nominant.

A lot of experts in the field of controlling and economic analysis believe that many financial indicators should be treated as nominants with a single nominal value or nominants with a recommended range of values, but the recommended range of values should be determined on the basis of empirical data concerning companies from a given industry and not only on the basis of theoretical values. The most frequent indicators included in this group, together with a proposal of the method of determining the recommended values, are presented in Table 2.

Table 1 Financial indicators-nominants and their recommended ranges of values

Group of indicators	Name of the indicator	Recommended range of values
Liquidity indicators	Current ratio	[1.2; 2]
	Quick ratio	[1; 1.2]
	Cash ratio	[0.1; 0.2]
Debt indicators	Debt ratio	[0.57; 0.67]
Activity indicators	Receivables turnover	[30 days; 60 days]
	Liabilities turnover	[30 days; 60 days]

Source Sierpińska and Jachna (1995, 2004), Hozer et al. (1997), Bednarski et al. (2003), Waśniewski and Skoczylas (2004), Tarczyński and Łuniewska (2004), Łuniewska and Tarczyński (2006), Gabrusewicz (2014)

Table 2 Financial indicators-nominants for which recommended values should be computed empirically (according to experts)

Name of the indicator	Method of computing recommended values
Current ratio	In case of treating these indicators as nominants with single nominal value, this value should be equal to median In case of treating these indicators as nominants with the recommended range of values, this range should be computed on the base of median and median absolute deviation or median and quartile deviation
Quick ratio	
Cash ratio	
Receivables turnover	
Liabilities turnover	
Inventory turnover	
Fixed assets cover ratio	
Debt margin	

Source Batóg and Skoczylas (2017)

From indicators-nominants with a recommended range of values, the current ratio and the debt ratio were selected for further research. The following reasons determined their selection:

- knowledge of theoretical recommended ranges of values,
- knowledge of the proposals of determination of the limits of the recommended ranges of values in an empirical way,
- both ratios can be regarded as asymmetric nominants, i.e. in case of current ratio, the higher values are favourable, and in case of debt ratio, smaller values are favourable.

2.2 *Selected Formulas of Transformation of Nominants with the Recommended Ranges of Values into Stimulants with the Range of [0; 1]*

The choice of the analysed transformations was based on the fact that after the transformation, the obtained stimulant values are normalized in the range of [0; 1]. This assumption excluded from consideration proposals that normalize the stimulant in the range [−1; 1] (e.g. Strahl and Walesiak 1997). Therefore, the paper will consider the proposals of Kukuła (2000) and Kowalewski (2006). Kukuła in his paper (2000) proposed three methods of transformation of nominants with a recommended range of values into stimulants normalized in the range [0; 1]. Equations 1–3 present these transformations.

$$x_{ij}^S = \begin{cases} \frac{1}{c_{1j}-a_j}\left(x_{ij}^N - a_j\right) & \text{for} \quad x_{ij}^N < c_{1j} \\ 1 & \text{for } c_{1j} \le x_{ij}^N \le c_{2j} \\ \frac{1}{c_{2j}-b_j}\left(x_{ij}^N - b_j\right) & \text{for} \quad x_{ij}^N > c_{2j} \end{cases} \tag{1}$$

$$x_{ij}^S = \begin{cases} \frac{-x_{ij}^2 + 2c_{1j}x_{ij}^N - a_j(2c_{1j} - a_j)}{(a_j - c_{1j})^2} & \text{for} \quad x_{ij}^N < c_{1j} \\ 1 & \text{for } c_{1j} \le x_{ij}^N \le c_{2j} \\ \frac{-x_{ij}^2 + 2c_{2j}x_{ij}^N - b_j(2c_{2j} - b_j)}{(b_j - c_{2j})^2} & \text{for} \quad x_{ij}^N > c_{2j} \end{cases} \tag{2}$$

$$x_{ij}^S = \begin{cases} \left(\frac{x_{ij}^N - a_j}{a_j - c_{1j}}\right)^2 & \text{for} \quad x_{ij}^N < c_{1j} \\ 1 & \text{for } c_{1j} \le x_{ij}^N \le c_{2j} \\ \left(\frac{x_{ij}^N - b_j}{b_j - c_{2j}}\right)^2 & \text{for} \quad x_{ij}^N > c_{2j} \end{cases} \tag{3}$$

where:

x_{ij}^S—value of stimulant in the range [0; 1] of jth indicator-nominant for ith company,
x_{ij}^N—value of jth indicator-nominant for ith company,
a_j—minimum value of jth indicator-nominant,
b_j—maximum value of jth indicator-nominant,
c_{1j}—lower limit of recommended range of values of jth indicator-nominant,
c_{2j}—upper limit of recommended range of values of jth indicator-nominant,
$i = 1, 2, \ldots, n$; n—number of companies,
$j = 1, 2, \ldots, m$; m—number of indicators-nominants.

The first method of transformation uses a segment-linear valuing function (Eq. 1) and is used when there is no preference for valuing a nominant outside the recommended range of values. The second method is based on the square segment-concave valuing function (Eq. 2), and its use is justified if the values of the nominant close to the lower and upper limits of the recommended range of values are evaluated only slightly worse than its values in the recommended range of values. The third method is based on the square segment-convex valuing function (Eq. 3), and its use is justified if even a small exceed of the recommended range of values is assessed as bad situation. It is also possible to use a transformation which is a combination of the second and third methods, i.e. values below the lower limit of the recommended range of values are described by the square concave function, and values above the upper limit of the recommended range of values are described by the square convex function or vice versa (Batóg 2003).

For the transformations described in Eqs. 1–3, the simulations of two cases were conducted depending on whether the distance between the minimum value of indicator and the lower limit of the recommended range of values was the same as the distance between the maximum value of indicator and the upper limit of the recommended range of values (symmetry with respect to the middle of the recommended range of values) or whether these distances were different (lack of symmetry with respect to the middle of the recommended range of values).

These simulations are shown in Figs. 1, 2 and 3. The left graphs refer to the situation of symmetry with respect to the middle of the recommended range of values,

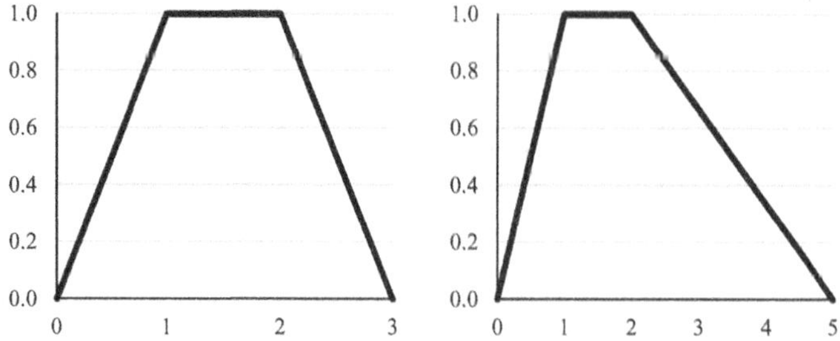

Fig. 1 Simulation of transformation according to Eq. 1—symmetry (left), lack of symmetry (right)

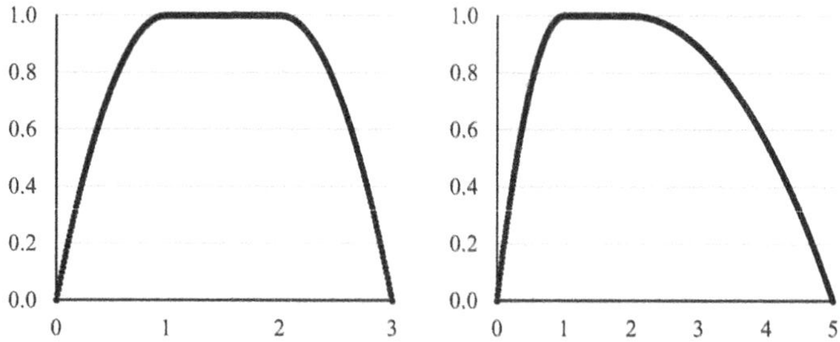

Fig. 2 Simulation of transformation according to Eq. 2—symmetry (left), lack of symmetry (right)

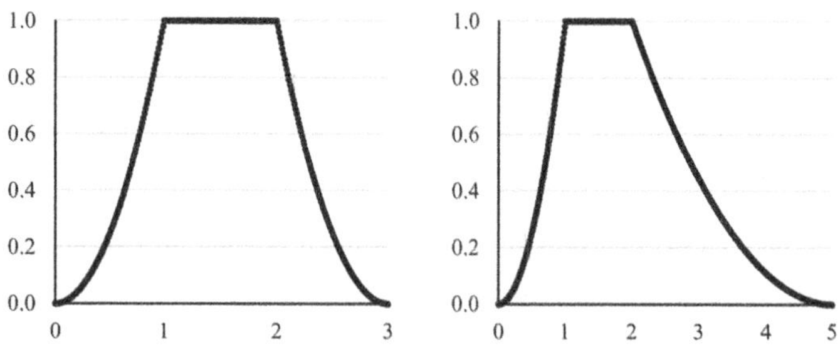

Fig. 3 Simulation of transformation according to Eq. 3—symmetry (left), lack of symmetry (right)

while the right ones refer to the one of the situations of the lack of symmetry—the distance between the minimum value of indicator and the lower limit of the recommended range of values is smaller than the distance between the maximum value of indicator and the upper limit of the recommended range of values. In all simulations, the range [1; 2] was selected as the recommended range of values.

On the base of Figs. 1, 2 and 3 (left graphs), it can be concluded that in case of symmetry, companies with values of indicators-nominants below and above the recommended range of values were assigned such values of stimulants which are the same for companies with values of indicators-nominants equally distant from the lower and upper limits of the recommended range of values. This in turn results in the fact that in terms of the values of this indicator, the companies are evaluated on the same level, i.e. they have the place in linear ordering (according to examined indicator-nominant).

However, in the lack of symmetry with respect to the middle of the recommended range of values (Figs. 1, 2 and 3, right graphs), the values of stimulant obtained for companies with the values of the indicators-nominants below and above the recommended range of values do not result from properties of the analysed indicator but depend on the minimum and maximum values of the analysed indicator for the surveyed companies. The greater the difference of the distance between the minimum value of indicator and the lower limit of the recommended range of values and the distance between the maximum value and the upper limit of the range of recommended values, the greater the inconsistency in the order of listed companies before and after transformation. In case of Eq. 1 (Fig. 1, right graph) when recommended range of values is [1; 2], the modified values for equally distant numbers from this interval are different—for example, for 0.5 and 2.5, originally we obtain 0.5 and 0.83, respectively. Then, the companies are not in the same place in linear ordering.

Another proposal for transformation of nominants into stimulants was presented by Kowalewski (2002, 2006). He introduced the notions of right-side and left-side asymmetric nominants. A left-side asymmetric nominant is when its values below the lower limit of the recommended range of values are better assessed than its values above the upper limit of the recommended range of values. A right-side asymmetric nominant is when its values above the upper limit of the recommended range of values are better assessed than its values below the lower limit of the recommended range of values. However, a nominant is symmetric when its values below the lower and upper limits of the recommended range of values are assessed equally—the closer to the limits the better, and the further away the limits, the worse. Equation 4 presents the general formula of the transformation of a nominant with the recommended range of values into stimulant proposed by Kowalewski (2006).

$$x_{ij}^{S} = \begin{cases} 1 & \text{for } x_{ij}^{N} \in \left[A_j, B_j\right] \\ \frac{-1}{k_r\left(x_{ij}^{N} - A_j - 1\right)} & \text{for } \quad x_{ij}^{N} < A_j \\ \frac{1}{k_l\left(x_{ij}^{N} - B_j + 1\right)} & \text{for } \quad x_{ij}^{N} > B_j \end{cases} \tag{4}$$

where:

x_{ij}^{S}—value of stimulant in the range [0; 1] of jth indicator-nominant for ith company,
x_{ij}^{N}—value of jth indicator-nominant for ith company,
$i = 1, 2, \ldots, n$; n—number of companies,
$j = 1, 2, \ldots, m$; m—number of indicators-nominants,
A_j—lower limit of recommended range of values of jth indicator-nominant,

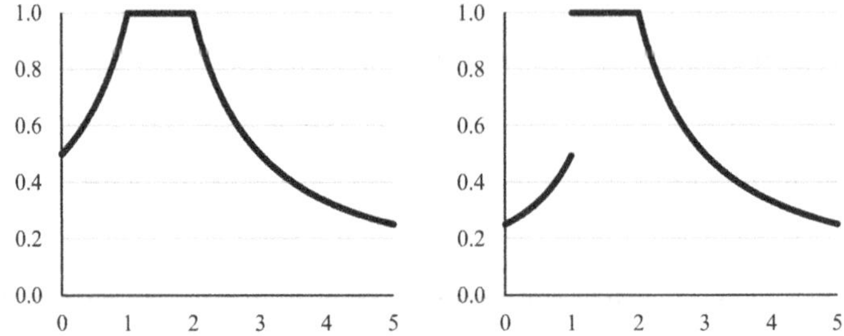

Fig. 4 Simulation of transformation according to Eq. 4—symmetric nominant (left), right-side asymmetric nominant (right)

B_j—upper limit of recommended range of values of jth indicator-nominant,
$k_r > 1$—coefficient of asymmetry for right-side asymmetric nominant ($k_l = 1$),
$k_l > 1$—coefficient of asymmetry for left-side asymmetric nominant ($k_r = 1$),
$k_r = k_l = 1$—coefficient of asymmetry for symmetric nominant.

Figure 4 presents simulations of transformations obtained on the basis of Eq. 4, where the left graph concerns a symmetric nominant, while the right one concerns right-side asymmetric nominant for $k_r = 2$.

Figure 4 shows that for the symmetric nominant, the consistency of the order of companies before and after the transformation is maintained. However, in the case of an asymmetric nominants, there is a gap on one side of the recommended range of values, which results in a discrepancy in the order of companies before and after the transformation. It is worth mentioning that in the discussed proposal for indicators-nominants, the lowest and the highest values after the transformation do not take the value of zero. Also of great importance are the constant k_r and k_l, for which there are no clear guidelines on how to determine them.

2.3 The Authors' Proposal of the Modification of Transformation of Nominants with the Recommended Range of Values into Stimulants

The modification of the transformation of nominants with the recommended range of values into stimulants in the range [0; 1] proposed in the paper concerns cases when the transformations take into account the minimum and the maximum value of the indicator-nominant for the group of companies, and there is no symmetry in relation to the middle of the recommended range of values. The Authors propose to extend the range of variability of the indicator-nominant by determining a new (artificial) minimum (Eq. 5) or a new (artificial) maximum (Eq. 6) depending on whether the distance between the minimum value of the indicator-nominant and the lower limit of the recommended range of values is smaller or larger than the distance between the maximum value of the indicator-nominant and the upper limit of the recommended range of values.

$$a_j^* = \begin{cases} a_j & \text{for } c_{1j} - a_j \geq b_j - c_{2j} \\ c_{1j} - \left(b_j - c_{2j}\right) & \text{for } c_{1j} - a_j < b_j - c_{2j} \end{cases} \tag{5}$$

$$b_j^* = \begin{cases} c_{2j} + \left(c_{1j} - a_j\right) & \text{for } c_{1j} - a_j \geq b_j - c_{2j} \\ b_j & \text{for } c_{1j} - a_j < b_j - c_{2j} \end{cases} \tag{6}$$

where:

a_j^*—minimum value of *j*th indicator-nominant after modification,
b_j^*—maximum value of *j*th indicator-nominant after modification,
a_j—minimum value of *j*th indicator-nominant before modification,
b_j—maximum value of *j*th indicator-nominant before modification,
c_{1j}—lower limit of the recommended range of values of *j*th indicator-nominant,
c_{2j}—upper limit of the recommended range of values of *j*th indicator-nominant.

The new minimum (or maximum) value should be used for the transformation of a nominant using Eqs. 1–3. Figure 5 shows the simulation of the transformation according to Eq. 1 assuming that the distance of the minimal value of the indicator-nominant from the lower limit of the recommended range of values is smaller than the distance of the maximum value from the upper limit of the recommended interval, i.e. taking into account the new minimum determined from Eq. 5.

Comparing Fig. 5 with the right graph in Fig. 1, it can be observed that the transformation with the new minimum presented in Fig. 5 ensures the consistency of the order of the companies before and after the transformation of the value of the indicator-nominants. In case when the recommended range of values is [1; 2], the modified values for equally distant numbers from this interval are the same—for example, for 0.5 and 2.5, originally we obtain 0.83 and 0.83, respectively. Then, the companies are in the same place in linear ordering.

The application of the proposed modification gives the same effect also for the transformations described in Eqs. 2 and 3.

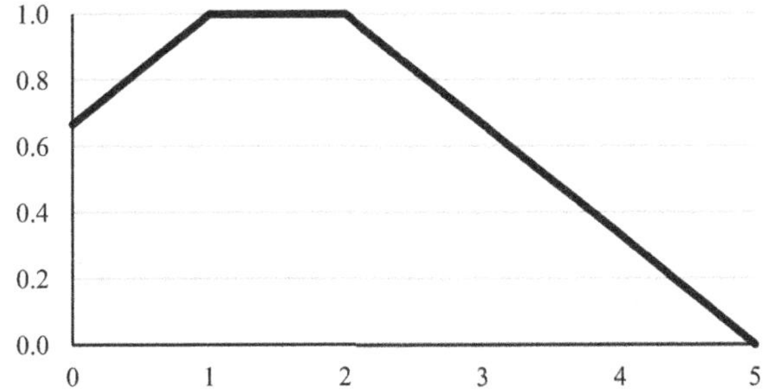

Fig. 5 Simulation of transformation according to Eq. 1 with modification of minimum

3 Application of Examined Transformations for Selected Indicators-Nominants—Empirical Example

As it was mentioned in the previous part of the paper, two indicators-nominants were selected for the research—current ratio and debt ratio. The values of these indicators concern the companies from the *Machinery Industry* sector listed on the Warsaw Stock Exchange in 2016.

The first step was to transform the indicators-nominant into stimulants using Eqs. 1–4 with the recommended range of values from the literature (Table 1). The recommended range for the current ratio is [1.2; 2], and the recommended range for the debt ratio is [0.57; 0.67]. Figures 6, 7, 8 and 9 present the results of these transformations for examined financial indicators.

In the next step, the Authors' proposal of modification of the minimum or the maximum value was used for the transformation of the indicators-nominants. For the current ratio, the distance between the minimum value of the indicator and the lower limit of the recommended range of values was smaller than the distance between the maximum value of the indicator and the upper limit of the recommended range of values, so the modification with Eq. 5 was applied. For the debt ratio, the distance between the minimum value of the indicator and the lower limit of the recommended range of values was greater than the distance between the maximum value of the

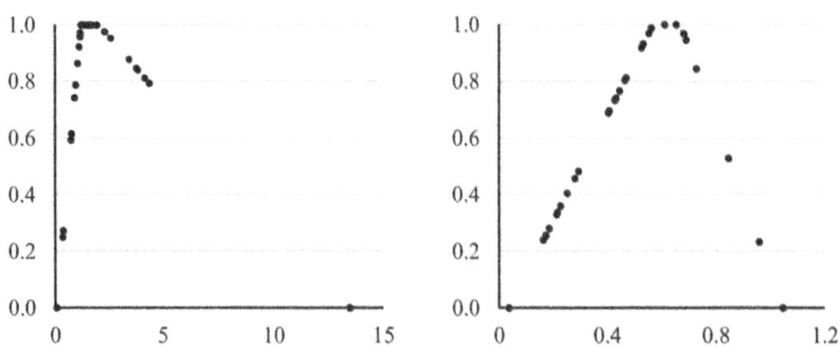

Fig. 6 Transformation according to Eq. 1—current ratio (left) and debt margin (right)

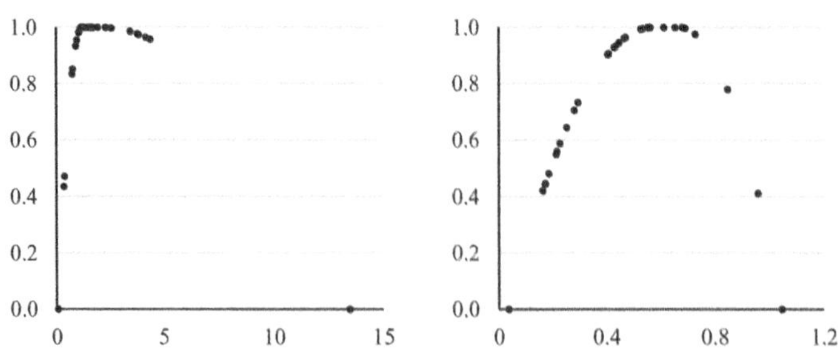

Fig. 7 Transformation according to Eq. 2—current ratio (left) and debt margin (right)

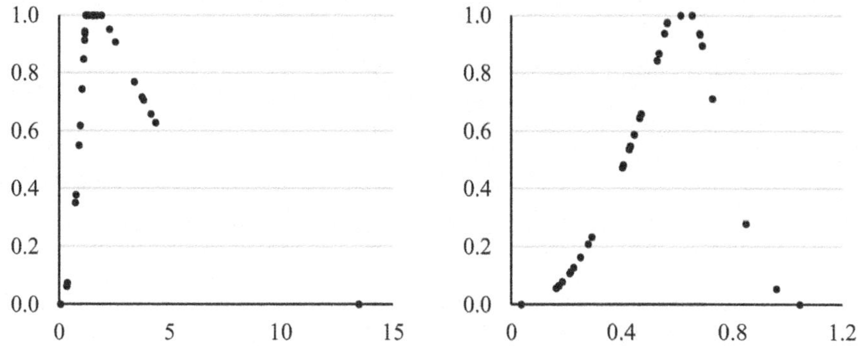

Fig. 8 Transformation according to Eq. 3—current ratio (left) and debt margin (right)

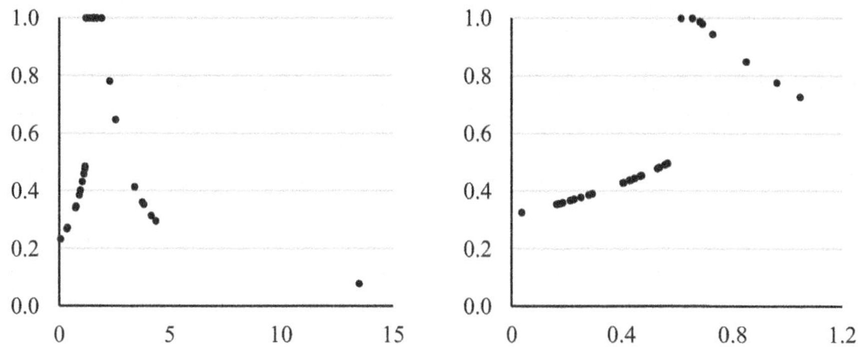

Fig. 9 Transformation according to Eq. 4—current ratio (left) and debt margin (right)

indicator and the upper limit of the recommended range of values, so the modification was applied by means of Eq. 6. The results of the modification of the transformation of Eq. 1 are presented in Fig. 10. The proposed modification can also be applied to the transformation given by Eqs. 2 and 3. In the case of Eq. 4, it is not possible to

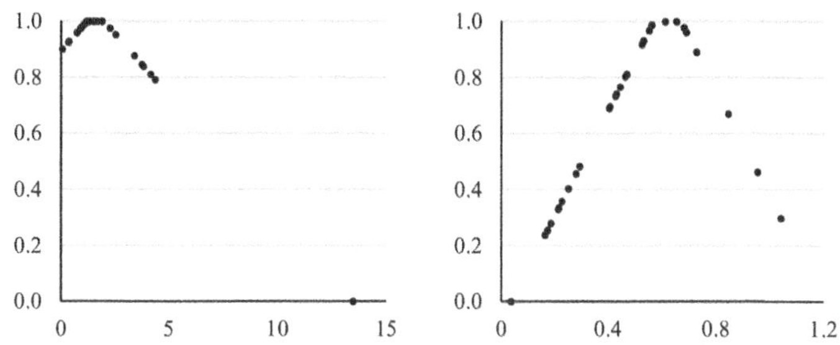

Fig. 10 Modified transformation from Eq. 1—current ratio (left) and debt margin (right)

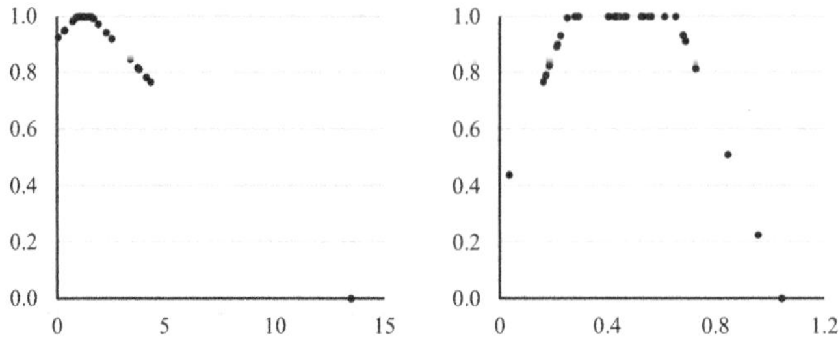

Fig. 11 Transformation given by Eq. 1 modified according to Eq. 5 with the empirical recommended range of values—current ratio (left) and debt margin (right)

apply the proposed modification, because in this transformation, the minimum and the maximum values of the indicator-nominant for the surveyed companies are not taken into account.

Figure 10 shows that despite the different distance between the minimum value of the indicator-nominant and the lower limit of the recommended range of values and the distance between the maximum value of the indicator-nominant and the upper limit of the recommended range of values, the symmetry of the values equally distant from the recommended range of values for both indicators was maintained after the transformation.

In the last part of the study, the transformation (Eq. 1) of the analysed financial indicators was modified according to Eq. 5 for the case when the recommended range of values is an empirical range determined as the median plus/minus median mean deviation.[1] The transformation using this approach is shown in Fig. 11.

Comparing Figs. 10 and 11, it can be observed that theoretical and empirical recommended ranges of values for both indicators-nominants differ from each other, but these differences concern not only the values of the lower and upper limits, but first of all the width of these ranges. This in turn results in the fact that the greater the width of empirical ranges, the more companies after transformation will obtain the value of the standardized stimulant at the level of 1.

The application of an empirical recommended range of values approach to research seems to be better than the adoption of theoretical ones—unchanging in time and space—because companies belonging to different sectors and examined at different periods are characterized by different operating conditions, which in turn has great impact on the results they achieve (Batóg and Skoczylas 2017).

4 Conclusions

On the basis of the simulations and studies carried out, the following conclusions concerning the selected formulas for the transformation of indicators-nominant with

[1] More on median absolute deviation could be found in Młodak (2006).

the recommended range of values into normalized stimulants in the range [0; 1] can be formulated:

- the transformation formulas (Eqs. 1–3) proposed by Kukuła are justified if the difference between the minimum and maximum value of the nominant is symmetric with respect to the middle of the recommended range of values, because then the values of the computed stimulant correspond to the real values of the indicator-nominant, which are outside the recommended range of values—e.g.: companies with current ratio at the level of 0.8 and 2.4 (with the recommended range of values from 1.2 to 2) after the transformation have the same values of the stimulant, so the order of the companies before and after the transformation of the indicator-nominant is consistent,
- if this difference is not symmetric, then the above regularity is not maintained, and then, a modification proposed by the Authors may be applied; it is based on extending the range of the indicator-nominant by determining a new minimum or a new maximum value depending on whether the distance of the minimum value of the indicator-nominant from the lower limit of the recommended range of values is smaller or bigger than the distance of the maximum value of the indicator-nominant from the upper limit of the recommended range of values,
- application of the modification proposed by Authors to the transformation formulas described in Eqs. 1–3 ensures that the order of companies before and after the transformation is consistent with the real values of the indicator-nominant,
- the proposed modification concerns cases when the transformations take into account the minimum and the maximum value of the indicator-nominant, and there is the lack of symmetry of these values in relation to the middle of the recommended range of values,
- the method of transformation of nominants into stimulants proposed by Kowalewski does not take into account the minimum and the maximum value of the indicator-nominant, so the variant with the proposed modification was not considered, and during its discussion, it was pointed out that the reason for the lack of the same order of companies before and after the transformation of the indicator-nominant may be the assumed level of the constant k in the case of asymmetric nominant; it turned out that the adoption of the constant k at a level slightly above 1 resulted in the fact that after the transformation of the current ratio (right-side asymmetric nominant), for a company with a lack of liquidity, the value of the stimulant was the same as the value for the company with the excessive liquidity,
- in the case of the transformation proposed by Kowalewski, the minimum and the maximum values of indicator-nominant are not transformed into zero for stimulant.

The last conclusion, which can be formulated on the basis of the conducted research, is not directly related to transformation formulas but concerns the method of determining the range of recommended values for indicator-nominant. According to the Authors, in order to link the results obtained by companies with their operating conditions, an empirical approach based on parameters that characterize the distribution of the indicator-nominant in the studied group of companies is more appropriate to determine the recommended range of values.

In further research on methods of transformation of nominant with a recommended range of values into standardized stimulant, the Authors want to focus on the left-side and right-side asymmetric nominants.

References

Batóg B, Skoczylas W (2017) Wykorzystanie taksonomicznego miernika rozwoju w ocenie sytuacji finansowej przedsiębiorstw. Dylematy zarządzania kosztami i dokonaniami. In: Kowalak R, Kowalewski M, Bednarek P (eds) Prace Naukowe Uniwersytetu Ekonomicznego we Wrocławiu 472, pp 387–397

Batóg J (2003) Klasyfikacja obiektów w przypadku agregacji danych. In: Metody ilościowe w ekonomii, Zeszyty Naukowe 365, Prace Katedry Ekonometrii i Statystyki 14, red. J. Hozera, Szczecin, pp 35–44

Bednarski L, Borowiecki R, Duraj J, Kurtys E, Waśniewski T, Wersty B (2003) Analiza ekonomiczna przedsiębiorstwa. Wydawnictwo Akademii Ekonomicznej im. Oskara Langego we Wrocławiu, Wrocław

Borys T (1978) Metody normowania cech w statystycznych badaniach porównawczych. Przegląd Statystyczny 2:227–239

Borys T (1984) Kategoria jakości w statystycznej analizie porównawczej. Prace Naukowe Akademii Ekonomicznej we Wrocławiu 284. Monografie i opracowania 23, Seria

Gabrusewicz W (2014) Podstawy analizy finansowej. PWE, Warszawa

Gatnar E, Walesiak M (eds) (2004) Metody statystycznej analizy wielowymiarowej w badaniach marketingowych. Wydawnictwo AE we Wrocławiu, Wrocław

Hellwig Z (1968) Zastosowanie metody taksonomicznej do typologicznego podziału krajów ze względu na poziom ich rozwoju oraz zasoby i strukturę wykwalifikowanych kadr. Przegląd Statystyczny 4, Warszawa, pp 307–327

Hozer J, Tarczyński W, Gazińska M, Wawrzyniak K, Batóg J (1997) Metody ilościowe w analizie finansowej przedsiębiorstwa. Główny Urząd Statystyczny, Warszawa

Kowalewski G (2002) Nominanty niesymetryczne w wielowymiarowej analizie sytuacji finansowej jednostek gospodarczych. Przegląd Statystyczny 2:123–132

Kowalewski G (2006) Jeszcze o nominantach w metodach porządkowania liniowego zbioru obiektów, Taksonomia 13. Klasyfikacja i analiza danych – teoria i zastosowania. Prace Naukowe Akademii Ekonomicznej 1126:519–528

Kukuła K (2000) Metoda unitaryzacji zerowanej. Wydawnictwo Naukowe PWN, Warszawa

Łuniewska M, Tarczyński W (2006) Metody wielowymiarowej analizy porównawczej na rynku kapitałowym. Wydawnictwo Naukowe PWN, Warszawa

Młodak A (2006) Analiza taksonomiczna w statystyce regionalnej. Difin, Warszawa

Sierpińska M, Jachna T (1995, 2004) Ocena przedsiębiorstwa według standardów światowych. Wydawnictwo Naukowe PWN, Warszawa

Sokołowski A, Markowska M (2017) Iteracyjna metoda liniowego porządkowania obiektów wielocechowych. Przegląd Statystyczny 2:153–162

Strahl D, Walesiak M (1997) Normalizacja zmiennych w skali przedziałowej i ilorazowej w referencyjnym systemie granicznym. Przegląd Statystyczny 1:69–77

Tarczyński W, Łuniewska M (2004) Dywersyfikacja ryzyka na polskim rynku kapitałowym. Wydawnictwo PLACET, Warszawa

Walesiak M (2011) Uogólniona miara odległości GDM w statystycznej analizie wielowymiarowej z wykorzystaniem programu R. Wydawnictwo UE we Wrocławiu, Wrocław

Waśniewski T, Skoczylas W (2004) Teoria i praktyka analizy finansowej w przedsiębiorstwie. Fundacja Rozwoju Rachunkowości w Polsce, Warszawa

Wójciak M (2003) Niesymetryczne metody wartościowania nominant. Taksonomia 10. Klasyfikacja i analiza danych – teoria i zastosowania. Prace Naukowe Akademii Ekonomicznej 988:519–528

Silhouette Index as Clustering Evaluation Tool

Andrzej Dudek

Abstract Silhouette index is commonly used in cluster analysis for finding the optimal number of clusters, as well as for final clustering validation and evaluation as a synthetic indicator allowing to measure the general quality of clustering (relative compactness and separability of clusters—see Walesiak and Gatnar in Statystyczna analiza danych z wykorzystaniem programu R. PWN, Warszawa, p. 420, 2009). Its advantage is low computational complexity and simple interpretation rules. Recently, some proposals have appeared to use this index directly as basis of clustering algorithms. The paper is a tryout of the evaluation of such approach. In the paper examples, when the "mechanical" use of the silhouette index leads to the results that do not correspond to the actual structure of the classes are shown, the recommendations on the principles of the correct application of the index are presented.

Keywords Cluster analysis · Cluster quality index · Silhouette index · Clustering trees

1 Introduction

Silhouette index in cluster analysis proposed by Rousseeuw (1987) defines for each object in dataset, the measure of how this object is similar to other objects from the same cluster (cohesion, compactness) in comparison with objects of other clusters (separation). For each object, the measure can obtain values from range $<-1, 1>$. The higher value means better matching to cluster to which it should be classified and lower fitness to other clusters.

The average silhouette index overall points is often treated as a clustering quality measure. It is very often used for finding the number of clusters in partitioned dataset. The silhouette index is defined in Eq. 1.

A. Dudek (✉)
Wrocław University of Economics and Business, Wrocław, Poland
e-mail: andrzej.dudek@ue.wroc.pl

K. Jajuga et al. (eds.), *Classification and Data Analysis*, Studies in Classification, Data Analysis, and Knowledge Organization,
https://doi.org/10.1007/978-3-030-52348-0_2

$$S(u) = \frac{1}{n}\sum_{i=1}^{n}\frac{b(i) - a(i)}{\max\{a(i); b(i)\}}, \quad S(u) \in [-1, \ 1] \qquad (1)$$

where

u—number of clusters,
n—number of objects in dataset,
$\{d_{ij}\}$—distance matrix,
$a(i) = \sum_{k \in \{P_r \setminus i\}} d_{ik}/(n_r - 1)$—average distance from object i to other objects belongings to cluster P_r (object i belongs to cluster P_r); n_r—number of objects in cluster P_r,
$b(i) = \min_{s \neq r}\{d_{iP_s}\}, d_{iP_s} = \sum_{k \in P_s} d_{ik}/n_s$—average distance from object i to other objects belongings to cluster P_s (object i does not belong to cluster P_s); n_s—number of objects in cluster P_s.

Silhouette index is commonly used in cluster analysis for finding the optimal number of clusters, as well as for final clustering validation and evaluation as a synthetic indicator allowing to measure general quality of clustering (relative compactness and separability of clusters (see. Walesiak and Gatnar 2009, p. 420). The synopsis of this routine is presented in Sect. 2 along with an example.

Recently, some proposals have appeared to use the silhouette index directly as the basis of clustering algorithms. This approach is described in Sect. 3, and its valuation of three dataset is presented in Sect. 4.

2 Finding Number of Clusters with Silhouette Index

Silhouette index can be used for finding the optimal number of clusters for the partitioned dataset. The procedure of finding the optimal number of clusters can be stated as:

1. Run clustering algorithm for different numbers of clusters (*clNo*) in range *j…k* (often *2…10*).
2. For each *clNo* calculate the silhouette index.
3. Choose optimal number of clusters as the one corresponding to maximal silhouette index calculated in each step.

The idea of this approach can be observed in Figs. 1, 2, 3, 4 and 5.

For given dataset (generated with *clusterGen* function from *clusterSim* R package with *model* = 8 parameter), the values of silhouette index for each cluster number initially set are labeled in Table 1. The partitioning around medoids (pam) method (Kaufman and Rousseeuw 1990) has been used as a clustering algorithm.

The optimal number of clusters for pam algorithm is thus equal to four (with corresponding average silhouette value equal to 0.852).

The source code of the procedure is included in Appendix 1.

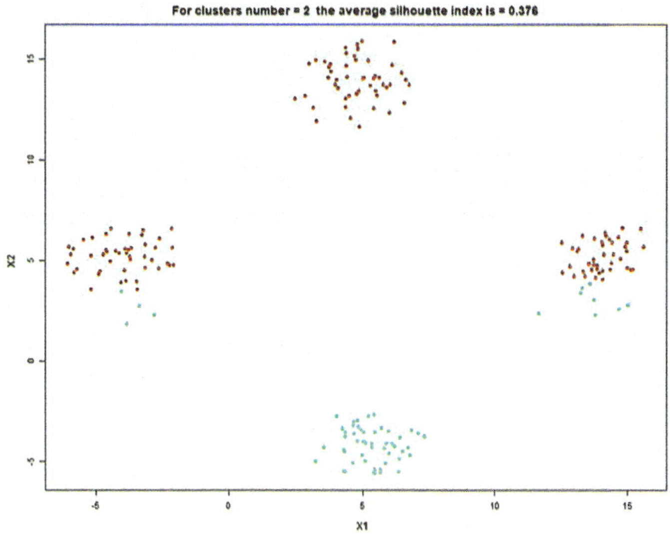

Fig. 1 Clustering procedure results for pam method for cluster number equal to two with calculated silhouette index value. *Source* Own's calculations

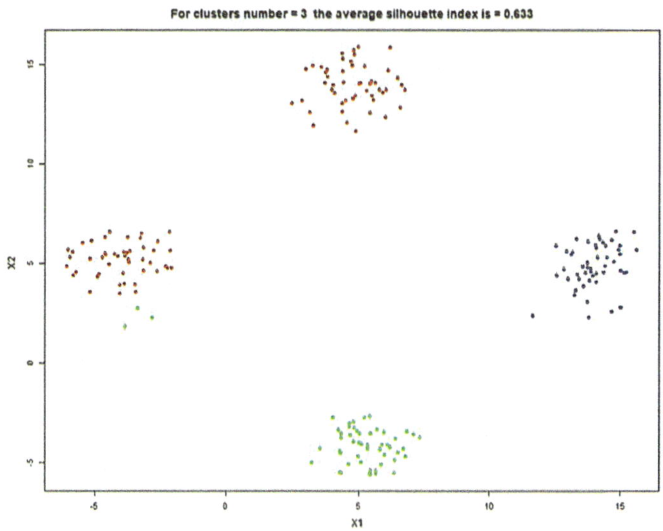

Fig. 2 Clustering procedure results for pam method for cluster number equal to three with calculated silhouette index value. *Source* Own's calculations

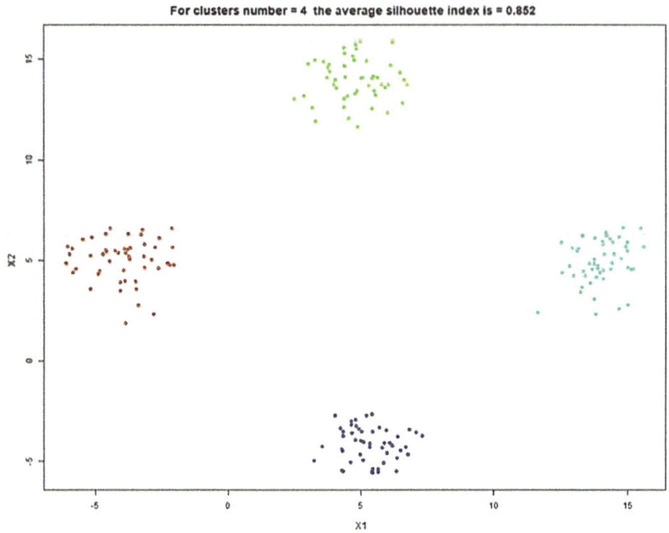

Fig. 3 Clustering procedure results for pam method for cluster number equal to four (optimal number) with calculated silhouette index value. *Source* Own's calculations

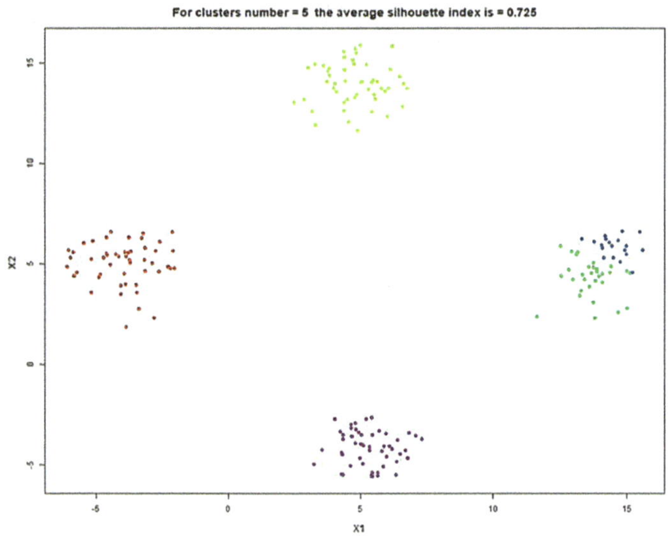

Fig. 4 Clustering procedure results for pam method for cluster number equal to five with calculated silhouette index value. *Source* Own's calculations

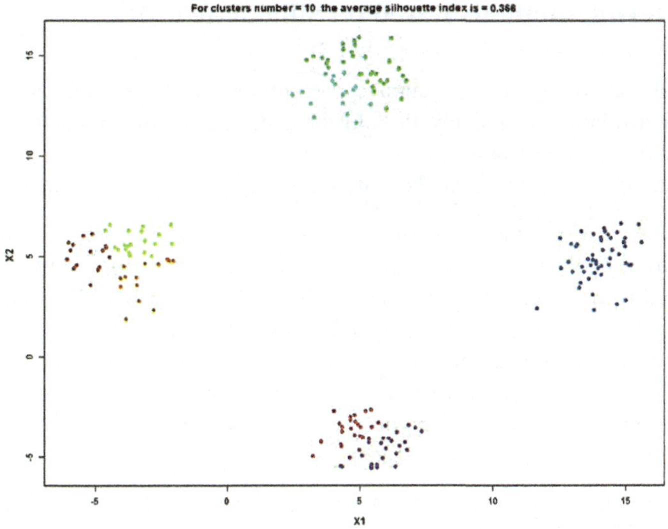

Fig. 5 Clustering procedure results for pam method for cluster number equal to ten with calculated silhouette index value. *Source* Own's calculations

Table 1 Silhouette index values for cluster number in 2…10

No. of clusters	Silhouette value
2	0.376
3	0.633
4	0.852
5	0.725
6	0.595
7	0.474
8	0.482
9	0.359
10	0.366

This approach is very popular in the literature of subject. Arbelaitz et al. (2013) have compared 38 cluster quality indices and found "Silhouette, Davies–Bouldin and Calinski–Harabasz at the top," Starczewski and Krzyżak (2015) stated "very good effectiveness of the *SILv2*-index." Migdał-Najman and Najman (2006) have noticed that the obtained cluster structure comparison method should be treated rather as a good start than the end of the research.

3 Clustering Methods Based on Silhouette Index

Kang et al. (2016) proposed recursive partitioning clustering tree (RPCT) algorithm which divides recursively the clustered dataset basing in each step on the maximization of silhouette index.

The idea of the algorithm can be stated in three steps:

Step 1. For each variable, calculate the silhouette statistics based on the results of partitioning the original data space into two subspaces (for each realization of the variable in object k (v_k), all object with $v_j <= v_k$ are assigned to first subspace and all objects with $v_j > v_k$ be assigned to second subspace).
Step 2. Find the set of partitioning points that maximize the silhouette gain.
Step 3. If a stop-recursion criterion is satisfied (there is no silhouette gain due to any split or the maximal number of splits is exceeded), the iteration ends. Otherwise, steps 1 and 2 are repeated.

The idea of the algorithm for dataset coming from multivariate normal distribution is shown in Figs. 6, 7, 8 and 9.

In first step, the split maximizing the silhouette index gain is corresponding to variable one and split point set to 2.014. This allows to maximize silhouette index to 0.513.

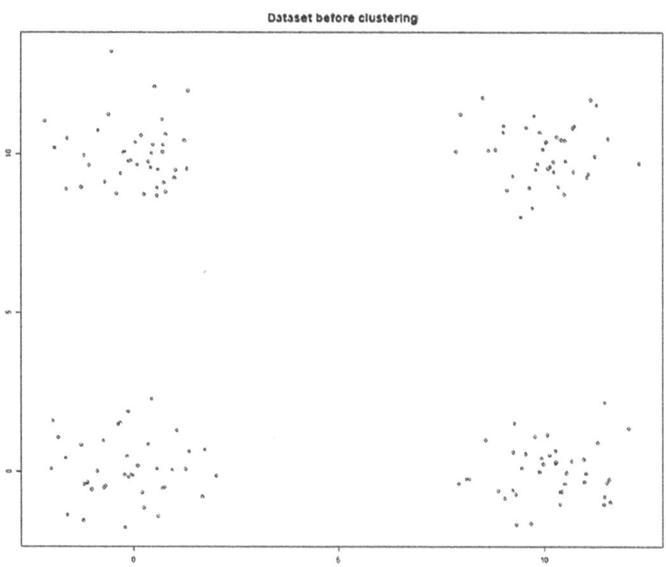

Fig. 6 RPCT method starting point. *Source* Own's calculations

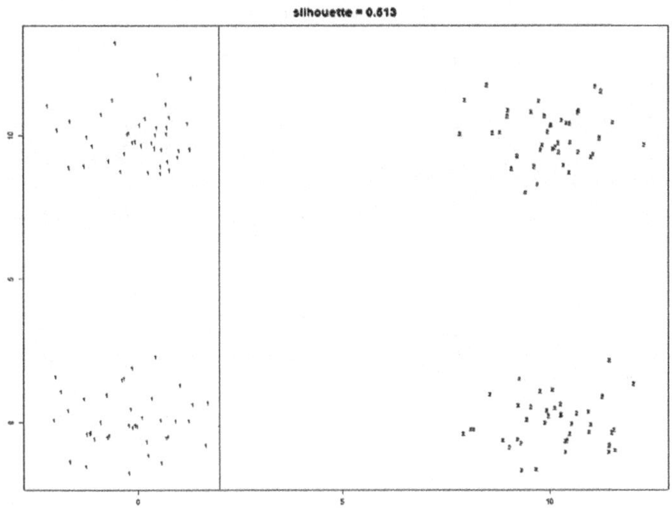

Fig. 7 RPCT method step one—one split. *Source* Own's calculations

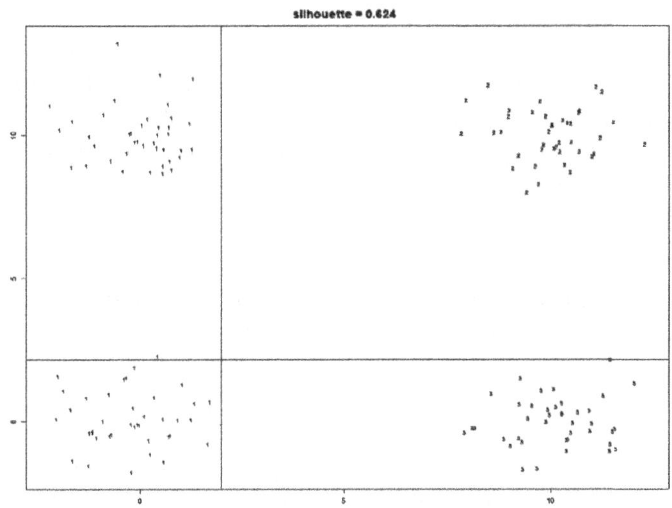

Fig. 8 RPCT method step two—two splits. *Source* Own's calculations

In second step, the split maximizing the silhouette index gain is corresponding to variable two (with first variable constrained by condition >= 2.014) and split point is set to 2.187. This allows to maximize silhouette index to 0.624 (with step gain equal to 0.111).

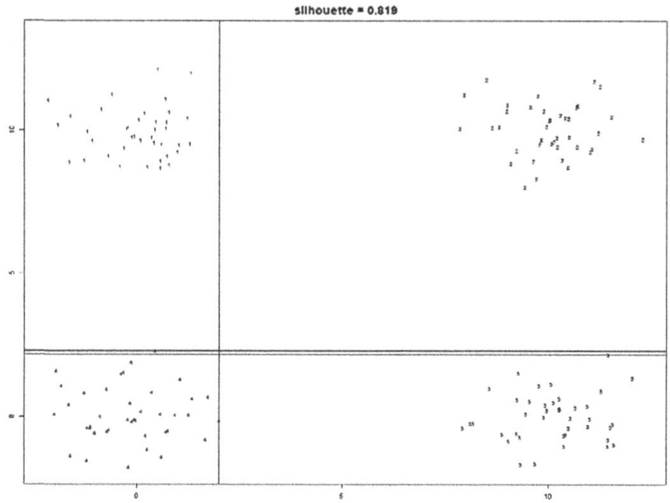

Fig. 9 RPCT method step three—three splits. *Source* Own's calculations

In third step, the split maximizing the silhouette index gain is corresponding to variable two (with first variable constrained by condition < 2.014) and split point is set to 2.293. Again, this split allows to maximize silhouette index to 0.819 (with step gain equal to 0.195).

In last step, no split that improves silhouette index average value can be found, so, the clustering algorithm converges.

4 The Performance of Clustering Methods Based on Silhouette Index

To evaluate the performance of RPCT algorithm, its implementation has been written in R language. The source code of this procedure is included in Appendix 2. For all evaluation datasets, adjusted rand index (ARI) (Hubert and Arabie 1985) comparing achieved clustering with real cluster structure has been calculated.

The first dataset has been generated by *cluster.Gen* function from *clusterSim* package (Walesiak and Dudek 2019) with parameter *model* = 13. The original cluster structure is shown in Fig. 10.

The results of clustering tree RPCT procedure are shown in Fig. 11. The final partitioning is related to one split only (two clusters).

- Split is corresponding to variable one and split point set to 0.045. The silhouette index value for this cluster structure is equal to 0.536.

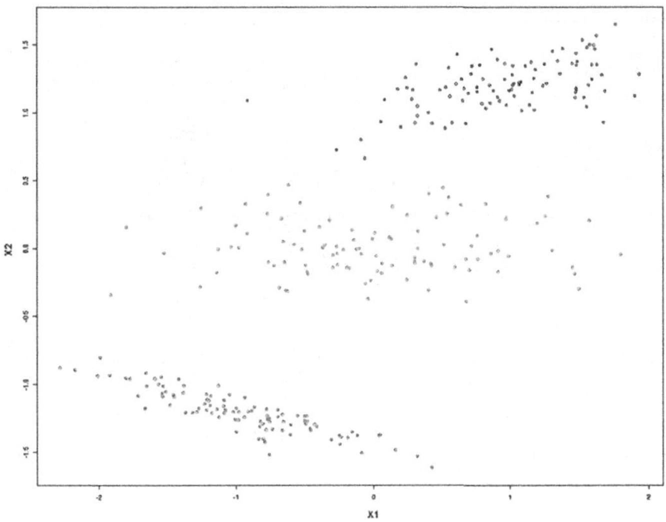

Fig. 10 First evaluation dataset—real clusters structure. *Source* Own's calculations

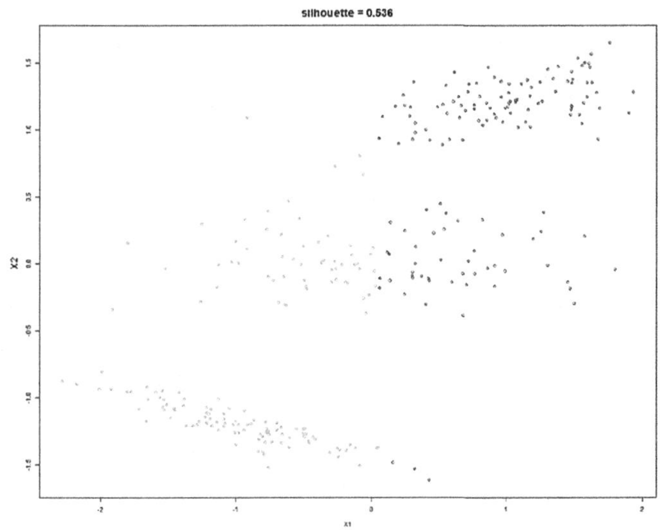

Fig. 11 First evaluation dataset—clustering tree algorithm results. *Source* Own's calculations

The adjusted Rand index (ARI) between RPCT algorithm results and actual clusters of this dataset is equal to 0.382, which cannot be recognized as satisfactory.

The second dataset comes from fundamental clustering problems suite (FCPS—https://github.com/Mthrun/FCPS)—chainlink dataset. The original cluster structure is shown in Fig. 12.

Fig. 12 Second evaluation dataset—real clusters structure. *Source* Own's calculations

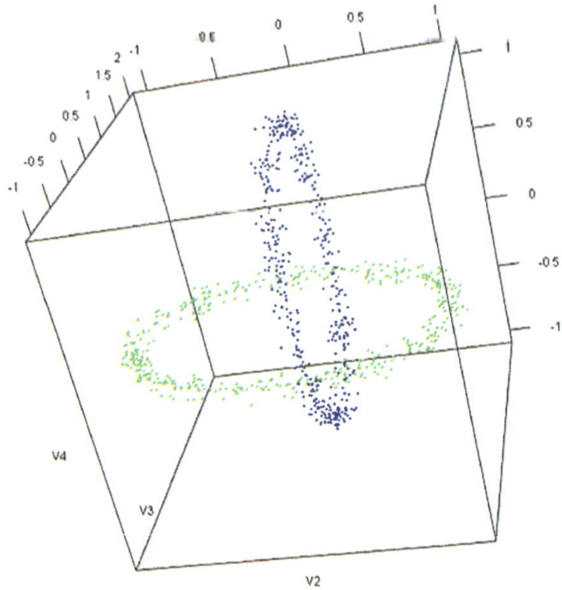

The results of clustering tree RPCT procedure on second dataset are shown in Fig. 13. The final partitioning is related to one split only (two clusters)

Fig. 13 Second evaluation dataset—clustering tree algorithm results. *Source* Own's calculations

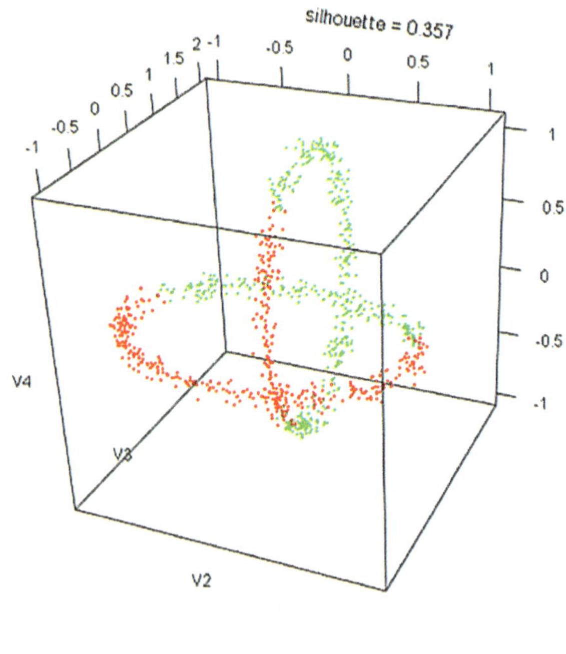

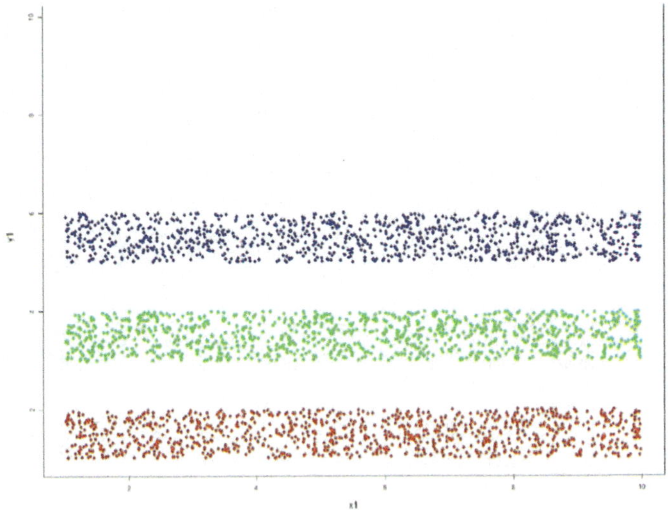

Fig. 14 Third evaluation dataset—real clusters structure. *Source* Own's calculations

- Split is corresponding to variable two and split point set to 0.357, the silhouette index value for this cluster structure is equal to 0.357.

The ARI index between RPCT algorithm results and actual clusters of this dataset is equal to 0.109, which can be treated as an information, that the real cluster structure has not been discovered by algorithm.

The third dataset is generated with three uniform distributions. The original cluster structure is shown in Fig. 14.

The results of clustering tree RPCT procedure are shown in Fig. 15. The final partitioning is related to two splits (three clusters)

- First split is corresponding to variable one and split point set to 4.01
- Second split is corresponding to variable one (with the same first variable constrained by condition >= 4.01), the split point set to 6.99 and silhouette index equal to 0.336.

The ARI index between RPCT algorithm results and actual clusters of this dataset is equal to 0.499, which again cannot be recognized as satisfactory.

5 Conclusions and Remarks

In the paper, the usage of silhouette index for finding number of clusters and for assessing the final clustering with clustering tree RPCT algorithm has been evaluated.

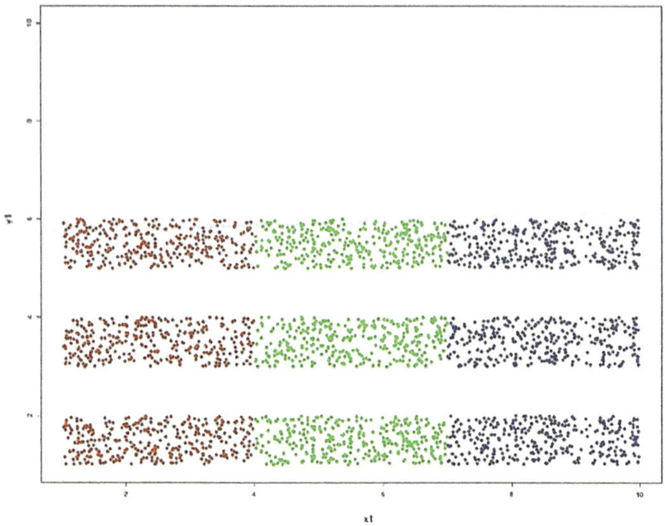

Fig. 15 Third evaluation dataset—clustering tree algorithm results. *Source* Own's calculations

The silhouette index and clustering methods based on maximizing the silhouette index such as clustering tree RPCT algorithm behave very well if there are Gaussian—based on a multidimensional normal distribution—type clusters in the analyzed dataset.

The silhouette index is rather inappropriate not only for atypical cluster shapes but even for clusters with elongated or inclined shapes.

Clustering methods based on maximizing the silhouette index such as clustering tree RPCT algorithm can give misleading results especially for non-typical clusters. All such cases are related with relatively small index value, so maybe such methods should be secured by adding a minimal value of index, below which the partitioning cannot be recognized as reliable/stable.

The index should rather be used to determine the optimal number of clusters than to assess the final clustering (especially when index value is relatively small).

Author is aware that all presented examples are basing on visual determination of actual cluster structure and latest tendencies in the literature of the subjects points that this is not necessarily the only point of view and the proper evaluation procedure should start from determining the "formal categorization principles" of clustering which may be (Henning 2015):

1. within-cluster dissimilarities should be small,
2. between-cluster dissimilarities should be large,
3. clusters should be fitted well by certain homogeneous probability models such as the Gaussian or a uniform distribution on a convex set, or by linear, time series or spatial process models,
4. members of a cluster should be well-represented by its centroid,

5. the dissimilarity matrix of the data should be well-represented by the clustering (i.e., by the ultrametric induced by a dendrogram, or by defining a binary metric "in same cluster/in different clusters"),
6. clusters should be stable,
7. clusters should correspond to connected areas in data space with high density.
8. the areas in data space corresponding to clusters should have certain characteristics (such as being convex or linear),
9. it should be possible to characterize the clusters using a small number of variables,
10. clusters should correspond well to an externally given partition or values of one or more variables that were not used for computing the clustering,
11. features should be approximately independent within clusters,
12. all clusters should have roughly the same size,
13. the number of clusters should be low.

but the Silhouette index and RCPT like methods itself are basing on concept that taken into account the visualization (longest distance in each dimension) and the minimal criterion that should be satisfied by those methods is to properly cluster the dataset for whom the structure is clearly visible.

The full analysis of the performance of each principle for datasets that have no visual structure requires further analysis.

Appendix 1: Source Code of Procedure of Finding Number of Cluster with Silhouette Index in R Language

```
library(clusterSim)

model<-cluster.Gen(model=10)
nrOfClusters<-2:10
silhouetteValues<-rep(NA,10)
d<-dist(model$data)
for(clusterNr in nrOfClusters){
  jpeg(paste("pam_nr_clusters",clusterNr,"jpeg",sep="."))
  cl<-pam(d,clusterNr)$clustering
  s<-index.S(d,cl)
  plot(model$data,col=grey.colors(10)[cl],pch=16,main=paste("For
clusters number =",clusterNr," the average silhouette index is
=",s))
  silhouetteValues[clusterNr]<-s
  graphics.off()
}
print(paste("the optimal cluster number by silhouette criterion is
",which.max(silhouetteValues)))
```

Appendix 2: Source Code of Clustering Tree (RPCT) Algorithm Implementation in R Language

```
calculateIndexForSplitPoint<-function(data,dist,clustering,nn,mm){
  newClusterNr<-max(clustering)+1
  for(i in 1:length(clustering)){
    if(clustering[i]==clustering[nn] && data[i,mm]>data[nn,mm]){
      clustering[i]<-newClusterNr
    }
  }
  return
(list(index=index.S(dist,clustering),clustering=clustering))
}

calculateMaxIndexSplit<-
function(data,dist,actualClustering,actualIndex){
  toReturn=list(index=actualIndex,clustering=actualClustering)
  m<-ncol(data)
  n<-nrow(data)
  for(i in 1:n){
    for(j in 1:m){
indexForSplit=calculateIndexForSplitPoint(data,dist,actualClusterin
g,i,j)
      if(  is.na(toReturn$index) || (!is.nan(indexForSplit$index)
&& indexForSplit$index>toReturn$index)){
        indexForSplit$variable<-j
        indexForSplit$splitValue<-data[i,j]
        toReturn=indexForSplit
      }
    }
  }
  return (toReturn)
}
```

```
cTree<-function(data,dist,maxClusterNr=10){
  splitLevel<-1;
  toReturn<-list(index=NA,clustering=rep(1,nrow(data)))
  #repeat process until no gain is achived or maximal number of
clusters appeared
  splits<-list()
  repeat{
    maxSplit<-
calculateMaxIndexSplit(data,dist,toReturn$clustering,toReturn$index
)
    if(!is.na(toReturn$index) && (maxSplit$index<=toReturn$index))
return (append(toReturn,list(splits=splits)))
    toReturn<-maxSplit
    splits[[splitLevel]]<-
list(variable=maxSplit$variable,splitValue=maxSplit$splitValue,spli
tIndex=maxSplit$index)
    splitLevel<-splitLevel+1;
    if(!is.na(toReturn$index) && splitLevel>=maxClusterNr) return
(append(toReturn,list(splits=splits)))
  }
}
```

References

Arbelaitz O, Gurrutxaga I, Muguerza J, PéRez JM, Perona I (2013) An extensive comparative study of cluster validity indices. Pattern Recogn 46(1):243–256

Henning Ch (2015) What are the true clusters? Pattern Recogn Lett 64:53–62

Hubert LJ, Arabie P (1985) Comparing partitions. J Classif 2:193–218

Kang JH, Park CH, Kim SB (2016) Recursive partitioning clustering tree algorithm. Pattern Anal Appl 19(2):355–367

Kaufman L, Rousseeuw PJ (1990) Finding groups in data: an introduction to cluster analysis. Wiley, New York

Migdał-Najman K, Najman K (2006) Wykorzystanie indeksu silhouette do ustalania optymalnej liczby skupień. Wiadomości Statystyczne 6:1–10

Rousseeuw PJ (1987) Silhouettes: a graphical aid to the interpretation and validation of cluster analysis. J Comput Appl Math 20:53–65

Starczewski A, Krzyżak A (2015) Performance evaluation of the Silhouette index. International conference on artificial intelligence and soft computing. Springer, Cham, pp 49–58

Walesiak M, Gatnar E (eds) (2009) Statystyczna analiza danych z wykorzystaniem programu R. PWN, Warszawa

Walesiak M, Dudek A (2019) clusterSim: searching for optimal clustering procedure for a data set. R package version 0.48-5

The Role of Discretization of Continuous Variables in Socioeconomic Classification Models on the Example of Logistic Regression Models and Artificial Neural Networks

Wioletta Grzenda

Abstract Logistic regression models and artificial neural networks require the use of appropriate quality data. One of the methods of improving the quality of raw data is the discretization of continuous variables. It can be a way to deal with outliers and influential observations and can be helpful when the assumptions of some of the models are not met. This paper shows that despite the fact that the discretization of continuous variables means that reduced information is used for the modeling, it can improve the classification accuracy of machine learning models. This is particularly important when searching for the best predictive model when a limited set of explanatory variables is available, as well as when analyzing large data sets. In addition, by selecting the methods used to discretize continuous variables we decide about the number and type of variables that are included in the model and, as a result, are subject to interpretation. The selection of cut-off points matching the purpose of the research can be made using supervised discretization methods. In this study, the data from the Generations and Gender Survey (GGS) for Poland was used. The status of respondents on the labor market was considered. For the considered data, the advantages of using supervised discretization of continuous variables based on the entropy criterion and the Gini criterion were pointed out. Importantly, discretization based on these methods provided predictive models of better classification accuracy than the models based on discretization procedure frequently applied in socioeconomic studies.

Keywords Discretization · Classification models · Logistic regression model · Artificial neural network

W. Grzenda (✉)
SGH Warsaw School of Economics, Collegium of Economic Analysis, Warsaw, Poland
e-mail: wgrzend@sgh.waw.pl

K. Jajuga et al. (eds.), *Classification and Data Analysis*, Studies in Classification, Data Analysis, and Knowledge Organization,
https://doi.org/10.1007/978-3-030-52348-0_3

1 Introduction

Understanding data and preparing it for modeling is a very labor-intensive and multifaceted data mining analysis stage that has a large impact on the quality of the models considered, including their predictive power. In this paper, we investigate logistic regression models and artificial neural networks. While the comparison of these two categories of models is presented *inter alia* in Dreiseitl and Ohno-Machado (2002), we focus on the impact of discretization of training data on the performance of these models.

Logistic regression models and artificial neural networks require data matching specified criteria. Empirical data for these models must not contain missing data; moreover, in the case of logistic regression, it is also required to meet the assumptions regarding the form of this model. The quality of logistic regression models and artificial neural networks may also significantly depend on the observations present in training data and variables included in them. In particular, the quality of these models can be affected by the method of introducing discrete and continuous variables into the model. Therefore, the transformation of individual variables is one of the key elements of data preprocessing. It also affects the computational cost of the algorithms. When properly applied, data transformation can simplify the process of learning and may improve the generalizability of the learned results (Kim and Lee 2004). One method of transforming variables is their discretization. Discretization of a continuous random variable as shown in this research can improve the quality of the predictive model and in some application areas, as pointed out in Drezner and Zerom (2016), are simply necessary. Moreover, there are some classifiers, for example, based on Bayesian methods, that can only deal with discrete variables (Flores et al. 2011).

An important aspect of the development of classification models, which is related to the transformation of variables, is the analysis of the impact of influential and outlier values present in the data on the models developed with these data. In socioeconomic research, a frequent example of a variable containing such observations is the variable describing the income of the respondent. The inclusion of influential observations in a predictive model may make the parameters of the model change significantly, while the occurrence of outliers may affect the value of sum of squared errors. It is worth emphasizing that outliers may or may not be influential (Larose 2006). Discretization of variables can be one of the ways of dealing with such observations.

In the case of logistic regression model, it is also required to meet the assumption of linearity of exogenous variables and log odds. In this model, it is assumed that the estimated odds ratio is constant for all possible values of the continuous random variable. This assumption according to Larose (2006) is not always met by the data, and naive use of the model may lead to erroneous conclusions, for example, to the conclusion that the variable under consideration is not statistically significant. As a solution to this problem, Larose (2006) proposes to categorize the continuous exogenous variable using the expert method. According to Stanisz (2016), there is a discussion in the literature about the inclusion of continuous variables in the logistic regression model, because apart from verifying the assumptions of the model, some researchers believe that including such variables makes it easier to understand the model. On the other hand, the categorization of a continuous variable and estimating for a given feature not one but several parameters is the added value of the model. According to other researchers, the assessment of only selected odds ratios is a loss of information. In addition, increasing the number of degrees of freedom causes the loss of power in assessing the significance of the estimated parameters.

In the case of classification, one of the related problems is also modeling imbalanced data sets, i.e., the data sets, which contain more samples from one class than from the remaining classes. Zhang et al. (2019) showed, using the example of a logistic regression model, that in the case of such data sets, the solution may be also to discretize continuous variables. They investigate four methods of discretizing variables such as distance, quantile, Gini, optimal binning and indicate that proper variable discretization can reduce the model bias.

As shown, discretization of continuous variables can be a solution to many problems in the construction of predictive models. Many different methods of discretizing continuous variables can be found in the literature (Cichosz 2000, Zhang et al. 2019). However, there are no clear conclusions as to which of these methods may result in the models with the best classification accuracy. In socioeconomic studies, the most common methods include discretization of continuous variables based on quartiles, equality of intervals, or expert knowledge. The identification of cut-off points by graphical methods, for example, based on a histogram, is also popular. However, when choosing a method of discretizing continuous variables in supervised learning models, one can also consider relationships between exogenous and endogenous variables. This study analyses the impact of unsupervised and supervised discretization methods applied to continuous variables on the classification accuracy in comparison with the impact of discretization typically applied in socioeconomic studies on this accuracy.

Each method of discretizing continuous variables requires determining cut-off points, which depend on the considered data and there are no general guidelines for the identification of these points (Kim and Lee 2004). The primary objective of this study is to show how the choice of different cut-off points of continuous variables affects the classification accuracy of the considered models. In particular, experiments performed in this study show that the use of discretization of continuous variables most commonly encountered in socioeconomic sciences does not always yield the best results. We propose to determine the cut-off points using a decision tree based on the entropy criterion and Gini index, and we compare this method to the use of quartiles and expert methods.

Moreover, in this paper attention is paid to the distribution of a target variable in each of the considered ranges depending on the applied division method, obtained ordinal variables and their interpretation.

2 Discretization of Continuous Variables

There are two main groups of discretization methods used for continuous variables: unsupervised methods also called uncontrolled and supervised methods (with/under supervision) also referred to as controlled methods (Witten et al. 2016).

In the case of supervised methods, the information about which class a given observation belongs to is used, while in the case of unsupervised methods this information is omitted, and the division is typically made into a predefined number of intervals regardless of the entire data set. The unsupervised methods in particular the division into intervals of equal width or frequency of occurrences they provide are not recommended for classification tasks (Witten et al. 2016). They cause the elimination of differences that could be used in machine learning, because modeled events can be distributed very unevenly in individual intervals. Therefore, when setting discretization intervals for a continuous explanatory variable, it is important to consider the distribution of a target variable. Otherwise, the feature for which the considered attributes are used will not be taken into account when selecting cut-off points.

One of the first and still used methods of discretization of continuous variables under supervision is the ChiMerge algorithm (Dougherty et al. 1995). This method is based on the use of chi-square statistics to select two intervals to connect that are most similar in terms of frequency distribution of classes for the observations contained in these intervals. Starting from the minimum ranges corresponding to the range of values of the attribute under consideration, the intervals are combined until the specified stop condition is reached. However, the most common technique in the literature is the entropy-based method called the method of class entropy discretization (Chmielewski and Grzymala-Busse 1996). Entropy is a measure of homogeneity, which takes a value of 0 when all observations in a group belong to the same class and a maximum value when the distribution of class membership in a given group is uniform. For the distribution of classes described by the vector $(p_1, p_2, \ldots, p_r)$, it is given by the formula:

$$-\sum_{i=1}^{r} p_i \log_2 p_i. \tag{1}$$

In this paper, the selection of cut-off points also took into account the Gini index, which is given by the following formula:

$$1 - \sum_{j=1}^{r} p_j^2 = 2 \sum_{j<k} p_j p_k. \tag{2}$$

In the case of the maximum diversity of classes in the group, this indicator takes the value of 0.5. In this paper, these measures were used to study the homogeneity of classes in groups into which observations were divided based on the cut-off points determined using decision tree models.

In Cichosz (2000), a division into local and global as well as naive (primitive) and advanced methods has also been made. The first division criterion differentiates discretization methods based on whether discretization of continuous features is carried out equally in the entire domain or in different ways depending on the subsets of this domain. In local methods, the division is made on the basis of a limited information. Moreover, in different application domains, feature discretization can be performed using different attribute values. Most often, it follows the principles of supervised discretization and is carried out in a manner strongly dependent on the machine learning algorithm used. Global methods relate to the discretization of a continuous feature independently of other features. The ranges obtained are global in the sense that they are the same for all observations. A more general definition of these terms can be found in Chmielewski and Grzymala-Busse (1996). Namely, discretization of only one variable is called local discretization in this work, and discretization concerning all continuous attributes at the same time is referred to as global discretization. These methods include both unsupervised and supervised approaches.

Naive methods according to Cichosz (2000) are the methods in which the distribution of attribute values and classes is not taken into account during discretization. In this type of method, an arbitrary number of intervals of equal width or frequency of occurrence is usually determined. Due to their simplicity, these methods are very often used in research, because they do not require extensive calculations. In the case of advanced methods, the division is made on the basis of information contained in the entire data set. In addition, appropriate measures are used to assess the quality of the division.

Some studies, e.g., Jankowski et al. (2014), distinguish also static and dynamic discretization. Static methods do not take into account the relationship between individual features, while dynamic methods consider cut-off points for many features at the same time. There are also other classifications of discretization methods for continuous variables, e.g., hierarchical and non-hierarchical. The question then arises, which method to choose and whether the discretization of continuous variables is always needed.

Discretization of continuous variables for some methods and data mining models is necessary due to the machine learning algorithms used, for example in the case of a naive Bayesian classifier, or in association analysis. In addition, it is necessary whenever calculations require the use of contingency tables (Tuv et al. 2009). Then the use of these methods is possible in machine learning tasks in which most such attributes occur (Cichosz 2000). In addition, it is used when better prediction results are obtained or if the algorithms used are faster (Witten et al. 2016), which results from replacing many values of a continuous feature with several values. In particular, decision tree models do not require discretization of continuous variables, but they work much slower on continuous attributes. At the same time, researchers use decision tree algorithms to discretize continuous variables, because they believe that the discrete form of variables facilitates the interpretation of modeling results.

Among the advantages of the discretization of continuous variables, Cichosz (2000) also indicates the increase in simplicity and readability of the hypotheses studied and the improvement in their accuracy. Hypotheses based on continuous attributes can be complex and illegible, while discretization of continuous features can yield their simpler versions.

In the case of artificial neural networks, both continuous and discrete variables can be included in the model, and the validity of the selected approach can be measured by the quality of the ultimate prediction. In logistic regression, the same approach can be used, but other assumptions have to be met by these models. A summary of the advantages and disadvantages of discretizing continuous variables in the logistic regression model can be found, among others, in Stanisz (2016). According to Stanisz (2016), the main disadvantages of the categorization of continuous variables introduced into the logistic regression model include the lack of clear guidelines as to the choice of cut-off points, the assumption that the relationship between the exogenous variable and the endogenous variable is fixed and problems with verifying this assumption, increasing the number of degrees of freedom by introducing several indicator variables, and thus the power loss when assessing the significance of parameter estimates, and considering only selected odds ratios, which is equivalent to the fact that only selected groups are compared with the reference group.

The treatment of discretized variables is also important—labels are commonly assigned to them and they are treated as nominal variables. Considering the fact that they result from the transformation of continuous variables, where it is significant, they should be declared as ordinal variables (Witten et al. 2016), because the ranges obtained are ordered. This is particularly important for decision tree models due to the number of possible divisions.

The choice of division points for the discretization of continuous variables remains an open issue. According to Provost and Fawcett (2013), it is necessary to verify all reasonable division points and choose the ones that provide the most information to the model. In Liu et al. (2002), three criteria for assessing the quality of discretization were distinguished. These include the number of ranges, the number of inconsistencies, and the accuracy of prediction. For the first criterion, the principle of minimizing the number of divisions applies. Inconsistency applies when observations of a certain pair are not distinguishable and have different decisions (classes) assigned to them. Discretization is not expected to increase the number of such cases.

In predictive modeling, the assessment of the stability of parameter assessments, and consequently, the resulting model are an important element of its evaluation. For this purpose, various measures are used to assess the quality of prediction of the classification model on the training set and validation (and/or testing set). Based on the values of these measures, a decision is made to use the model for future observations (Witten et al. 2016). Many measures and methods are used to assess the accuracy of the predictions, and the main ones are the proportion of incorrect classifications, the ROC curve for classification and the mean square error for regression tasks, respectively. In this paper, these criteria were used to assess the quality of the model after and before discretization performed with individual methods. At the same time, the impact of taking into account the dependence between the discretized explanatory variable and the target variable when choosing cut-off points on prediction accuracy was analyzed.

3 Empirical Examples

To perform tests described in this study, the data from the Generations and Gender Survey panel survey (GGS) for Poland conducted as part of the Generations and Gender Program (GGP) was used. The data comes from the second half of 2014. In addition, missing data that has not changed over time was imputed using the previous round of research carried out in 2010–2011. The survey was conducted on a random sample of respondents aged 18–79. From the entire data set, persons who reported to belong to one of the following groups were selected:

- employed in the public sector,
- employed in the private sector,
- self-employed out of agriculture sector,
- unemployed.

In this way, a sample of 5741 observations was obtained and used for modeling. The target variable *work* was defined as a binary variable taking the value of 1, if the person declared that he was working and 0 otherwise. The set of qualitative explanatory variables taken into account in this study is given in Table 1. In addition, two continuous features, i.e., *age* and *income,* were considered. The former one describes the age of the respondents at the time of the survey (min = 18, max = 79, average = 41.44), whereas the latter one denotes the average monthly total net income in PLN (including overtime income) in the last 12 months (min = 60, max = 24,000, average = 2184.14). The variable describing income had 52.71% of missing values, as both employed and unemployed are considered in the study. Therefore, a binary variable in the model, which takes the value 1 in the absence of this information and 0 otherwise was also included in modeling. Other variables contained no missing data except for a few incorrectly entered values—these observations were removed before the sample of 5741 observations was established.

In the next stage, discretization of continuous variables was performed in each experiment using one of the following methods:

1. into three intervals using a decision tree based on the entropy criterion,
2. into four intervals using a decision tree based on the entropy criterion,
3. into four intervals using the decision tree based on the Gini index,
4. into four intervals using quartiles,
5. into four intervals using the expert method.

In this research, the division of continuous variables into equal ranges was not considered, because it would cause large differences in the number of observations contained in each of them. For example, the oldest age group obtained with such a division would contain less than 2% of observations. A similar situation applies to the variable describing income.

To determine the cut-off points in this paper, we propose to use a decision tree. The three degenerated decision tree models were developed with only one feature included in each of them. Multiple divisions needed to perform discretization were determined by building a tree including up to three or four branches, depending on the number of desired cut-off points. Moreover, either the Gini index or entropy criterion has been chosen to guide the process of partitioning the input space when constructing the decision trees.

In the expert method, in the case of the *age* variable, the cut-off points were selected on the basis of a frequent practice in demography studies in Poland of mapping raw values to one of three discrete values based on cut-off points listed in the Expert column of Table 2, while for the *income* variable the first cut-off point was selected based on the minimum salary, and the second one was at the level of the average income based on the CSO data for the considered period. Table 2 presents the obtained cut-off points for each of the variables for the discretization methods considered. It is worth emphasizing that in the case of the *age* variable, the cut-off points given by methods (2)–(4) typically differ significantly from the ones commonly found in the literature. This may mean that young people are now entering the labor market earlier or more frequently combine work and study. Tables 3 and 4 show the distribution of the target variable in each of the distinguished intervals

Table 1 Target variable and qualitative explanatory variables

Variable	Description	Levels	Proportion (%)
Work	Labor market status work	0 = unemployment	15.66
		1 = employment	84.34
Sex	Respondent's gender	1 = woman	47.03
		2 = man	52.97
Marital status	Marital status of the respondent	1 = unmarried, separated or divorced, a widower, a widow	37.59
		2 = married	62.41
Child	Having a child	0 = no	72.64
		1 = yes	27.36
Education status	The level of education of the respondent at the time of the survey	1 = higher	23.08
		2 = undergraduate or engineering and post-secondary	13.57
		3 = secondary professional	20.82
		4 = secondary general	9.14

(continued)

Table 1 (continued)

Variable	Description	Levels	Proportion (%)
		5 = basic vocational	25.62
		6 = primary school, incomplete basic and without education	7.77
Financial situation	Financial situation of the respondent's household	1 = poor or no response	24.77
		2 = rather poor	25.69
		3 = rather good	34.09
		4 = good and very good	15.45
Health	General assessment of the respondent's health	1 = very good	22.38
		2 = rather good	52.36
		3 = rather poor	21.62
		4 = poor or very poor	3.64
Health problems	Does the respondent have long-term health problems or chronic diseases	0 = no	77.44
		1 = yes	22.56
Living with birth parents	Flat with parent/biological parents	0 = no	78.09
		1 = yes	21.91
Satisfaction	Satisfaction with the respondent's current situation on the labor market	1 = poor	12.65
		2 = average	1.18
		3 = good or very good	0.26
		4 = no response	85.91
Region	Region of residence	1 = central (łódzkie, mazowieckie)	16.88
		2 = south (małopolskie, śląskie)	17.63
		3 = east (lubelskie, podkarpackie, świętokrzyskie, podlaskie)	20.29
		4 = northwest (wielkopolskie, zachodniopomorskie, lubuskie)	16.46
		5 = southwest (dolnośląskie, opolskie)	13.01
		6 = north (kujawsko-pomorskie, warmińsko-mazurskie, pomorskie)	15.73

Table 2 Cut-off points depending on the discretization method selected

Cut-off point	Entropy (1)	Entropy (2)	Gini (3)	Quartile (4)	Expert (5)
Age variable					
1	25	20	20	32	25
2	50	30	25	41	44
3	–	47	47	52	60
Income variable					
1	933	695	605	1350	1630
2	1190	933	845	1970	4000
3	–	1190	1090	2500	8000

Table 3 Distribution of the target variable in each of the ranges depending on the division method for the *age* variable (in %)

Range number	The entropy criterion (2)		Expert method (5)	
	0 (unemployed)	1 (working)	0 (unemployed)	1 (working)
1	45.24	54.76	31.93	68.07
2	20.41	79.59	11.40	88.60
3	10.30	89.70	16.24	83.76
4	17.06	82.94	14.06	85.94

Table 4 Distribution of the target variable in each of the ranges depending on the chosen method of division for the *income* variable (in %)

Range number	The entropy criterion (2)		Quartile method (5)	
	0 (unemployed)	1 (working)	0 (unemployed)	1 (working)
1	69.48	30.52	25.79	74.21
2	42.31	57.69	2.61	97.39
3	12.90	87.10	1.45	98.55
4	1.82	98.18	1.03	98.97

depending on the method used for the discretization of continuous variables on the example of entropy and expert methods, respectively.

Before modeling started, the data set was randomly divided into the training set (70%) and the test set (30%). In the next stage, logistic regression models were estimated and artificial neural networks were trained on the training data set prepared in this way. In the case of artificial neural networks, a standard multilayer perceptron network with one hidden layer was used. Levenberg–Marquardt gradient algorithm was used to train the network (Masters 1995). First, the models for which the variable *age* was included in the original version, and then the models developed with the data transformed with each of the five discretization methods were considered. The *income* variable due to the occurrence of missing data was imputed using a decision tree. The values of selected measures used to assess the quality of predictions of these models are presented in Tables 5, 6, 7 and 8. The index of the discretization method was added; after the model name as presented in Table 2, the lack of the number means taking into account the continuous variable without discretization.

Table 5 Values of selected measures and criteria used to assess the quality of the models with *age* variable before and after discretization; the case of logistic regression models

Model	Misclassification rate		Mean square error		Area under ROC curve	
	Training	Testing	Training	Testing	Training	Testing
Regression	0.0184	0.0255	0.0149	0.0189	0.995	0.991
Regression1	0.0206	0.0266	0.0155	0.0187	0.993	**0.992**
Regression2	0.0196	**0.0185**	0.0150	**0.0170**	0.992	0.991
Regression3	0.0156	0.0197	0.0134	**0.0170**	0.994	**0.992**
Regression4	0.0212	0.0249	0.0161	0.0194	0.994	0.991
Regression5	0.0221	0.0272	0.0215	0.0265	0.940	0.925

The value for the best model in terms of a given performance measure is shown in bold

Table 6 Values of selected measures and criteria used to assess the quality of models with variable *age* before and after discretization; the case of artificial neural network models

Model	Misclassification rate		Mean square error		Area under ROC curve	
	Training	Testing	Training	Testing	Training	Testing
Networks	0.0139	0.0243	0.0103	0.0208	0.998	0.990
Networks1	0.0181	0.0261	0.0127	**0.0193**	0.997	0.991
Networks2	0.0149	**0.0237**	0.0110	0.0197	0.998	0.987
Networks3	0.0136	0.0243	0.0110	0.0198	0.997	0.989
Networks4	0.0174	0.0272	0.0118	0.0234	0.998	0.988
Networks5	0.0171	0.0272	0.0118	0.0212	0.997	**0.992**

The value for the best model in terms of a given performance measure is shown in bold

Table 7 Values of selected measures and criteria of assessing the quality of models with *income* variable before and after discretization; the case of logistic regression models

Model	Misclassifications rate		Mean square error		Area under ROC curve	
	Training	Testing	Training	Testing	Training	Testing
Regression	0.0219	0.0272	0.0178	0.0215	0.992	0.991
Regression1	0.0219	**0.0261**	0.0164	0.0205	0.994	0.991
Regression2	0.0219	0.0272	0.0178	0.0215	0.992	0.991
Regression3	0.0219	0.0272	0.0178	0.0215	0.992	0.991
Regression4	0.0206	0.0266	0.0159	**0.0200**	0.994	**0.992**
Regression5	0.0206	0.0266	0.0161	0.0201	0.994	0.990

The value for the best model in terms of a given performance measure is shown in bold

Table 8 Values of selected measures and criteria of assessing the quality of models with *income* variable before and after discretization; the case of artificial neural network models

Model	Misclassifications rate		Mean square error		Area under ROC curve	
	Training	Testing	Training	Testing	Training	Testing
Networks	0.0156	0.0313	0.0122	0.0250	0.997	0.990
Networks1	0.0154	0.0249	0.0128	**0.0196**	0.996	**0.992**
Networks2	0.0154	0.0266	0.0126	0.0205	0.996	0.989
Networks3	0.0159	**0.0243**	0.0131	0.0202	0.996	0.991
Networks4	0.0119	0.0266	0.0100	0.0232	0.997	0.991
Networks5	0.0144	0.0278	0.0121	0.0221	0.997	**0.992**

The value for the best model in terms of a given performance measure is shown in bold

For example, Regression1 is a logistic regression model that includes a continuous variable after division into three intervals using a decision tree based on the entropy criterion.

In the first stage of the study, *age* variable was considered. The classification accuracy of the considered models was investigated on the test set. In addition, the values of the measures under consideration are also given on the training set, which can be used to assess the stability of constructed models. Based on the results presented in Table 5, it can be concluded that in the case of the most popular model used in socioeconomic sciences—logistic regression—the best results in terms of predictive power were obtained for supervised discretization methods. The best accuracy of the prediction was obtained for the model in which the *age* variable was taken into account by dividing it earlier into four intervals using a decision tree based on the entropy criterion. On the other hand, definitely the lowest predictive power was obtained when this variable was included in the model after its division into four intervals using the expert method based on a review of the literature. At the same time, in the case of artificial neural networks (Table 6), after using the expert method of division for this variable, high accuracy predictions were obtained, as shown by the area under the ROC curve (AUC). Nevertheless, they are very similar to the results obtained when using this variable after previously dividing it into three intervals using a decision tree based on the entropy criterion. In addition, including this variable in the artificial neural network model after performing division using the entropy criterion resulted in the lowest mean square error values. The lowest proportion of incorrect predictions was obtained for the neural network model in which the continuous variable was divided into four intervals using a decision tree based on the entropy criterion.

The second of the continuous variables considered in this study is the *income* variable. This variable includes significant proportion of missing data, so the results obtained are affected by the imputation method. The logistic regression model for which the lowest mean square error value was obtained turned out to be the model in which the *income* variable after division into four intervals using quartiles was included (Table 7). For the ROC curve, very similar results were obtained for each of the considered models. Moreover, for the criterion of the proportion of incorrect classifications, the logistic regression model, in which the *income* variable was taken into account after dividing its values into three ranges using a decision tree based on the entropy criterion, proved to be the best model in predictive terms. In the case of artificial neural networks, the best model quality in terms of mean square error was obtained for the model in which the *income* variable was included as a discrete variable using the division into three intervals determined with a decision tree based on the entropy criterion (Table 8). However, the lowest proportion of incorrect classifications was attained by the model in which the division of the *income* variable into four intervals using decision tree based on the Gini index was taken into account (Table 8).

However, the values of the considered measures and criteria of assessing the quality of models do not differ significantly. The most important conclusion resulting from this study is that for the data set under consideration and selected continuous variables, the best value was not obtained for any of the considered performance measures while including a continuous variable without prior discretization. In all the cases, the best predictive power was observed when model development was preceded with the discretization of continuous variables.

The main purpose of this paper was to assess the impact of the discretization of continuous variables on the classification accuracy; however, the logistic regression model considered in the study is also a common tool to describe socioeconomic phenomena. Therefore, this model was estimated and interpreted taking into account the *age* and *income* variables after discretization into four ranges based on the entropy criterion. The results for logistic regression model have been shown in Table 9. We presented the odds ratio estimates only for these variables for which parameter estimation was statistically significant for at least one level.

It was *inter alia* received, that people up to and including 20 years old had 99.1% less chance of having a job than people over 47 years old. People aged between 20 and 30 had a 89% lower chance compared to those in the oldest age group. People aged between 30 and 47 have 11.5% higher chance to have a job compared to those over 47. Respondents who declared that their average monthly income was less than

Table 9 Odds ratio estimates for logistic regression model

Effect (reference level)	Level	Odds ratio
Age (over 47 years old)	From 18 to 20 years old From 21 to 30 years old From 31 to 47 years old	0.009 0.110 1.115
Income (over 1190 PLN)	To 695 From 695 PLN to 933 PLN From 933 PLN to 1190 PLN	0.079 0.401 0.454
Sex (man)	Woman	1.054
Marital status (married)	Unmarried, separated or divorced, a widower, a widow	0.801
Child (yes)	No	0.976
Education status (primary school, incomplete basic and without education)	Higher Undergraduate or engineering and post-secondary	4.041 1.617
	Secondary professional Secondary general Basic vocational	0.895 0.831 1.517
Financial situation (good and very good)	Poor or no response Rather poor Rather good	0.386 0.557 1.009
Health (poor or very poor)	Very good Rather good Rather poor	0.598 0.772 0.870
Health problems (yes)	No	1.462
Living with birth parents (yes)	No	1.799
Region (north)	Central South East Northwest Southwest	0.416 0.373 0.343 1.098 0.422

or equal to PLN 695 had 92% less chance of having a job than people who declared income above PLN 1190. On the other hand, people with income from PLN 695 to PLN 933 and from PLN 933 to PLN 1190 had about 69% less chance of having a job compared to people who declared income above PLN 1190. The interpretation of the results for the variable describing the average monthly total net income when modeling the declared status on the labor market should be made cautiously, as the cause-and-effect relationship can be bidirectional here. However, taking into account the primary purpose of this work—the classification accuracy—it was decided to consider this variable in the model.

People with at least tertiary education had more than four times higher chances of having a job compared to people with the lowest education status. People with undergraduate or engineering education had a 62% greater chance compared to respondents with the basic vocational education. Persons with secondary education, both vocational and general, had about 15% less chance of having a job than persons with primary or lower secondary education. There were no major differences in the opportunities for having a job based on gender. Single people were less likely to have a job compared to married people. Similarly, people with children had a slightly higher chance of having a job compared to people without children. When it comes to the impact of health on having a job, people with health problems were less likely to have a job compared to people who did not declare such problems. People who do not live with biological parents were 80% more likely to have a job compared to people who lived with their parents. In the case of young people, it may be related to problems with becoming independent, another reason may be taking care of parents or financial support from parents. For the variable describing the respondent's region of residence, two levels turned out to be statistically insignificant, and the results obtained for the other levels of this variable are similar to each other. It can be said that people living in regions other than the northern ones more often declared that they were unemployed.

Last but not least, decision tree model was used to interpret the results obtained using artificial neural networks. It was obtained that the most important features differentiating the respondents in terms of their situation on the labor market, were the features describing the average monthly total net income, and to a lower extent age and education.

4 Summary and Conclusions

This study shows that the methods of discretization of continuous variables most frequently used in socioeconomic sciences based on quartiles or expert knowledge often do not result in the best solutions. The choice of discretization method for continuous variables should depend on the type of research problem being considered and the type and amount of data available. At the same time, at modeling stage, when choosing a method for discretizing continuous variables, it is very important to take into account the relationship between exogenous and endogenous variables. For the considered data and the target binary variable, models with the best predictive power were most often obtained using controlled methods of discretization of continuous variables based on the entropy criterion and the Gini criterion. However, the worst predictive power was obtained when a continuous variable was included in the model without its prior discretization. The differences in predictive power, though not large, show that discretization, when properly applied, may even improve both the predictive power and interpretability of a model.

The issues presented in this paper are considered mainly in the literature on machine learning, where the main goal is to obtain the best accuracy of prediction. However, regardless of the field of application, the topic is important, because the method of discretization of continuous variables has an impact on not only the classification accuracy, but above all on what parameters and how many of them are estimated in the model. Discretization of continuous exogenous variables, taking into account the values taken by the endogenous variable and replacing initially continuous variables with discretized variables may result in more accurate conclusions than the use of naive methods (Witten et al. 2016). This is the case, for example, with the continuous variable describing the income of respondents and the potential consideration of its change by PLN 1.

In this paper, a logistic regression model was considered, which is one of the most commonly used models in socioeconomic research. Still, very rarely how discrete values for originally continuous exogenous variables were adopted is described. In this paper, artificial neural networks have also been considered, which are increasingly used in this field of science. Based on the results presented in this paper and based on other studies (Jankowski et al. 2014), it can be concluded that there is no one best method of discretization of continuous variables, just as there is no one best method of data imputation. The selection of the best method is always closely related to the empirical data being considered, and not just the model being considered. In this paper, in contrast to the results presented in Rosenfeld et al. (2018), it was shown that for logistic regression, the worst predictive quality was obtained when a continuous variable was included in the model without its prior discretization.

In many questionnaires, the categorization of continuous features is often based on a literature review. This has its advantages, because it makes it possible to obtain comparable results. The best solution would be to include continuous features in the questionnaires, but in some cases discretization of continuous features is the only way to get answers to the so-called sensitive questions. However, taking into account demographic changes currently taking place in Europe, for example, defining cut-off points for continuous features based on the analysis of data obtained in pilot studies is worth considering.

The main purpose of the paper was to indicate the possibilities offered by controlled methods of discretization of continuous variables in socioeconomic researches, but let us emphasize that the results obtained and the benefits of initial data processing are strictly dependent on the considered data set. One more aspect of future works is how to combine continuous variables with their discretized versions. The authors of one of the recent studies (Rosenfeld et al. 2018) suggest that discrete features should not replace continuous ones, but should also be included in a model.

References

Chmielewski MR, Grzymala-Busse JW (1996) Global discretization of continuous attributes as preprocessing for machine learning. Int J Approx Reason 15(4):319–331

Cichosz P (2000) Systemy uczące się. Wydawnictwo Naukowo-Techniczne, Warszawa

Dougherty J, Kohavi R, Sahami M (1995) Supervised and unsupervised discretization of continuous features. In: Prieditis A, Russell S (eds) Machine learning proceedings 1995. Morgan Kaufmann, San Francisco, pp 194–202

Dreiseitl S, Ohno-Machado L (2002) Logistic regression and artificial neural network classification models: a methodology review. J Biomed Inform 35(5–6):352–359

Drezner Z, Zerom D (2016) A simple and effective discretization of a continuous random variable. Commun Stat Simul Comput 45(10):3798–3810

Flores MJ, Gámez JA, Martínez AM, Puerta JM (2011) Handling numeric attributes when comparing Bayesian network classifiers: does the discretization method matter? Appl Intell 34(3):372–385

Generations and Gender Programme (GGP). http://www.ggp-i.org/. Accessed 2 Sept 2019

Jankowski C, Borowik G, Kowalski K (2014) Dyskretyzacja danych numerycznych metodami przekształceń boolowskich. Przegląd Telekomunikacyjny i Wiadomości Telekomunikacyjne 10:1334–1342

Kim KJ, Lee WB (2004) Stock market prediction using artificial neural networks with optimal feature transformation. Neural Comput Appl 13(3):255–260

Larose DT (2006) Data mining methods and models. Wiley, Hoboken, NJ

Liu H, Hussain F, Tan CL, Dash M (2002) Discretization: an enabling technique. Data Min Knowl Discov 6(4):393–423

Masters T (1995) Advanced algorithms for neural networks: a C++ sourcebook. Wiley, New York

Provost F, Fawcett T (2013) Data science for business: what you need to know about data mining and data-analytic thinking. O'Reilly Media, USA

Rosenfeld A, Illuz R, Gottesman D et al (2018) Using discretization for extending the set of predictive features. EURASIP J Adv Signal Process 2018:7. https://doi.org/10.1186/s13634-018-0528-x Accessed 2 Sept 2019

Tuv E, Borisov A, Runger G et al (2009) Feature selection with ensembles, artificial variables, and redundancy elimination. J Mach Learn Res 10:1341–1366

Stanisz A (2016) Modele regresji logistycznej: zastosowania w medycynie, naukach przyrodniczych i społecznych. StatSoft Polska, Kraków

Witten IH, Frank E, Hall MA et al (2016) Data Mining: practical machine learning tools and techniques. Morgan Kaufmann, USA

Zhang L, Ray H, Priestley J, Tan S (2019) A descriptive study of variable discretization and cost-sensitive logistic regression on imbalanced credit data. J Appl Stat: 1–14

Intuitionistic Fuzzy Synthetic Measure for Ordinal Data

Bartłomiej Jefmański

Abstract The paper presents an intuitionistic fuzzy synthetic measure for ordinal data based on Hellwig's pattern of the development method. The intuitionistic fuzzy synthetic measure allows for a comparative analysis of objects due to the complex phenomenon described by ordinal measurement scales. It also allows for taking into account the uncertainty in comparing objects expressed in the form of neutral points on the ordinal measurement scales. The proposed approach is a part of the research into ordinal data using the fuzzy set theory. The method of the construction of the proposed synthetic measure was presented on the example of the subjective quality of life research of the residents of the communes of the Kraina Łęgów Odrzańskich region in Poland.

Keywords Intuitionistic fuzzy sets · Synthetic measures · Ordinal data · Comparative analysis

1 Introduction

Synthetic measures (composite indexes) measure complex socioeconomic phenomena (quality, well-being, poverty, sustainability, innovation) that cannot be covered by a single variable. One of the approaches to the construction of synthetic measures is the pattern methods based on the appropriate distances between the assessed objects and the pattern object containing the most favorable values of partial variables. A pioneering development in this area was the proposal for a pattern of development measure proposed by Hellwig (1967), also called in the literature as the Wrocław taxonomic method (e.g., Mazziotta and Pareto 2017).

Complex socioeconomic phenomena for which synthetic measures are built are often described by variables measured on ordinal measurement scales (e.g., quality of life, quality of services and products). Comparative analysis of objects due to this type of complex phenomena is possible using at least two approaches. The first of

B. Jefmański (✉)
Wrocław University of Economics and Business, Wrocław, Poland
e-mail: bartlomiej.jefmanski@ue.wroc.pl

K. Jajuga et al. (eds.), *Classification and Data Analysis*, Studies in Classification, Data Analysis, and Knowledge Organization,
https://doi.org/10.1007/978-3-030-52348-0_4

these consists in enhancement ordinal measurement scales and treating the results as from metric (interval and/or ratio) scales. The second approach involves the use of the fuzzy set theory and the presentation of measurement results in the form of fuzzy sets. There are two ways to distinguish here: fuzzy scales and fuzzy conversion scales (e.g., de Sáa et al. 2013). In the first case, the respondent assesses a given variable using the fuzzy number. In the second case, the measurement results on the ordinal scale are usually transformed into fuzzy numbers (usually triangular or trapezoidal fuzzy numbers).

The main goal of the paper is to present a new concept for the construction of a synthetic measure for objects described by ordinal variables using intuitionistic fuzzy sets. The methodology of Hellwig's pattern of development was used in the study. The proposed synthetic measure enables comparative analysis of objects due to a complex phenomenon. It also allows for taking into account the degree of measurement uncertainty expressed in neutral categories of ordinal measurement scales.

2 Fuzzy Sets Theory

The fuzzy sets theory proposed by Zadeh (1965) is a generalization of classical theory of sets. This generalization consists that, in classical variant elements belong to particular set or not; however, in the case of fuzzy sets there is a possibility of partial elements' membership to a set.

Definition 1 A fuzzy set $\tilde{A}$ in $X = \{x\}$ space marked as $\tilde{A} \subseteq X$ is defined by the set of pairs $\tilde{A} = \{(x, \mu_{\tilde{A}}(x)) \mid x \in X\} \forall x \in X$, where $\mu_{\tilde{A}} : X \to [0, 1]$ means the membership to $\tilde{A}$ fuzzy set function, which assigns to each $x \in X$ element its membership level to $\tilde{A}$ fuzzy set (Zimmermann 2001, pp. 11–12).

Definition 2 A triangular fuzzy number $\tilde{A}^t = (a;\ b;\ c)$ is a special case of fuzzy set whose membership function is defined as follows:

$$\mu(x) = \begin{cases} \frac{x-a}{b-a}, \ if\ a \le x < b \\ \frac{c-x}{c-b}, \ if\ b \le x \le c\ , \\ 0, \ otherwise \end{cases} \tag{1}$$

where b indicates the center value, and a, c indicate the lower and upper limit values, respectively.

Definition 3 An Euclidean distance between two triangular fuzzy numbers $\tilde{A}_1^t = (a_1,\ b_1,\ c_1)$ and $\tilde{A}_2^t = (a_2,\ b_2,\ c_2)$ is expressed by the following formula:

$$d(\tilde{A}_1^t, \tilde{A}_2^t) = \sqrt{\frac{(a_1 - a_2)^2 + (b_1 - b_2)^2 + (c_1 - c_2)^2}{3}}. \tag{2}$$

3 Intuitionistic Fuzzy Sets Theory

The intuitionistic fuzzy sets theory proposed by Atanassov (1986) is a generalization of the classical fuzzy sets theory proposed by Zadeh (1965). The extension consists in repealing the assumption that if the object's degree of membership to the fuzzy set is equal a then the degree of non-membership must be equal to $(1 - a)$.

Definition 4 An intuitionistic fuzzy set A' in space X is defined as Atanassov (1999, pp. 1–2):

$$A' = \{< x, \mu_{A'}(x), \nu_{A'}(x) > |x \in X\}, \tag{3}$$

where functions:

$$\mu_{A'} : X \to [0, 1],$$

and:

$$\nu_{A'} : X \to [0, 1],$$

respectively determine the level of membership and non-membership in set A' where for each $x \in X$:

$$0 \le \mu_{A'}(x) + \nu_{A'}(x) \le 1.$$

Definition 5 A level of uncertainty of whether element $x \in X$ is a member of the intuitionistic fuzzy set A' is defined by the following formula:

$$\pi_{A'}(x) = 1 - \mu_{A'}(x) - \nu_{A'}(x). \tag{4}$$

Definition 6 A triangular intuitionistic fuzzy number $A''^t = [(a, b, c); w_{A''}], [(d, b, f); u_{A''}]$ is a special intuitionistic fuzzy set whose membership and non-membership functions are defined as follows (Chaira 2019, p. 74):

$$\mu_{A''}(x) = \begin{cases} \frac{x-a}{b-a} w_{A''}, \ a \le x \le b \\ w_{A''}, \quad x = b \\ \frac{c-x}{c-b} w_{A''}, \ b \le x \le c \\ 0, \quad x < a \text{ or } x > c \end{cases} \tag{5}$$

$$\nu_{A''}(x) = \begin{cases} \frac{b-x+u_{A''}(x-d)}{b-d}, \ d \le x \le b \\ u_{A''}, \quad x = b \\ \frac{x-b+u_{A''}(f-x)}{f-b}, \ b \le x \le f \\ 1, \quad x < d \text{ or } x > f \end{cases} \tag{6}$$

where the values $w_{A''}$ and $u_{A''}$ represent the maximum membership degree and the minimum non-membership degree such that they satisfy the conditions: $0 \le w_{A''} \le 1, 0 \le u_{A''} \le 1$ and $0 \le w_{A''} + u_{A''} \le 1$.

Definition 7 An Euclidean distance between two intuitionistic fuzzy sets A' and B' is expressed by the following formula (Szmidt and Kacprzyk 2000, p. 514):

$$d(A', B') = \sqrt{\frac{1}{2}\sum_{j=1}^{m} (\mu_{A'}(x_i) - \mu_{B'}(x_i))^2 + (\nu_{A'}(x_i) - \nu_{B'}(x_i))^2 + (\pi_{A'}(x_i) - \pi_{B'}(x_i))^2}. \tag{7}$$

4 Transformation of Ordinal Data into Intuitionistic Fuzzy Sets

Let us assume that we have a set of objects $O = \{O_1, O_2, \ldots, O_n\}, i = (1, 2, \ldots, n)$ characterized by a set of variables $X = \{X_1, X_2, \ldots, X_m\}$, $j = (1, 2, \ldots, m)$, which were evaluated by respondents. Each variable was measured using the ordinal measurement scale consisting of k scale points in the form of word categories (e.g., very low, definitely important, etc.). Through x_{ij} we denote the i-th category of the variable X_j whereby $i = (1, \ldots, u, l, \ldots, k)$. Let us also assume that the respondent may omit the assessment of a selected variable and/or has the option of answering outside the accepted measuring scale in the form of "it's hard to say", "I have no opinion".

Let us assume that A'_j is an intuitionistic fuzzy set containing right-sided categories from 1 to k expressing the evaluation of the j-th variable. The transformation of the variable X_j measurement results into an intuitionistic fuzzy set A'_j, assuming the above, requires determination $\mu_{A'}(X_j)$, $\nu_{A'}(X_j)$ and $\pi_{A'}(X_j)$. Formally, the degree of membership and non-membership can be obtained by calculating the following equations, respectively:

$$\mu_{A'}(X_j) = \frac{\sum_{i=l}^{k} N_{ij}}{N} \tag{8}$$

$$\nu_{A'}(X_j) = \frac{\sum_{i=1}^{u} N_{ij}}{N} \tag{9}$$

where N_{ij}—the number of respondents who, when assessing the *j*-variable, chose the *i*-category on the ordinal measurement scale.

The degree of uncertainty $\pi_{A'}(X_j)$ as to the membership of a variable X_j to an intuitionistic fuzzy set A'_j will be the sum of the number of missing responses and/or the number of neutral responses. We formally obtain its value in accordance with definition 5 by calculating the following equation:

$$\pi_{A'}(X_j) = 1 - \mu_{A'}(X_j) - \nu_{A'}(X_j). \tag{10}$$

5 Intuitionistic Fuzzy Synthetic Measure

Let us assume again that we have a set of objects $O = \{O_1, O_2, \ldots, O_n\}$, $i = (1, 2, \ldots, n)$, characterized by a set of ordinal variables $X = \{X_1, X_2, \ldots, X_m\}$, $j = (1, 2, \ldots, m)$, expressed in the form of intuitionistic fuzzy sets summarized in the data table:

$$W = \begin{bmatrix} <\mu_{11}, \nu_{11}, \pi_{11}> & <\mu_{12}, \nu_{12}, \pi_{12}> & \ldots & <\mu_{1m}, \nu_{1m}, \pi_{1m}> \\ <\mu_{21}, \nu_{21}, \pi_{21}> & <\mu_{22}, \nu_{22}, \pi_{22}> & \ldots & <\mu_{2m}, \nu_{2m}, \pi_{2m}> \\ \ldots & \ldots & \ldots & \\ <\mu_{n1}, \nu_{n1}, \pi_{n1}> & <\mu_{n2}, \nu_{n2}, \pi_{n2}> & \ldots & <\mu_{nm}, \nu_{nm}, \pi_{nm}> \end{bmatrix} \tag{11}$$

where $<\mu_{ij}, \nu_{ij}, \pi_{ij}>$—intuitionistic fuzzy set for variable X_j $j = (1, 2, \ldots, m)$ and object O_i $i = (1, 2, \ldots, n)$.

The construction of the intuitionistic fuzzy synthetic measure for such a case takes place in three stages:

Stage 1 Determining the pattern object with the "most favorable" values for each variable. Depending on the variable preference function, the coordinates of the pattern object will have the values:

$$O^+ = (<\mu_1^+, \nu_1^+, \pi_1^+>, <\mu_2^+, \nu_2^+, \pi_2^+>, \ldots, <\mu_m^+, \nu_m^+, \pi_m^+>), \tag{12}$$

where

$$< \mu_j^+, \nu_j^+, \pi_j^+ > = \begin{cases} < \max\{\mu_j\}, \min\{\nu_j\}, 1 - \max\{\mu_j\} - \min\{\nu_j\} > & \text{when } X_j \text{ is a stimulant} \\ < \min\{\mu_j\}, \max\{\nu_j\}, 1 - \min\{\mu_j\} - \max\{\nu_j\} > & \text{when } X_j \text{ is a destimulant} \end{cases}$$

X_j variable is a stimulant, when for every two of its observations x_{ij}^S, x_{kj}^S referring to objects O_i, O_k take $x_{ij}^S > x_{kj}^S \Rightarrow O_i \succ O_k$ ($\succ$ means O_i object domination over O_k object). Variable X_j is a destimulant, when for every two of its observations x_{ij}^D, x_{kj}^D referring to objects O_i, O_k take $x_{ij}^D > x_{kj}^D \Rightarrow O_i \prec O_k$ ($\prec$ means O_k object domination over O_i object) (Walesiak 2016, p. 11).

On account of the ordinal measurement scale in the proposed method, there is no need for the normalization of variables.

Stage 2 Determining the distance of individual objects from the pattern object using the Euclidean distance measure for intuitionistic fuzzy sets:

$$d_{iO^+} = \sqrt{\frac{1}{2}\sum_{j=1}^{m} (\mu_i(X_j) - \mu_{O^+}(X_j))^2 + (\nu_i(X_j) - \nu_{O^+}(X_j))^2 + (\pi_i(X_j) - \pi_{O^+}(X_j))^2} \tag{13}$$

Stage 3 Calculation of the intuitionistic synthetic measure according to the formula:

$$S_i = 1 - \frac{d_{iO^+}}{d_0}, \tag{14}$$

where $d_0 = \bar{d}_0 + 2S(d_0)$, $\bar{d}_0 = \frac{1}{n}\sum_{i=1}^{n} d_{iO^+}$, $S(d_0) = \left[\frac{1}{n}\sum_{i=1}^{n} (d_{iO^+} - \bar{d}_0)^2\right]^{\frac{1}{2}}$.

The measure takes values from the interval [0, 1]. The higher values of the measure, the less the object is away from the pattern object.

6 Empirical Example

The method of constructing an intuitionistic fuzzy synthetic measure is presented on the example of a study on the subjective quality of life of residents of 11 communes of the Kraina Łęgów Odrzańskich region in Poland. The aim of the study was to develop a ranking of communes in terms of the subjective quality of life of residents. The research sample contained 641 respondents. Each respondent assessed the inhabited commune in terms of 8 variables: X_1—satisfaction with health care, X_2—satisfaction with public safety, X_3—satisfaction with access to services, products and transport accessibility, X_4—satisfaction with the cultural and sports offer, X_5—satisfaction with the educational offer, X_6—satisfaction with the professional and material situation, X_7—satisfaction with the condition of the environment, X_8—satisfaction with social relations. The variables were evaluated using a 4-degree ordinal scale with the following points: 1—"very unsatisfied", 2—"unsatisfied", 3—"satisfied", 4—"very satisfied". In addition, each of the respondents had the opportunity to give a neutral answer in the form of "hard to say".

For the purposes of constructing an intuitionistic fuzzy synthetic measure, an intuitionistic fuzzy set A'_j containing right-hand categories from the applied ordinal measurement scale ("satisfied" and "very satisfied") expressing the evaluation of the jth variable was determined for each of the variables. Transformations of the measurement results to intuitionistic fuzzy sets were carried out using the methodology presented in the point 4. The results of the evaluation of individual variables expressed in the form of intuitionistic fuzzy sets for each of the analyzed communes are presented in Tables 1 and 2.

The data set transformed according to the methodology proposed in the paper is presented in the form of the data table:

$$W = \begin{bmatrix} < 0.645; 0.065; 0.290 > & < 0.742; 0.065; 0.193 > & \dots & < 0.742; 0.000; 0.258 > \\ < 0.563; 0.063; 0.374 > & < 0.458; 0.188; 0.354 > & \dots & < 0.833; 0.000; 0.167 > \\ \dots & \dots & & \dots \\ < 0.689; 0.122; 0.189 > & < 0.633; 0.156; 0.211 > & \dots & < 0.778; 0.033; 0.189 > \end{bmatrix}$$

Stage 1. In the presented example, the coordinates of the pattern object have been determined with the application of the method of transformation of variables to the forms of intuitionistic fuzzy sets presented in Chapter 4. Hence, if the variable X_j is a stimulant, then the j-th coordinate of the pattern object (12) assumes the form of the intuitionistic fuzzy set $< 1, 0, 0 >$. In the case the variable X_j is an destimulant, then the jth coordinate of the pattern object assumes the form of the intuitionistic fuzzy set $< 0, 1, 0 >$. Since all variables are stimulant, the following coordinates of the pattern object were adopted:

$$O^+ = (< 1, 0, 0 >, < 1, 0, 0 >, < 1, 0, 0 >, < 1, 0, 0 >, \\ < 1, 0, 0 >, < 1, 0, 0 >, < 1, 0, 0 >, < 1, 0, 0 >).$$

Stage 2. Distances of communes from the pattern object are presented in Table 3.

Stage 3. In order to calculate the value of the intuitionistic fuzzy synthetic measure in accordance with Eq. (14) partial calculations were made and the results are presented in Table 4.

Table 1 Evaluation of variables in the form of intuitionistic fuzzy sets

Communes	Parameters of the intuitionistic fuzzy sets	X_1	X_2	X_3	X_4
A	$\mu_{A'_j}(X_j)$	0.645	0.742	0.645	0.355
	$\nu_{A'_j}(X_j)$	0.065	0.065	0.129	0.258
	$\pi_{A'_j}(X_j)$	0.290	0.193	0.226	0.387
B	$\mu_{A'_j}(X_j)$	0.563	0.458	0.625	0.334
	$\nu_{A'_j}(X_j)$	0.063	0.188	0.188	0.208
	$\pi_{A'_j}(X_j)$	0.374	0.354	0.187	0.458
C	$\mu_{A'_j}(X_j)$	0.571	0.333	0.405	0.333
	$\nu_{A'_j}(X_j)$	0.143	0.143	0.167	0.262
	$\pi_{A'_j}(X_j)$	0.286	0.524	0.428	0.405
D	$\mu_{A'_j}(X_j)$	0.574	0.426	0.556	0.296
	$\nu_{A'_j}(X_j)$	0.204	0.315	0.130	0.463
	$\pi_{A'_j}(X_j)$	0.222	0.259	0.314	0.241
E	$\mu_{A'_j}(X_j)$	0.603	0.492	0.747	0.572
	$\nu_{A'_j}(X_j)$	0.159	0.095	0.063	0.111
	$\pi_{A'_j}(X_j)$	0.238	0.413	0.190	0.317
F	$\mu_{A'_j}(X_j)$	0.703	0.453	0.719	0.266
	$\nu_{A'_j}(X_j)$	0.125	0.141	0.047	0.172
	$\pi_{A'_j}(X_j)$	0.172	0.406	0.234	0.562
G	$\mu_{A'_j}(X_j)$	0.574	0.544	0.603	0.338
	$\nu_{A'_j}(X_j)$	0.132	0.088	0.103	0.250
	$\pi_{A'_j}(X_j)$	0.294	0.368	0.294	0.412
H	$\mu_{A'_j}(X_j)$	0.702	0.449	0.609	0.414
	$\nu_{A'_j}(X_j)$	0.126	0.264	0.207	0.322
	$\pi_{A'_j}(X_j)$	0.172	0.287	0.184	0.264
I	$\mu_{A'_j}(X_j)$	0.678	0.526	0.559	0.424
	$\nu_{A'_j}(X_j)$	0.102	0.169	0.136	0.288
	$\pi_{A'_j}(X_j)$	0.220	0.305	0.305	0.288
J	$\mu_{A'_j}(X_j)$	0.471	0.549	0.392	0.333
	$\nu_{A'_j}(X_j)$	0.294	0.235	0.333	0.255
	$\pi_{A'_j}(X_j)$	0.235	0.216	0.275	0.412

(continued)

Table 1 (continued)

Communes	Parameters of the intuitionistic fuzzy sets	X_1	X_2	X_3	X_4
K	$\mu_{A'_j}(X_j)$	0.689	0.633	0.600	0.356
	$\nu_{A'_j}(X_j)$	0.122	0.156	0.044	0.233
	$\pi_{A'_j}(X_j)$	0.189	0.211	0.356	0.411

The method of calculating the intuitionistic fuzzy synthetic measure is presented on the example of the commune A:

$$S_A = 1 - \frac{d_{AO^+}}{d_0} = 1 - \frac{1.085}{1.704 + 2 \times 0.509} = 0.601$$

The values of the intuitionistic fuzzy synthetic measure for the other communes are presented in Table 5.

In the presented empirical example, none of the communes achieved high value of the intuitionistic fuzzy synthetic measure. There is also a clear division of communes into two groups with very similar values of the synthetic measure. The first group includes communes (A–E) for which the values of the synthetic measure belong to the interval (0.5–0.65) which means the average level of subjective quality of life of the residents. The second group includes communes (F–K) with very low values of synthetic measure which means a low level of subjective quality of life of the residents.

7 A Comparative Analysis

The proposed method of the construction of the intuitionistic fuzzy synthetic variable was compared with two other approaches indicated in the initial part of the paper. The first approach consists in the enhancement of the measurement scale and of the construction of the synthetic measure in accordance with the method proposed by Hellwig (1967). Let us mark such an approach as Hellwig's synthetic measure (HSM). The other approach consists in the construction of Hellwig's synthetic measure with the application of triangular fuzzy numbers, and it will be marked as FHSM (fuzzy Hellwig's synthetic measure) (Jefmański and Dudek 2016). In order to achieve this goal, one implemented the fuzzy conversion scale proposed by Lubiano et al. (2016). The parameters of triangular fuzzy numbers assigned to the four points of the scale are given in Table 6.

The graphic form of the fuzzy conversion scale is given in Fig. 1.

Before starting the comparative analysis of the selected methods, it should be emphasized that in both the approaches (HSM and FHSM), the category "hard to say" was not taken into account because in the research it did not constitute a point of the measurement scale. Because the number of "hard to say" responses in the case of

Table 2 Evaluation of variables in the form of intuitionistic fuzzy sets

Communes	Parameters of the intuitionistic fuzzy sets	X_5	X_6	X_7	X_8
A	$\mu_{A'_j}(X_j)$	0.710	0.484	0.484	0.742
	$\nu_{A'_j}(X_j)$	0.065	0.129	0.032	0.000
	$\pi_{A'_j}(X_j)$	0.225	0.387	0.484	0.258
B	$\mu_{A'_j}(X_j)$	0.729	0.354	0.479	0.833
	$\nu_{A'_j}(X_j)$	0.042	0.167	0.146	0.000
	$\pi_{A'_j}(X_j)$	0.229	0.479	0.375	0.167
C	$\mu_{A'_j}(X_j)$	0.714	0.381	0.667	0.786
	$\nu_{A'_j}(X_j)$	0.048	0.214	0.071	0.024
	$\pi_{A'_j}(X_j)$	0.238	0.405	0.262	0.190
D	$\mu_{A'_j}(X_j)$	0.722	0.500	0.537	0.778
	$\nu_{A'_j}(X_j)$	0.074	0.315	0.241	0.037
	$\pi_{A'_j}(X_j)$	0.204	0.185	0.222	0.185
E	$\mu_{A'_j}(X_j)$	0.603	0.476	0.619	0.778
	$\nu_{A'_j}(X_j)$	0.048	0.111	0.095	0.000
	$\pi_{A'_j}(X_j)$	0.349	0.413	0.286	0.222
F	$\mu_{A'_j}(X_j)$	0.625	0.563	0.594	0.797
	$\nu_{A'_j}(X_j)$	0.016	0.094	0.063	0.000
	$\pi_{A'_j}(X_j)$	0.359	0.343	0.343	0.203
G	$\mu_{A'_j}(X_j)$	0.603	0.397	0.279	0.721
	$\nu_{A'_j}(X_j)$	0.118	0.118	0.265	0.015
	$\pi_{A'_j}(X_j)$	0.279	0.485	0.456	0.264
H	$\mu_{A'_j}(X_j)$	0.713	0.483	0.472	0.839
	$\nu_{A'_j}(X_j)$	0.069	0.195	0.241	0.023
	$\pi_{A'_j}(X_j)$	0.218	0.322	0.287	0.138
I	$\mu_{A'_j}(X_j)$	0.559	0.356	0.746	0.780
	$\nu_{A'_j}(X_j)$	0.102	0.119	0.034	0.017
	$\pi_{A'_j}(X_j)$	0.339	0.525	0.220	0.203
J	$\mu_{A'_j}(X_j)$	0.529	0.333	0.412	0.667
	$\nu_{A'_j}(X_j)$	0.176	0.333	0.196	0.039
	$\pi_{A'_j}(X_j)$	0.295	0.334	0.392	0.294

(continued)

Table 2 (continued)

Communes	Parameters of the intuitionistic fuzzy sets	X_5	X_6	X_7	X_8
K	$\mu_{A'_j}(X_j)$	0.656	0.444	0.644	0.778
	$\nu_{A'_j}(X_j)$	0.111	0.156	0.089	0.033
	$\pi_{A'_j}(X_j)$	0.233	0.400	0.267	0.189

Table 3 Distances of communes from the pattern object

Communes	d_{iO^+}
A	1.085
B	1.223
C	1.278
D	1.175
E	1.037
F	2.315
G	2.158
H	2.099
I	2.221
J	1.917
K	2.231
$\bar{d}_0$	1.704

Table 4 Results of the partial calculations for the intuitionistic fuzzy synthetic measure (14)

Communes	$d_{iO^+} - \bar{d}_0$	$(d_{iO^+} - \bar{d}_0)^2$
A	−0.618	0.382
B	−0.481	0.231
C	−0.425	0.181
D	−0.528	0.279
E	−0.666	0.444
F	0.612	0.374
G	0.454	0.206
H	0.395	0.156
I	0.517	0.267
J	0.214	0.046
K	0.528	0.279
$\sum_{i=1}^{n}(d_{iO^+} - \bar{d}_0)^2$		2.846
$S(d_0) = \left[\frac{1}{n}\sum_{i=1}^{n}(d_{iO^+} - \bar{d}_0)^2\right]^{\frac{1}{2}}$		0.509

Table 5 Values of the intuitionistic fuzzy synthetic measure for communes

Communes	S_i	Ranking position
A	0.601	2
B	0.551	4
C	0.530	5
D	0.568	3
E	0.619	1
F	0.149	11
G	0.207	8
H	0.229	7
I	0.184	9
J	0.295	6
K	0.179	10

Table 6 Parameters of triangular fuzzy numbers representing the categories in the 4-degree ordinal scale of measurement

Categories	Lower limit value (a)	Center value (b)	Upper limit value (c)
Very unsatisfied	0	0	3.333
Unsatisfied	0	3.333	6.666
Satisfied	3.333	6.666	10
Very satisfied	6.666	10	10

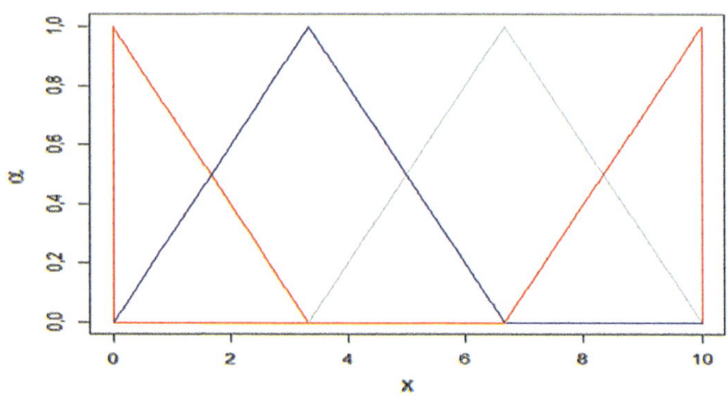

Fig. 1 Graphic interpretation of the fuzzy conversion scale

certain variables is relatively high, a large amount of information are not taken into account by the methods selected for the comparative analysis. The proposed method makes it possible to take this information into account (in the form of the degree of uncertainty), but a direct comparison of the results with other methods is in such a case more difficult, and imposes caution while interpreting them.

In the case of the first method selected for the comparative analysis, it is necessary to average the ratings of each of the variables on the level of the analyzed communes. Average values of the variables are given in Tables 7 and 8.

The coordinates of the pattern object were determined on the basis of the numbers assigned to the extreme categories of the 4-degree ordinal scale. Due to the fact that each of the variables possesses the characteristics of a stimulant, the pattern coordinates were determined in the following way:

$$O^+ = (4,\ 4,\ 4,\ 4,\ 4,\ 4,\ 4,\ 4). \tag{16}$$

By applying the Euclidean distance measure one calculated the distance of each commune from the pattern object. The distances from the pattern object, the values

Table 7 Average values of the variables for HSM method

Communes	X_1	X_2	X_3	X_4
A	2.955	2.960	2.917	2.632
B	3.444	2.909	2.692	2.111
C	2.867	2.650	2.833	2.520
D	2.750	2.745	2.770	2.759
E	2.792	2.865	3.098	2.907
F	2.943	2.816	3.102	2.679
G	2.875	2.977	2.938	2.525
H	2.889	2.629	2.845	2.531
I	2.936	2.917	2.941	2.948
J	2.641	2.825	2.568	2.533
K	3.123	2.775	3.069	2.642

Table 8 Average values of the variables for HSM method

Communes	X_5	X_6	X_7	X_8
A	3.167	2.789	2.938	3.174
B	3.154	2.714	2.833	3.200
C	3.125	2.680	2.968	3.206
D	2.848	2.789	2.846	2.920
E	3.073	2.919	2.889	3.082
F	3.220	2.976	2.905	3.157
G	2.980	2.829	2.459	3.180
H	3.118	2.712	2.661	3.147
I	3.016	3.020	3.081	3.104
J	2.889	2.471	2.710	3.111
K	2.913	2.704	2.955	3.096

Table 9 Values of the HSM for communes

Communes	Distance from the pattern object	S_i	Ranking position
A	3.032	0.177	4
B	3.339	0.094	8
C	3.296	0.106	6
D	3.389	0.081	9
E	2.977	0.192	3
F	2.939	0.202	2
G	3.328	0.097	7
H	3.400	0.077	10
I	2.847	0.227	1
J	3.669	0.005	11
K	3.123	0.153	5

of the synthetic measure HSM, and the positions in the ranking for each of the communes are presented in Table 9.

In the case of the second method, it was necessary to replace the measurement results by fuzzy numbers with the parameters specified in Table 6. Having at the disposal, the measurements result in the form of triangular fuzzy numbers, one averaged the ratings of the variables on the level of each of the communes. Average values of the variables in the form of triangular fuzzy numbers are given in Table 10.

The coordinates of the pattern object were determined on the basis of the triangular fuzzy numbers assigned to the extreme categories of the 4-degree ordinal scale. The coordinates of the pattern object in the form of triangular fuzzy numbers were determined in the following way:

$$O^{+} = ((6.667;\ 10;\ 10);\ (6.667;\ 10;\ 10);\ \ldots\ (6.667;\ 10;\ 10)) \qquad (17)$$

By applying the Euclidean distance measure for triangular fuzzy numbers, one calculated the distance of each commune from the pattern object. The distances from the pattern object, the values of the synthetic measure FHSM, and the positions in the ranking for each of the communes are presented in Table 11.

In order to facilitate the comparative analysis, the results obtained for the tree methods of the construction of the synthetic measures are given in Table 12 and in Fig. 2.

The conducted comparative analysis demonstrated clear differences in the results obtained with the application of the three methods of the construction of synthetic measures. The values of the intuitionistic fuzzy synthetic measure possess a decisively larger range of variability with reference to the remaining two approaches. The smallest range of variability and the lowest values of synthetic measures were observed in the case of the approach based on the enhanced measurement scale and classical Hellwig's method in the construction of the synthetic measure (HSM). The

Table 10 Average values of variables in the form of triangular fuzzy numbers

Communes	Variables	Lower limit value (a)	Center value (b)	Upper limit value (c)
A	X_1	3.790	7.078	9.452
	X_2	2.676	5.916	9.249
	X_3	3.563	6.897	9.770
	X_4	2.264	5.472	8.554
	X_5	3.043	6.377	9.517
	X_6	2.531	5.679	8.951
	X_7	3.182	6.515	9.596
	X_8	3.653	6.987	9.863
B	X_1	3.778	7.111	9.667
	X_2	2.473	5.699	8.925
	X_3	2.991	6.069	8.974
	X_4	2.308	5.128	8.205
	X_5	3.693	6.937	9.730
	X_6	2.266	5.333	8.667
	X_7	2.555	5.889	9.222
	X_8	4.000	7.334	10.000
C	X_1	3.000	6.222	9.222
	X_2	2.500	5.500	8.667
	X_3	2.778	6.111	9.028
	X_4	1.866	5.067	8.400
	X_5	3.750	7.084	9.792
	X_6	2.266	5.600	8.800
	X_7	3.226	6.559	9.677
	X_8	4.019	7.353	9.902
D	X_1	2.778	5.953	8.968
	X_2	1.916	5.250	8.583
	X_3	2.883	6.126	9.279
	X_4	1.463	4.390	7.561
	X_5	3.566	6.899	9.690
	X_6	2.121	5.455	8.712
	X_7	2.381	5.714	8.968
	X_8	3.560	6.894	9.849
E	X_1	2.778	5.972	9.167
	X_2	2.973	6.216	9.369
	X_3	3.660	6.994	9.739
	X_4	3.023	6.357	9.457

(continued)

Table 10 (continued)

Communes	Variables	Lower limit value (a)	Center value (b)	Upper limit value (c)
	X_5	3.577	6.911	9.756
	X_6	3.063	6.397	9.369
	X_7	2.963	6.297	9.556
	X_8	3.605	6.939	10.000
F	X_1	3.144	6.478	9.497
	X_2	2.719	6.053	9.211
	X_3	3.673	7.007	9.796
	X_4	2.500	5.595	8.452
	X_5	4.065	7.399	9.919
	X_6	3.333	6.588	9.444
	X_7	3.095	6.349	9.603
	X_8	3.856	7.190	10.000
G	X_1	3.125	6.250	9.167
	X_2	3.256	6.589	9.535
	X_3	3.125	6.459	9.514
	X_4	2.167	5.083	8.167
	X_5	3.333	6.599	9.388
	X_6	2.952	6.095	9.048
	X_7	1.802	4.865	8.108
	X_8	3.933	7.267	9.933
H	X_1	3.102	6.297	9.352
	X_2	2.258	5.430	8.602
	X_3	2.911	6.150	9.061
	X_4	2.031	5.104	8.281
	X_5	3.774	7.059	9.657
	X_6	2.542	5.706	8.870
	X_7	2.312	5.538	8.764
	X_8	3.822	7.156	9.911
I	X_1	3.768	6.884	9.348
	X_2	2.601	5.854	9.106
	X_3	3.089	6.342	9.268
	X_4	2.301	5.476	8.492
	X_5	3.419	6.667	9.402
	X_6	2.738	5.834	8.929
	X_7	3.478	6.812	9.855
	X_8	3.901	7.234	9.929

(continued)

Table 10 (continued)

Communes	Variables	Lower limit value (a)	Center value (b)	Upper limit value (c)
J	X_1	2.479	5.470	8.376
	X_2	3.000	6.083	8.750
	X_3	2.072	5.225	8.288
	X_4	2.222	5.111	8.111
	X_5	3.241	6.296	8.889
	X_6	2.059	4.902	7.843
	X_7	2.580	5.699	8.710
	X_8	3.889	7.037	9.630
K	X_1	3.790	7.078	9.452
	X_2	2.676	5.916	9.249
	X_3	3.563	6.897	9.770
	X_4	2.264	5.472	8.554
	X_5	3.043	6.377	9.517
	X_6	2.531	5.679	8.951
	X_7	3.182	6.515	9.596
	X_8	3.653	6.987	9.863

Table 11 Values of the FHSM for communes

Communes	Distance from the pattern object	S_i	Ranking position
A	67.780	0.334	3
B	79.959	0.214	6
C	81.295	0.201	7
D	97.829	0.039	10
E	65.767	0.354	2
F	64.211	0.369	1
G	81.827	0.196	8
H	85.937	0.155	9
I	71.067	0.302	4
J	99.061	0.026	11
K	72.535	0.287	5

values of the synthetic measure were within the range <0.005; 0.227>. The application of the method based on triangular fuzzy numbers (FSM) resulted in an increase in the variability range as well as an increase in the values of synthetic measures for the analyzed objects. In the case of the proposed method IFSM, the highest values of the synthetic measure exceeded the level 0.6. The proposed measure differentiates the analyzed objects to the highest degree.

Table 12 Results of comparative analysis for the tree methods

Communes	IFSM		HSM		FSM	
	S_i	Ranking position	S_i	Ranking position	S_i	Ranking position
A	0.601	2	0.177	4	0.334	3
B	0.551	4	0.094	8	0.214	6
C	0.530	5	0.106	6	0.201	7
D	0.568	3	0.081	9	0.039	10
E	0.619	1	0.192	3	0.354	2
F	0.149	11	0.202	2	0.369	1
G	0.207	8	0.097	7	0.196	8
H	0.229	7	0.077	10	0.155	9
I	0.184	9	0.227	1	0.302	4
J	0.295	6	0.005	11	0.026	11
K	0.179	10	0.153	5	0.287	5

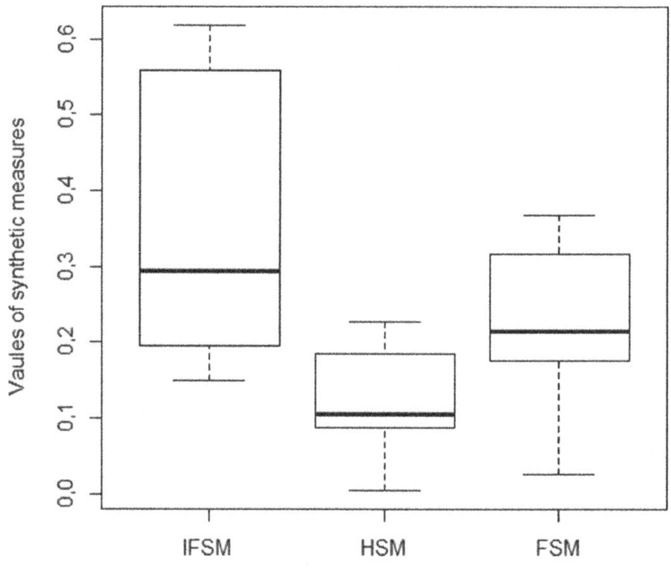

Fig. 2 Box-plots for values of the synthetic measures

Analyzing the results given in Table 12, one may also notice the impact of the three applied methods of the construction of synthetic measures on the position of objects in the obtained rankings. Kendall's tau-b measure was implemented in the evaluation of the significance of differences in the obtained rankings, and the results are given in Table 13.

Table 13 Values of Kendall's tau-b measures for three rankings

	IFSM	HSM	FSM
IFSM	1	–	–
HSM	−0.127	1	–
FSM	−0.055	0.782[a]	1

[a]Correlation significant at the 0.01 level

The obtained results demonstrate that the choice of the method of the construction of synthetic measures had an impact both on the values of the measurements and on the order of objects in the rankings. The greatest discrepancies were obtained in the rankings of objects obtained with the implementation of the methods IFSM and HSM. A high similarity of results was obtained in the case of the methods based on the fuzzy sets IFSM and FSM; but in the case of some objects, there exist great discrepancies as far as the rankings positions are concerned (e.g., D and F).

8 Conclusions

Complex socioeconomic phenomena are often described by ordinal data. Comparative analysis of objects (e.g., countries, cities) in such a situation is difficult due to the limited number of arithmetic operations that can be made on ordinal measurement scales. Therefore, the paper proposes an intuitionistic fuzzy synthetic measure that after prior transformation of ordinal data into intuitionistic fuzzy sets enables comparative analysis of objects due to complex phenomena described by ordinal data. The proposed intuitionistic fuzzy synthetic measure also allows for the inclusion in comparative analyzes of neutral categories of ordinal measurement scales which are characteristic for the so-called Likert measuring scales.

The comparative analysis carried out in the paper demonstrated a higher variability of the value of the intuitionistic fuzzy synthetic measure with reference to the remaining two methods of the construction of synthetic variables. The most similar results of ordering the communes were observed in the case of these methods of the construction of synthetic measures which take advantage of the classical and intuitionistic fuzzy sets. However, it should be emphasized that, on account of the ordinal scale of measurement which was implemented in the analysis, it was not justified to take into account the category "hard to say" in the method based on triangular fuzzy numbers (FSM) as well as in the method enhancing the measurement scale (HSM).

The proposed intuitionistic fuzzy synthetic measure can be particularly useful in the analysis of data sets available by public statistics (e.g., Eurostat) where in the case of subjective indicators describing phenomena measurement results are given in the aggregate form (e.g., in the form of the percentage of respondents who selected specific statements from the measurement scale). It is also an alternative to synthetic measures constructed on the basis of fuzzy conversion scales where the transformation of ordinal data into fuzzy numbers is subjective and most often without justification for fuzzy number parameters.

References

Atanassov KT (1986) Intuitionistic fuzzy sets. Fuzzy Sets Syst 20(1):87–96

Atanassov KT (1999) Intuitionistic fuzzy sets. Springer, Berlin Heidelberg

Chaira T (2019) Fuzzy sets and its extension. Intuitionistic fuzzy set. Wiley Inc, New Jersey

de Sáa SR, Gil MA, Garcia MTL, Lubiano MA (2013) Fuzzy rating vs fuzzy conversion scales: an empirical comparison through the MSE. In: Kruse R et al (eds) Synergies of soft computing and statistics AISC 190. Springer, Berlin, Heidelberg, pp 135–143

Hellwig Z (1967) Procedure of evaluating high level manpower data and typology of countries by means of the taxonomic method. COM/WS/91 Warsaw (unpublished UNESCO working paper)

Jefmański B, Dudek A (2016) Syntetyczna miara rozwoju Hellwiga dla trójkątnych liczb rozmytych. In: Appenzeller D (ed) Matematyka i informatyka na usługach ekonomii. Wybrane problemy modelowania i prognozowania zjawisk gospodarczych, Wydawnictwo Uniwersytetu Ekonomicznego w Poznaniu, Poznań, pp 29–40

Lubiano MA, de Sáa SR, Montenegro M, Sinova B, Gil MA (2016) Descriptive analysis of responses to items in questionnaires. Why not using a fuzzy rating scale? Inf Sci 360:131–148

Mazziotta M, Pareto A (2017) Synthesis of Indicators: The Composite Indicators Approach. In: Maggino F (ed) Complexity in society: from indicators construction to their synthesis. Social indicators research series, vol 70, Springer, Cham, pp 159–191

Szmidt E, Kacprzyk J (2000) Distances between intuitionistic fuzzy sets. Fuzzy Sets Syst 114(3):505–518

Walesiak M (2016) Visualization of linear ordering results for metric data with the application of multidimensional scaling. Econometrics 2(52):9–21

Zadeh LA (1965) Fuzzy sets. Inf Control 8:338–353

Zimmermann JH (2001) Fuzzy set theory and its applications, 4th edn. Kluwer Academic Publishers, Boston

Improving Classification Accuracy of Ensemble Learning for Symbolic Data Trough Neural Networks' Feature Extraction

Marcin Pełka

Abstract The key element that has a major impact on the modeling process is the method selection and selection of variables (information that will be used in the model). One of the approaches that allow to improve model's accuracy is the selection of variables and the second one is the transformation of variables. The paper presents a procedure that combines these two approaches—extracting variables from neural networks (multilayer perceptron for symbolic data) as the method of variable selection for the purposes of ensemble learning for symbolic data. The main aim of the paper is to analyze the usefulness of the proposed approach for the prediction power of the ensemble model. In the empirical part, a symbolic data set describing a thousand German borrowers is used.

Keywords Symbolic data analysis · Hybrid models · Ensemble learning · Credit scoring

1 Introduction

Generally speaking, a credit score is a model-based estimate of the probability that a particular borrower will show some undesirable behavior in the future (see, e.g., Lessmann et al. 2015).

Existing studies have incorporated the use of many machine learning (data-mining) techniques and algorithms for credit scoring analysis such as discriminant analysis, neural networks, support vector machines, decision trees, logistic regression, fuzzy logic, genetic algorithms, Bayesian networks, hybrid methods, ensemble learning approach, and survival analysis (Munkhdalai et al. 2019, Louzada et al. 2016, Lessmann et al. 2015, Leo et al. 2019).

When analyzing results for classical data, we can say in general that neural networks, logistic regression, support vector machines, and fuzzy logic reach better

M. Pełka (✉)
Department of Econometrics and Computer Science, Wroclaw University of Economics and Business, Wrocław, Poland
e-mail: marcin.pelka@ue.wroc.pl

K. Jajuga et al. (eds.), *Classification and Data Analysis*, Studies in Classification, Data Analysis, and Knowledge Organization,
https://doi.org/10.1007/978-3-030-52348-0_5

results (in terms of model error and other measures) than other methods (Munkhdalai et al. 2019, Louzada et al. 2016, Lessmann et al. 2015, Leo et al. 2019).

When considering symbolic data, only Dudek (2013) and Pełka (2018) presented results using symbolic data sets. Both papers use decision trees for symbolic data, random forests, symbolic kernel discriminant analysis *k*-nearest neighbor. The data sets in both cases were divided into a training set (928 objects) and test set (27 objects). The random forest model reached the lowest error (the only measure provided in both papers) that was equal to 5.556%.

The key element that has a major impact on the modeling process is the method selection and selection of variables (information that will be used in the model). One of the approaches that allows to improve model's accuracy is the selection of variables and the second one is the transformation of variables. The paper presents a procedure that combines these two approaches—extracting variables from neural networks (multilayer perceptron for symbolic data) as the method of variable selection for the purposes of ensemble learning for symbolic data. This paper uses artificial neural networks as they allow to transform original (input variables) into new variables via nonlinear variable transformation in hidden layers. New variables are selected during the construction of the neural network in such a way that new variables have stronger discriminative power. The paper by Mori et al. has proposed to use the last layer of neural network as the new variables for the support vector machine method for face recognition (Mori et al. 2005).

The main aim of the paper is to analyze the usefulness of the proposed approach for the prediction power of the ensemble model. In the empirical part, a symbolic data set describing a thousand German borrowers is used.

2 Symbolic Data and Variable Selection from Neural Networks

In classical data, situation objects are described by single-valued (numerical or categorical) variables. This allows to represent each object as a vector of qualitative or quantitative measurements where each column represents a single variable. Such data representation is too restrictive to represent more complex data. If we want to take into consideration the uncertainty and variability of the data, we must assume sets of categories or intervals (with frequencies or weights in some cases). Such kind of data representation has been studied in the symbolic data analysis (SDA).

Each symbolic object can be described by Bock and Diday (2000), Billard and Diday (2006), Diday and Noirhomme-Fraiture (2008), Noirhomme-Fraiture and Brito (2011):

1. Quantitative (numerical) variables:
    - numerical single-valued,
    - numerical multi-valued,

- interval variables,
- histogram variables.

2. Qualitative (categorical) variables:

- categorical single-valued,
- categorical multi-valued,
- categorical modal.

Examples of symbolic variables with their realizations are given in Table 1.

As the objects in the symbolic data analysis are described by non-classical variables, we can describe any type of phenomena in a more detailed way. However, symbolic data representation requires to apply special distance measures, methods, and algorithms that can deal with complex data.

Predictive power (model quality) depends strictly on data that was used in modeling process. The key element is to provide the most important variables for the model. There are many data preprocessing methods that can solve this issue. The following part of the paper will present method of variable selection via using neural networks (multilayer perceptron for symbolic data).

An artificial neural network, in general, is an interconnected group of nodes that are inspired by a simplification of neurons in a human brain. Figure 1 presents a simplified example of a neural network. Each circular node represents an artificial

Table 1 Examples of symbolic variables with realizations

Symbolic variable	Realizations	Variable type
1	2	3
Price of a new car (in PLN)	<27,000, 38,000>; <35,000, 50,000> <20,000, 30,000>; <25,000, 37,000>	Interval-valued (non-disjoint intervals)
Engine's capacity (in ccm)	<1000, 1200>; <1300, 1400> <1500, 1800>; <1900, 2200>	Interval-valued (disjoint intervals)
Chosen car color	{red, black, green, blue}	Categorical multivalued
Preferred car	{Toyota (0.3); Volvo (0.7)} {Audi (0.6), VW (0.4), Skoda (0.05)}	Categorical modal
Distance travelled	<10, 20 > (0.65); <21, 30 > (0.35)	Histogram
Sex	{M; F}	Nominal
Number of customers	(0, 1, 2, 3, …)	Ratio

Source Own elaboration

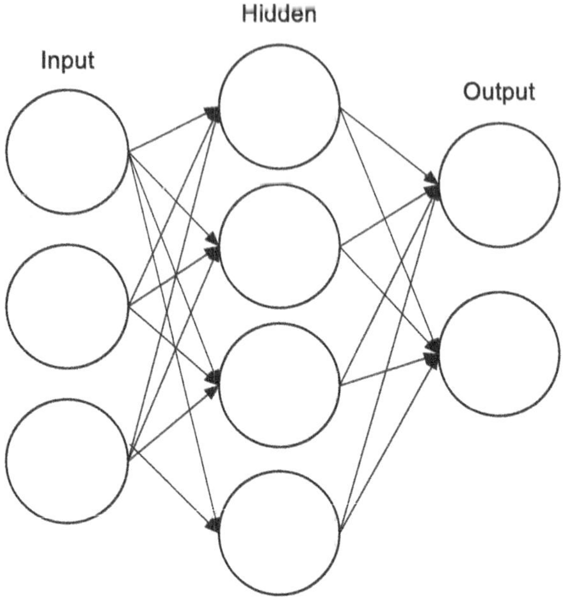

Fig. 1 A simplified model of a neural network. *Source* Own elaboration

neuron, and an arrow represents a connection from the output of one neuron to the input of another one.

In the case of symbolic data, only a solution on how to apply multilayer perceptron has been proposed (Diday and Noirhomme-Fraiture 2008).

In the case of a multilayer perceptron for symbolic data, the transformation of symbolic variables into numerical values is needed. Categorical single-valued variables with unordered m categories are replaced with m binary variables. Categorical single-valued variables with m ordered categories are replaced with a single numerical variable, where each category is replaced by rank. Interval-valued symbolic variables are replaced by a mean (midpoint) and the interval's length (or a logarithm of length if possible). Categorical multi-valued variables are coded as m numerical variables, but several 1 are allowed for a given variable. Modal variables are a generalization of categorical multi-valued variables. Again m numerical variables are used but instead, 1 and 0 weights from modal variables are used (Diday and Noirhomme-Fraiture 2008).

Alternative approaches for interval-valued symbolic data (interval-valued inputs in general) have been proposed by Šíma (1995), Simoff (1996) and Beheshti et al. (1998). The basic idea of these proposals is to apply interval arithmetic as an extension of standard arithmetic for interval-valued data. The main advantage is that we take into account whole interval-valued variable. However, Rossi and Conan-Guez (2002) have showed that the recording (coding) approach provides better results than interval arithmetic approach.

When we have symbolic data transformed into numerical values then classical, well-known multilayer perceptron model is used (Haykin 1998).

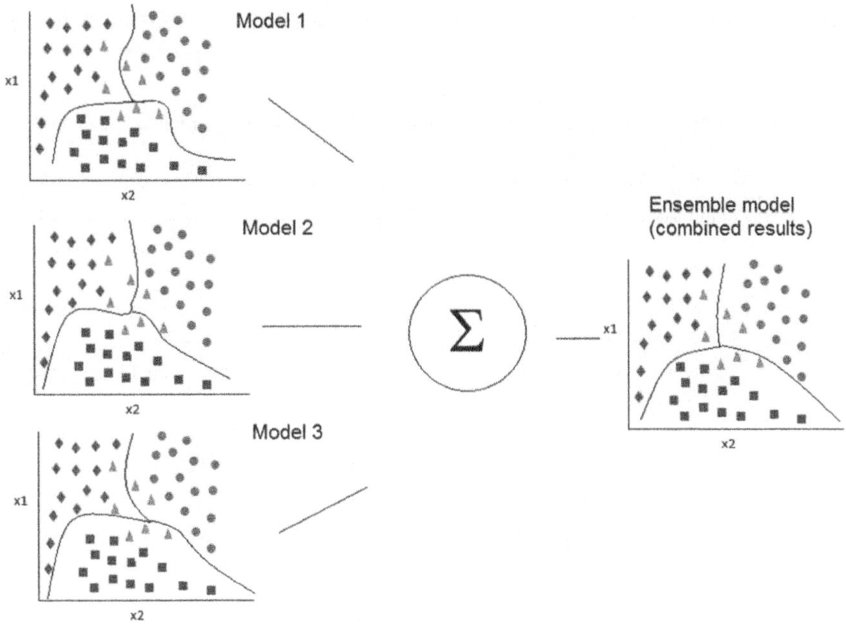

Fig. 2 General idea of ensemble learning approach. *Source* Own elaboration

Ensemble techniques, that combine results provided by different base models into one single (aggregated) model (see Fig. 2), are a useful tool in discrimination and regression tasks (Polikar 2006). In general different models, e.g., decision trees, SVM's, logistic regression models, etc., same models with different initial parameters, same models with subsamples of data, same models with different variable subsets are used (Polikar 2006). This idea can be also applied in the case of clustering.

The most important reason to use ensemble systems is the problem of model selection. Usually, many different models can be applied and each of them can lead to quite different results and typically there is no "best" one. What is more important ensemble models reach better results than any of the models that are part of the ensemble. However, there is no guarantee that the ensemble's average performance will be the best one (Fumera and Roli 2005).

Ensemble-based classifiers can be also useful when dealing with problems of too much or too little data in the data set. When we deal with the problem of big data sets (too much data) then using subsets of the entire data sets are smaller, easier to learn. When we have too little data ensemble systems, like bootstrapping, can be used.

Another problem, where ensemble systems can be a solution, is the problem of nonlinear, complex data sets. The ensemble system divides the complex problem into smaller, easier to learn, subsets ("divide-and-conquer" strategy).

Also, paper by Diettrich (2000) presents three other reasons to use ensembles: statistical, computational, and representational. The computational reason is just

Table 2 Procedure that uses multilayer perceptron for symbolic data as the method for variable extraction

Step no.	Description
1	Divide the data set into learning L and test part T (also validation part V can be obtained at this step)
2	Transform symbolic variables into numerical values
3	Set the variants of neural network (multilayer perceptron) parameters (number of hidden layers, number of elements in each layer). Build models for each variant and use the one where classification error for validation set V is lowest
4	Extract the last hidden layer of the multilayer perceptron and use its neurons as new variables and build new data set L_{net} for next step
5	Use the new data set L_{net} in the ensemble learning or single modeling technique (e.g., SVM, decision tree, etc.)
6	Transform symbolic data from test set T into numerical values and transform it by using the last hidden layer of the multilayer perceptron—like in step 4
7	Calculate the accuracy of the classification for the test set T

Source Own elaboration based on Trzęsiok (2018)

a model selection problem. The statistical reason is the lack of adequate data to properly represent the data distribution. The representational reason is the "divide-and-conquer" strategy.

The general procedure that uses multilayer perceptron for symbolic data as the method for variable extraction is shown in Table 2.

In this paper, extracted variables will be used to build a classical decision tree (decision tree for classical variables). There is no point using a symbolic decision tree as we no longer have symbolic variables after extracting them from a multilayer perceptron.

A decision tree is a well-known tool that uses a tree-like model of decisions. The main objectives of a decision tree are Safavian and Landgrebe (1991):

1. To classify correctly as much of the training sample as possible.
2. Generalize beyond the training sample, so that unseen samples could be classified with as high of accuracy as possible.
3. Be easy to update as more training sample becomes available.
4. To have as simple structure as possible.

The classical decision tree in this paper will be build using `rpart` package and `rpart` function of R software (Therneau et al. 2019). This function allows building regression or classification trees with different parameters (e.g., weights, minimum number of observations that must exist in a node in order for a split to be attempted) (Therneau et al. 2019).

In the empirical part of the paper, data set containing one thousand German credit borrowers will be used and the presented procedure will be applied which is used.

Table 3 Variables describing credit burrowers

No.	Variable	Variable type
1	Cluster membership: 1—credit paid without problems, 2—credit paid with some issues	Nominal
2	Duration of a credit	Symbolic interval-valued
3	Information about former credits	Multi-valued
4	Credit's purpose	Multi-valued
5	Amount of a credit	Symbolic interval-valued
6	Savings	Symbolic interval-valued
7	Employment (duration)	Symbolic interval-valued
8	Installment	Symbolic interval-valued
9	Sex	Multi-valued[a]
10	Guarantors	Multi-valued
11	Most valuable assets	Multi-valued
12	Age	Symbolic interval-valued
13	Information about other credits	Multi-valued
14	Flat	Multi-valued[a]
15	Former credits in the same bank	Multi-valued[a]
16	Occupation	Multi-valued[a]
17	Foreigner	Multi-valued[a]

[a]Means that multi-valued variable has only one category
Source Own elaboration

3 Results of Credit Scoring

For purposes of credit scoring, data set containing one thousand German credit burrowers was applied. The symbolic data table for this data set was prepared by Dudek (2013). In this data set, we have first-order symbolic objects[1] that are described by seventeen symbolic variables (see Table 3). This data set was divided into three subsets: 400 learning set, 300 validation set, and 300 test set.

For the original data set, single models were build—symbolic decision tree (SDT)—obtained with `symbolicDA` (Dudek et al. 2019) package for R software, multilayer perceptron (MLP), and multilayer perceptron with the last hidden layer used as the new data set for decision tree for classical data (MLP-cDT).

As mentioned before in the case of a multilayer perceptron for symbolic data, the transformation of symbolic variables into numerical values is needed. All interval-valued symbolic variables were replaced by their mean and the length of a variable (as some of them had length equal to 0 and the logarithm could not be applied). All multivalued variables were coded to m binary variables (where m is the number

[1]It means we have single units (burrowers in this case) that are described by symbolic variables.

of categories). As a result instead of 17 initial symbolic variables, the data set was described by 52 classical variables.

Besides that, the mentioned models were used in the ensemble approach (with 50 base models in each case), where the majority voting rule was applied (it is the simplest rule to assign objects into final groups, clusters from many models. The object is assigned to that group, cluster, where most of the models assign it).

Decision tree for symbolic data has been proposed in Bock and Diday (eds.) (2000) and is based on the construction of binary questions for choosing the best split of a decision tree. The criterion that is used split nodes of the decision tree is calculated as follows:

$$W_j(t, c) = \log \prod_{k=1}^{n} [p_k(l)P_l(s) + p_k(r)P_r(s)], \tag{1}$$

where $j = 1, \ldots, m$—variable number, t—node number, c—cutting value, $p_k(l)$—probability that kth object will be assigned to the left node, $p_k(r) = 1 - p_k(l)$—probability that kth object will be assigned to the right node, P_l(s) and P_r(s)—conditional probability that in the left (right) node a cluster s that the k-th object belongs to will be observed.

The symbolic decision tree is built as follows (Bock and Diday (eds.) 2000):

1. Construction of contingency table for nominal, ordinal, and symbolic multivalued variables. In the case of symbolic multivalued variables, this table contains information on how many times each category was observed within symbolic objects. For interval-valued symbolic, we calculate means for all possible combinations of upper and lower bounds of these variables. Classical ratio and interval data are treated as symbolic interval-valued data.
2. For symbolic interval-valued variable, we calculate all possible midpoints (means). These midpoints will be cutting values (thresholds) c. If considered cutting value c is within symbolic variable, then $p_k(l) = \frac{c - \underline{v}_{kj}}{\bar{v}_{kj} - \underline{v}_{kj}}$. If it is lower than the lower bound of a symbolic variable, then $p_k(l) = 0$ and if it is greater than the upper bound of a symbolic variable then $p_k(l) = 1$ (where $\underline{v}_{kj}$—is the lower bound of j-th variable in k-th object, $\bar{v}_{kj}$—is the upper bound of j-th variable in k-th object).
3. Stopping criteria for W (W^*) and the minimal size of a node to be split n^* have to be set.
4. Criterion value W is calculated for all possible cutting values c.
5. The highest W value is selected for each variable.
6. The highest W value is selected for all variable if it meets the condition $W_j(t, c) > W^*$. The node is cut under the condition its size (number of objects) is bigger than n^*.
7. Steps 4–6 are repeated until terminal nodes are obtained. Cutting values that were used before are not considered in further analysis.
8. The rate of correct predictions is calculated for the final decision tree.

For all models and approaches, most important measures were calculated—accuracy, sensitivity, specificity, and model's error. The simple decision tree for this data set is shown in Fig. 3.

The numerical values in Fig. 3 represent probabilities of observing cluster 1 and cluster 2 in a particular node. When taking into consideration, the savings of a customer (first split of a decision tree) we have a probability equal to 83.4% of observing second cluster in the right node (and only 16.6% for the first cluster in this node). We have the probability equal to 65.2% of observing the second cluster in the left node (and 34.8% for the first cluster in this node).

The most important variable that has influence on the credit scoring decision is savings, credit's duration, duration of employment and credit amount.

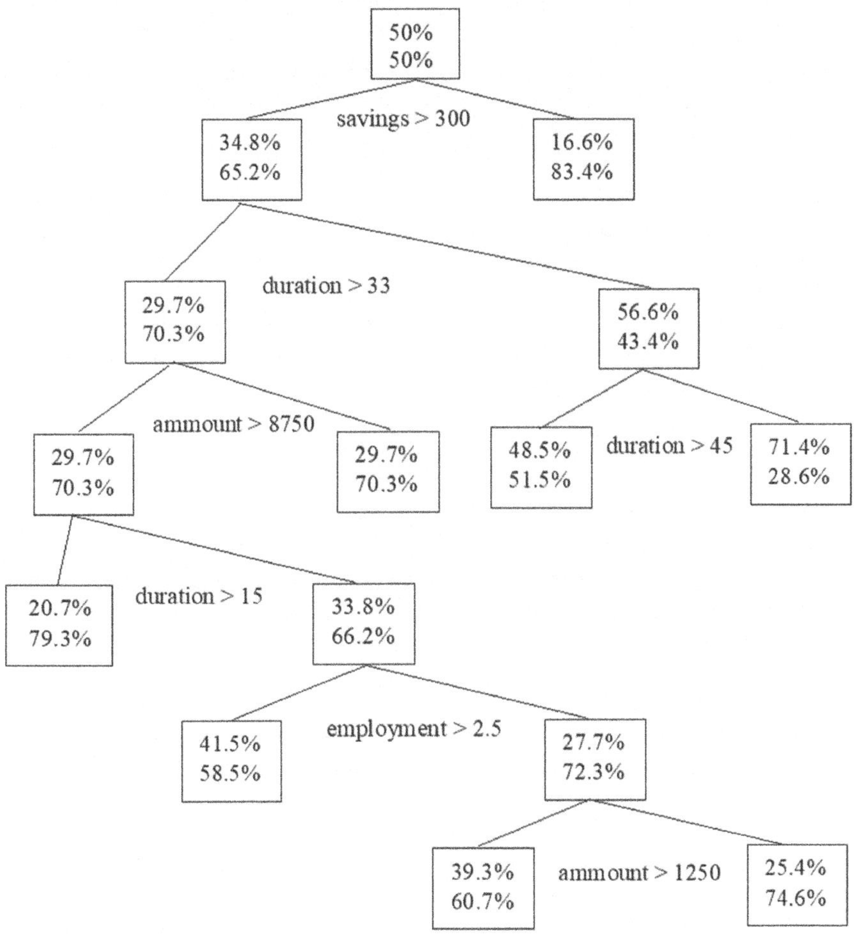

Fig. 3 Decision tree for credit data. *Source* Own elaboration obtained with R software

Table 4 Results of the analysis obtained for test data set

Measure	Method		
	SDT[a]	MLP[b]	MLP-cDT[c]
Single model			
Accuracy	0.89	0.90	0.910
Sensitivity	0.79	0.79	0.913
Specificity	0.945	0.95	0.907
Error	0.11	0.10	0.090
Ensemble model—50 models in each case			
Accuracy	0.95	0.93	0.937
Sensitivity	0.885	0.86	0.904
Specificity	0.95	0.97	0.972
Error	0.05	0.07	0.063

[a]symbolic decision tree, [b]multilayer perceptron for symbolic data, [c]multilayer perceptron for symbolic data combined with a classical decision tree
Source Own elaboration

For the further calculation multilayer perceptron with three hidden layers and 25, 14, 10 neurons in each layer respectively. This reduced initial 52 numeric variables into 10 numerical variables. This neural network reached error equal to 0.10. The last layer of this network was used as the new data set for the purposes of ensemble learning, where the same models were used with different initial parameters. The results obtained for the test set are shown in Table 4.

4 Final Remarks

When looking at the final results, it can be said that extracting variables from multilayer perceptron for symbolic data does not provide such error improvement as it would be expected. The main reason may be the initial transformation of symbolic data into numerical values.

The proposed approach allows transforming initial symbolic data set into new, smaller, variable data set. What is more, these variables are no longer symbolic—so also any classical data analysis method can be applied. The main drawback of the proposed approach is the need of transformation of symbolic data to classical variables. Like any type of transformation, this may lead to some information loss. Another problem is the size of the transformed data set, as it is much bigger than the initial symbolic data set.

The major impact of the data transformation of symbolic to classical data lies in the possible information loss and making the initial big symbolic data set even bigger, so computer software may have problems while training the neural network.

The proposed approach could be improved by using different models on the transformed data set, and on the other hand, it could be combined with the application of the methods that use the initial data set (e.g., symbolic decision trees combined with classical discriminant analysis build on the transformed data set).

The proposed extraction of variables from initial symbolic data set might be also useful when presenting symbolic objects, described by different symbolic variable types, on a two-dimensional map via applying multidimensional scaling.

References

Billard L, Diday E (2006) Symbolic data analysis. Conceptual statistics and data mining. Wiley, Chichester

Bock H-H, Diday E (eds) (2000) Analysis of symbolic data. Explanatory methods for extracting statistical information from complex data. Springer, Berlin-Heidelberg

Beheshti M, Berrached A, de Korvin A, Hu C, Sirisaengtaksin O (1998) On interval weighted three-layer neural networks. In: Proceedings of the 31st annual simulation symposium. IEEE Computer Society Press, Los Alamos, CA, pp 188–194

Diday E, Noirhomme-Fraiture M (2008) Symbolic data analysis and the SODAS software. Wiley, Chichester

Diettrich T (2000) Ensemble systems in machine learning. In: International workshop on multiple classifier systems, Lecture notes in computer science, vol 1857, pp 1–15. Springer

Dudek A, Pełka M, Wilk J, Walesiak M (2019) The `symbolicDA` package for R software. url: https://cran.r-project.org/package=symbolicDA

Dudek A (2013) Metody analizy danych symbolicznych w badaniach ekonomicznych. Wydawnictwo Uniwersytetu Ekonomicznego we Wrocławiu, Wrocław

Fumera G, Roli F (2005) A theoretical and experimental analysis of linear combiners for multiple classifier systems. IEEE Trans Pattern Anal 27(6):942–956

Haykin S (1998) Neural networks. A comprehensive foundation. Prentice Hall, New Jersey

Leo M, Shama S, Maddulety K (2019) Machine learning in banking risk management. Risks 7(1):1–22

Lessmann S, Seow H-V, Baesens B, Thomas L (2015) Benchmarking state-of-the-art classification algorithms for credit scoring: An update of research. Eur J Oper Res 247(1):124–136

Louzada F, Ara A, Fernandes G (2016) Classification methods applied to credit scoring: A systematic review and overall comparison. Surv Oper Res Man Sci 21(2):117–134

Mori K, Matsugu M, Suzuki T (2005) Face recognition using SVM Fed with intermediate output of CNN for face detection. In: IAPR conference on machine vision applications, pp 410–413, 16–18 May 2005

Munkhdalai L, Munkhdalai T, Namsrai O-E, Lee J, Ryu K (2019) An empirical comparison of machine-learning methods on bank client credit assessments. Sustainability. https://doi.org/10.3390/su11030699

Noirhomme-Fraiture M, Brito P (2011) Far beyond the classical data models: symbolic data analysis. Stat Ana D Min 4(2):157–170

Polikar R (2006) Ensemble based systems in decision making. IEEE Circ Sys Mag 6(3):21–45

Pełka M (2018) Podejście wielomodelowe analizy danych symbolicznych w ocenie zdolności kredytowej osób fizycznych. Prace Naukowe Uniwersytetu Ekonomicznego we Wrocławiu 507:200–207

Safavian S, Landgrebe D (1991) A survey of decision tree classifier methodology. IEEE Trans Syst Man Cybern 21(3):660–674

Šíma J (1995) Neural expert systems. Neural Netw 8(2):261–271

Simoff S (1996) Handling uncertainty in neural networks: an interval approach. In: IEEE International conference on neural networks, vol 3, pp 535–549
Rossi F, Conan-Guez B (2002) Multilayer perceptron on interval data. In: Jajuga K, Sokołowski A, Bock H-H (eds) Classification, clustering and data analysis. Berlin, Springer, pp 427–434
Therneau T, Atkinson B, Ripley B (2019) The `rpart` package for R software. url: https://cran.r-project.org/package=rpart
Trzęsiok M (2018) Wzmacnianie zdolności predykcyjnych modeli dyskryminacyjnych przez wyodrębnianie zmiennych objaśniających z sieci neuronowych. Prace Naukowe Uniwersytetu Ekonomicznego we Wrocławiu 508:227–236

Applications in Finance

Inequality Restricted Least Squares (IRLS) Model of Real Estate Prices

Mariusz Doszyń

Abstract The aim of the paper is developing an econometric model that may support the process of real estate mass appraisal. The research hypothesis assumes that a model with restrictions enables a more precise determination of the impact of real property attributes on the prices than an analogous model without restrictions. The so-called Szczecin algorithm of real estate mass appraisal serves as a starting point for the model determination. A unitary price of undeveloped land real properties designated for low-rise residential development constitutes an explained variable. A set of explanatory variables is comprised of the following real estate attributes: surface area, plot physical properties, utilities, transport availability, real estate neighbourhood. The impact of a location was considered through dummy variables adopted for city surveying sections. All the variables were introduced into the model taking into account the measurement scales best suited for each of them. Two types of restrictions, (1) non-negativity of an attribute impact and (2) monotonicity of an attribute impact, will be imposed on the model parameters. These restrictions refer to the parameters with variables (attributes) other than surface area, which is measured in m^2. The procedure of estimation of a model with restrictions will be discussed. The model will be verified with the use of a real transaction database from the Szczecin real estate market concerning undeveloped land real estate designated for low-rise residential development.

Keywords Mass appraisal · Econometrics · Inequality restricted least squares (IRLS) · Multicollinearity

1 Introduction

A well-constructed econometric model may seem to be an attractive tool for real estate appraisal. It refers chiefly to mass appraisal, for which econometric models constitute a natural choice. Econometric models are also used in individual valuations.

M. Doszyń (✉)
University of Szczecin, Szczecin, Poland
e-mail: mariusz.doszyn@usz.edu.pl

K. Jajuga et al. (eds.), *Classification and Data Analysis*, Studies in Classification, Data Analysis, and Knowledge Organization,
https://doi.org/10.1007/978-3-030-52348-0_6

The usefulness of econometric models in real estate appraisal needs to be considered in two spheres, i.e. from theoretical and empirical points of view.

Applying econometric models in valuation requires the satisfaction of numerous formal assumptions. Namely, the assumptions of a classic linear model; see, e.g., Pawłowski (1981). Those assumptions are rarely satisfied in sciences other than experimental ones.

In econometric appraisal models, explanatory variables comprise real estate attributes, including a location, which also accounts for the factors attributable to the demand side. Real estate attributes are typically random qualitative variables measured on an ordinal scale or a nominal scale. In this study, a unitary real estate price constitutes the explained variable and the database contains information regarding market transactions. Therefore, an econometrician has virtually no influence on which real properties are included in the database, and it is "determined" by the market.

Real estate constitutes a specific type of goods demonstrating low repeatability of qualities that result, for instance, from exceptional location, which is demonstrated in the differentiation of random deviation variance. Such specificity of real estate results in heteroscedasticity of an error term.

In many cases, real properties form continuous areas, and this means that similar real properties are adjacent to one another. The neighbourhood of similar real estate may entail the spatial correlation of an error term.

Another important issue decreasing the effectiveness of estimators of the econometric model parameters involves the (statistical) multicollinearity of attributes. It is yet one more specific quality of real estate. The real estate featuring a better state of attribute A typically demonstrates more favourable states of attributes B, C, etc. For instance, real properties built in a more favourable location usually enjoy better transport availability or better neighbourhood. Strong multicollinearity may increase the variance of estimators to such a degree that the estimations of parameters may have an incorrect sign.

The article presents an attempt at resolving this issue. Restrictions in the form of inequalities were imposed on the econometric model parameters, which should increase the effectiveness of estimators. The restrictions arise from the a priori knowledge of the model parameters. It concerns the non-negative impact of attributes (other than surface area). The second type of restrictions enforces the monotonicity of an attribute impact, which means that a better state of an attribute has an impact not smaller than the one preceding it.

Recapitulating, the objective of the paper is developing an econometric model that may support the process of real estate mass appraisal. The research hypothesis assumes that a model with restrictions enables a more precise determination of the impact of real property attributes on their prices than an analogous model without restrictions.

2 Literature Review

The econometric model of real estate appraisal proposed in the article was developed on the basis of the so-called Szczecin algorithm of real estate mass appraisal. The algorithm is described, e.g. in Hozer et al. (1999). Selected problems related to its econometric specification were described in Doszyń and Hozer (2017).

The possibilities of applying econometric models in appraisal are discussed in the works of, e.g., Benjamin et al. (2004), Isakson (1998), Dell (2017). Econometric models are also frequently considered in the context of mass appraisal; see, e.g., Pagourtzi et al. (2003), Kauko and D'amato (2008), Jahanshiri et al. (2011), McCluskey et al. (2013).

Econometric models featuring spatial effects are often proposed for mass appraisals (Fik et al. 2003). In the models of spatial econometrics, spatial effects are treated as a proxy variable for a location.

In presented paper, econometric models with restrictions in the form of inequalities imposed on parameters are proposed for the purpose of equalizing the negative effects of the multicollinearity of explanatory variables. The restrictions concern the parameters of attribute impact on a unitary real estate price. In the restrictions, the a priori information on the parameters is taken into account. A description of such types of models can be found in Judge and Takayama (1966), Lovell and Prescott (1970), Liew (1976). More advanced tests verifying the hypotheses of the veracity of restrictions in the form of inequalities can be found in Wolak (1989), Grömping (2010).

Judge and Takayama (1966) demonstrate the fact that regression models with the a priori knowledge about parameters can be presented as a square programming task. They analyse the properties of estimators in a regression model with restrictions in the form of inequalities for various regression equations. In the presented example, they use solutions proposed for probability estimation in a Markov process.

On the basis of the studies conducted by Lovell and Prescott (1970), the researchers find that in Ordinary Least Squares (OLS) models, variables with incorrect signs of parameter estimates, resulting from the strong multicollinearity of explanatory variables, need to be rejected. Rejection of this type of variables leads to the biasedness of estimators, but it significantly increases their efficiency. Furthermore, leaving collinear variables disturbs the values of t-Student statistics (by elevating the standard errors of parameters estimation).

Liew (1976) finds that Inequality Restricted Least Squares (IRLS) models enable maintaining the consistency of the obtained results with the theory. He discusses in detail the rules of IRLS model estimation, and he further explains a variance and co-variance matrix of IRLS estimators. The properties of estimators are considered for small and large samples. In a simulation test, using Monte Carlo methods, the biasedness and effectiveness of IRLS and OLS model estimators are compared.

In Wolak (1989), the tests of validity of linear restrictions in the form inequalities in linear econometric models are developed. Those tests are also generalized for the case of linear models of co-dependent equations. The proposed tests were illustrated with examples.

Models featuring restrictions in the form of inequalities, in the context of mass appraisals, are analysed in Pace and Gilley (1990). The authors point out to the fact that in mass appraisal models the multicollinearity of explanatory variables may lead to estimates inconsistent with the theory. Through a series of Monte Carlo experiments, they demonstrated that the estimators of models with restrictions in the form of inequalities are far more effective than in models without any restrictions.

In Grömping (2010), the rules related to the conclusions drawn through the use of restrictions in the form of inequalities were described along with an ic.infer software package in R programming language, which enables the construction of this type of models. The software package additionally enables the verification of the hypotheses regarding the validity of constraints in the form of inequalities. The calculations for this article were conducted with the above-mentioned software package.

3 Research Methodology

The econometric model presented in the study was constructed on the basis of the so-called Szczecin algorithm of real estate mass appraisal. A detailed description of the algorithm is featured in, e.g., Hozer et al. (1999).

The algorithm can be noted as follows:

$$w_{ji} = wwr_j \cdot pow_i \cdot w_{baz} \prod_{k=1}^{K} \prod_{p=1}^{k_p} (1 + A_{kpi}) \tag{1}$$

where

w_{ji}—market value of ith real property in jth location attractiveness zone,
wwr_j—market value coefficient in jth location attractiveness zone ($j = 1, 2, \ldots, J$),
J—number of location attractiveness zones,
pow_i—surface area of ith real property,
w_{baz}—estimated value of 1 m^2 of real property of the worst states of attributes in the least attractive location zone,
A_{kpi}—impact of p state of k attribute for ith real property determined by property surveyors $(k = 1, 2, \ldots, K; p = 1, 2, \ldots, k_p)$,
K—number of attributes,
k_p—number of states of k attribute.

The real estate value defined for representative real properties by property surveyors is a dependent variable in the algorithm. In the econometric model considered in this paper, a transaction price of real estate constitutes the explained variable.

Location attractiveness zones constitute continuous areas defined by experts, which feature a similar impact of a location. They are spatial units for which the so-called market value coefficients are determined. They ought to be homogeneous in terms of real properties similarity.

In the econometric model, no location attractiveness zones will be featured. A city surveying unit will serve as a spatial unit.

The following operations were performed on algorithm (1) in order to construct the econometric model:

(a) In place of a real estate value (w_{ji}), real estate transaction price $\left(c_{ji}\right)$ was substituted, where subscript $j = 1, 2, \ldots, J$. now refers to surveying sections, and not to location attractiveness zones (as was the case in the algorithm).
(b) Equation (1) was divided by surface area $\left(\text{pow}_i\right)$ and logarithmized.
(c) With the exception of surface area, which also constitutes an attribute, the remaining attributes are qualitative variables measured on an ordinal scale. The attributes were introduced into the set of explanatory variables as dummy variables for all states of an attribute, except for the worst one. The econometric model will feature a constant term; therefore, the worst states of attributes are omitted in order to avoid strict collinearity of explanatory variables.
(d) In the econometric model, the impact of a location will be accounted for by adding dummy variables for individual city surveying sections. For the avoidance of any strict collinearity, a dummy variable for the surveying section featuring the lowest unitary real estate prices will be omitted. It means that the impact of a location in the remaining surveying sections ought to be non-negative. The parameters with dummy variables for surveying sections are an equivalent of market value coefficients in algorithm (1).

After the performance of the above-mentioned operations, the following model hypothesis was obtained:

$$\ln\left(\frac{c_{ji}}{pow_i}\right) = \ln c_{baz} + \alpha_p pow_i + \sum_{k=1}^{K}\sum_{p=2}^{k_p} \alpha_{kp} x_{kpi} + \sum_{j=2}^{J} \beta_j oe_{ji} + u_i \quad (2)$$

where

c_{ji}—transaction price of ith real property in jth surveying section,
pow_i—surface area of ith real property,
$\ln c_{baz}$—logarithmized base price (constant term),
K—number of attributes ($k = 1, 2, \ldots, K$),
k_p—number of states of k attribute $\left(p = 2, \ldots, k_p\right)$,
α_{kp}—impact of p state of k attribute,
x_{kpi}—0–1 variable equal to one for p state of k attribute,
β_j—impact of location in jth surveying section,

oe_{ji}—0–1 variable equal to one for jth surveying section ($j = 2, 3, \ldots, J$),
J—number of surveying sections,
u_i—error term.

For a large number of attributes and their states, restrictions can be conveniently presented in the form of a matrix. The notation of model (2) in a matrix form is as follows:

$$\mathbf{C} = \mathbf{X}\boldsymbol{\alpha} + \mathbf{S}\boldsymbol{\beta} + \mathbf{u} \tag{3}$$

where

C—vector of logarithmized unitary prices of real estate ($N \times 1$),
N—number of real estate,
X—matrix of 0–1 variables for the states of attributes ($N \times g$),
g—number of explanatory variables (with surface area but with exclusion of location attractiveness zones),
$\boldsymbol{\alpha}$—parameter vector of the impact of attributes states ($g \times 1$),
S—matrix of 0–1 variables for surveying sections ($N \times J - 1$),
$\boldsymbol{\beta}$—parameter vector of the impact of location in surveying sections ($J - 1 \times 1$),
u—vector of error terms ($N \times 1$).

Two types of restrictions in the form of inequalities were taken into consideration:

(a) $\alpha_{kp} \geq 0$—non-negative impact of attributes states,
(b) $\alpha_{k,p+1} \geq \alpha_{kp}$—monotonicity of attributes impact.

The restrictions refer to parameters α_{kp}. According to the a priori knowledge, attributes (other than surface area) should elevate a unitary real estate price, hence the assumption of their non-negative impact. Monotonicity of the impact of a given attribute means that a better state of attribute ($p + 1$) increases a unitary real estate price by at least the same amount as state (p) preceding it.

In model (3), the squared sum of residuals is minimized:

$$\hat{\mathbf{u}}^{\mathrm{T}}\hat{\mathbf{u}} = \left(\mathbf{C} - \mathbf{X}\hat{\boldsymbol{\alpha}} - \mathbf{S}\hat{\boldsymbol{\beta}}\right)^{\mathrm{T}}\left(\mathbf{C} - \mathbf{X}\hat{\boldsymbol{\alpha}} - \mathbf{S}\hat{\boldsymbol{\beta}}\right) \rightarrow \min \tag{4}$$

while the following restrictions must be satisfied

$$\mathbf{Z}\hat{\boldsymbol{\alpha}}\prime \geq 0 \tag{5}$$

where

$\hat{\boldsymbol{\alpha}}\prime$—restricted vector of estimators of the impact of attributes states,
Z—matrix of the restrictions imposed on vector $\hat{\boldsymbol{\alpha}}\prime$ estimators,
$\hat{\mathbf{u}}$—vector of residuals.

Vector $\hat{\boldsymbol{\alpha}}\prime$ contains only estimators with restrictions. It does not include constant term and estimator next to surface area. Impact of surface area on unitary real estate

Table 1 Real estate attributes and their states

Attribute	Attribute state/symbol
Surface area	Measured in m^2—pow_i
Utilities	None incomplete—uz_1 complete—uz_2
Neighbourhood	Unfavourable average—o_1 favourable—o_2
Transport availability	Unfavourable average—dk_1 favourable—dk_2
Physical properties	Unfavourable average—cf_1 favourable—cf_2

price is often negative. The number of rows in matrix **Z** corresponds to the number of restrictions, while the number of columns is equal to the number of elements of vector $\hat{\boldsymbol{\alpha}}\prime$. Relationships (4) and (5) form the quadratic programming problem.

Real estate attributes, their states and symbols are presented in Table 1.

For the attributes presented in accordance with the order presented in Table 1 (with the exception of surface area), a matrix notation of restrictions is as follows:

$$\mathbf{Z}\hat{\boldsymbol{\alpha}}\prime = \begin{bmatrix} 1 & 0 & 0 & 0 & 0 & 0 & 0 & 0 \\ -1 & 1 & 0 & 0 & 0 & 0 & 0 & 0 \\ 0 & 0 & 1 & 0 & 0 & 0 & 0 & 0 \\ 0 & 0 & -1 & 1 & 0 & 0 & 0 & 0 \\ 0 & 0 & 0 & 0 & 1 & 0 & 0 & 0 \\ 0 & 0 & 0 & 0 & -1 & 1 & 0 & 0 \\ 0 & 0 & 0 & 0 & 0 & 0 & 1 & 0 \\ 0 & 0 & 0 & 0 & 0 & 0 & -1 & 1 \end{bmatrix} \begin{bmatrix} \hat{\alpha}_{uz1} \\ \hat{\alpha}_{uz2} \\ \hat{\alpha}_{o1} \\ \hat{\alpha}_{o2} \\ \hat{\alpha}_{dk1} \\ \hat{\alpha}_{dk2} \\ \hat{\alpha}_{cf1} \\ \hat{\alpha}_{cf2} \end{bmatrix} \geq \begin{bmatrix} 0 \\ 0 \\ 0 \\ 0 \\ 0 \\ 0 \\ 0 \\ 0 \end{bmatrix} \quad (6)$$

The matrix is featured in (5). The restrictions included in the above matrix mean that each average state of an attribute has a non-negative impact on a unitary real estate price. Moreover, the impact of each favourable state is not smaller than the impact of the state preceding it, i.e. an average state.

4 Data and Research Results

Two versions of model (2) were estimated, a model without restrictions (OLS—Ordinary Least Squares) and a model with restrictions in the form of inequalities

(IRLS—Inequality Restricted Least Squares). The calculations were performed with the ic.infer software package, which is available in R programming language.

The database featured information on 94 transactions of undeveloped land designated for low-rise residential development in Szczecin. The prices were determined for 2018 year.

Table 2 contains descriptive statistics of real estate unitary prices, as well as their attributes. On the grounds of the table provided, it is possible to conclude that the unitary real prices oscillated between 133.70 and 400.80 PLN/m^2, with the median equal to 273.57 PLN/m^2. Real estate surface areas featured in the database were ranging between 187 and 2896 m^2, while the median of real estate surface area was 860.50 m^2. Quartile deviation (Q) and positional coefficient of variation $\left(V_Q\right)$ demonstrate very low variability of price and the surface area. The variability of surface area is only slightly higher.

Attributes other than surface area, i.e. the ones measured on an ordinal scale, were encoded in the following manner: the worst variant—1, average variant—2, favourable variant—3. In Table 3, the shares of attribute states other than surface area are presented.

In case of all attributes, there was usually one dominant state. As many as 79.8% of real estate has complete utility, and 53.2% real estate has unfavourable neighbourhood. For transport availability and physical properties, the average variant was dominant. It was respectively 85.1% and 78.7%.

The results of OLS and IRLS model estimation are presented in Table 4. It features the estimations of parameters and p-values for the OLS model, whereas for the IRLS model only the estimations of parameters were provided. It results from the fact that in the models with restrictions in the form of inequalities, the distributions typically used

Table 2 Descriptive statistics determined for unit prices and surface area

Statistics	Price 1 m^2	Surface area
Min	133.70	187.00
$Q_{1.4}$	239.80	684.75
M	273.57	860.50
$Q_{3.4}$	299.55	1214.75
Max	400.80	2896.00
Q	29.88	265.00
V_Q (%)	10.92	30.80

Table 3 Shares of the attribute states measured on ordinal scale

Attribute state	Utilities	Neighbourhood	Transport availability	Physical properties
1	0.096	0.532	0.085	0.074
2	0.106	0.277	0.851	0.787
3	0.798	0.191	0.064	0.138

Table 4 Results of the estimation of a model without restrictions (OLS) and a model with restrictions (IRLS)

Variable	Estimations of parameters (OLS)	p-value (OLS)	Estimations of parameters (IRLS)
$const$	4.6607	0.0000*	4.6705
pow	−0.0002	0.0001*	−0.0002
u_{z1}	0.0092	0.9305	0.0000
u_{z2}	−0.0076	0.9147	0.0000
o_1	0.2346	0.0032*	0.2412
o_2	0.4528	0.0004*	0.4459
dk_1	0.0105	0.8326	0.0000
dk_2	−0.0487	0.5402	0.0000
cf_1	0.1595	0.0383*	0.1445
cf_2	0.2588	0.0069*	0.2441
oe_1	0.5769	0.0272*	0.5138
oe_2	0.6678	0.0013*	0.6246
oe_3	0.6976	0.0011*	0.6467
oe_4	0.8025	0.0001*	0.7815
oe_5	0.5444	0.0196*	0.5428
oe_6	0.4861	0.0526	0.4238
oe_7	1.0288	0.0005*	0.9772
oe_8	0.8522	0.0002*	0.8493
oe_9	0.3078	0.1718	0.2983
oe_{10}	0.8427	0.0011*	0.7774
oe_{11}	0.9398	0.0001*	0.9385
oe_{12}	0.7102	0.0003*	0.6971
oe_{13}	0.7223	0.0007*	0.7028
oe_{14}	0.6666	0.0010*	0.6535
oe_{15}	1.0694	0.0000*	1.0049

*p-value lower than 0.05

in an econometric model, such as t—Student distribution, χ^2, F, JB distributions, etc., do not apply. Therefore, it is not possible to apply the usual diagnostic tests of an econometric model. In OLS model, with respect to Jarque–Bera test and Breusch–Pagan test, normality of residuals was rejected, but residuals were homoscedastic (significance level 0.05).

With regard to the signs of parameter estimations for OLS and IRLS models, the attributes (except for surface area) and variables for surveying sections are defined in such as way so that their impact is non-negative. This is the case for the IRLS model, in which only the impact of surface area was negative, a result being consistent with the assumptions (real properties of greater surface area obtain lower unitary prices). In the OLS model, the impact of utilities and transport availability is incorrect. In the case of these two attributes, a favourable attribute state exerted a negative impact, which is inconsistent with the theoretical assumptions. In line with the adopted premise, the impact of attributes ought to be non-negative and monotonous. The estimations of parameters for utilities and transport availability do not satisfy those two assumptions, including also the assumption of a monotonous impact of an attribute.

However, it can be observed that in the OLS model, parameters with dummy variables for utilities and transport availability do not significantly differ from zero (significance level of 0.05). In the IRLS model, the estimations of parameters with dummy variables for utilities and transport availability are equal to zero, which, in the light of the results obtained for the OSL model, is a correct result. The IRLS model equalizes to zero the influence of those variables that occurred to be statistically insignificant in the OLS model. It does not need to be so in each case. In the models under consideration, the imposition of restrictions is equivalent to eliminating the variables that differ from zero to an insignificant degree. It is consistent with the proposal analysed in Lovell and Prescott (1970).

In the case of the remaining attributes, i.e. neighbourhood and physical properties, the estimations of parameters in the OLS and IRLS models were similar and consistent with the a priori assumptions. Parameter estimations with dummy variables adopted for those attributes were non-negative, and the condition of the monotonous impact of those attributes is also satisfied. It means that in the case of those two attributes, a favourable state increased a unitary price more than an average variant did.

Overall, it can be concluded that among all attributes, neighbourhood and physical properties exerted the greatest influence on unitary prices, and surface area demonstrated the least influence. Utilities and transport availability did not affect real property unitary prices in a statistically significant manner.

Nevertheless, the parameter estimations with dummy variables adopted for surveying sections reveal that, after all, it was the location attribute that exerted the most decisive impact on prices. The parameters next to those variables were, with the exceptions of two cases, significantly different from zero and they assumed fairly high, positive values. Only the parameters with oe_6 and oe_9 variables were statistically insignificant. Parameter estimations with variables adopted for surveying sections ought to be positive, which is the case, since a surveying section featuring the lowest real estate unitary prices was omitted.

Once the restrictions were imposed, the coefficient of determination decreased slightly from 0.644 (in the OLS model) to 0.641 (in the IRLS model), which proves the validity of the restrictions. Therefore, it means that the imposition of restrictions is "consistent" with the data.

On account of the fact that conventional diagnostic tests cannot be conducted for the IRLS model, the above two models will be compared in terms of accuracy of the appraisals generated with them, bearing in mind that in this case appraisals are theoretical prices determined on the grounds of the analysed models.

At this juncture, it can be added that the author considered performing a division of a set of observations into a training data set, which served for the models estimation, and into a test data set, in which the accuracy of appraisals was supposed to be examined. However, the number of observations was too small to conduct this sort of a division. In particular, it concerns a number of observations in individual surveying sections. A low number of observations in certain surveying sections rendered this kind of division (into a training and test data sets) impossible. Therefore, theoretic unitary prices of real estate were defined for all the observations in the data set.

The following appraisal errors were determined:

(a) Percentage Error (PE):

$$PE_i = \frac{c_i - \hat{c}_i}{c_i} \tag{7}$$

where

c_i—actual unitary price of ith real estate,

$\hat{c}_i$—theoretical unitary price of ith real estate (obtained on the basis of the OLS or IRLS model),

(b) Absolute Percentage Error (APE):

$$APE_i = \frac{\left|c_i - \hat{c}_i\right|}{c_i} \tag{8}$$

c) Share of appraisals with $APE_i < 10\%$ ($S_{APE<10\%}$):

$$S_{APE<10\%} = \frac{m}{n} \tag{9}$$

where

m—number of appraisals with $APE_i < 10\%$,

n—number of appraisals.

Measure (9) is nonparametric, it is a share of appraisals with APE below 10%. From the point of view of appraisal accuracy, it can be assumed that the valuations burdened with an error greater than ±10% are unfavourable; therefore, for each of the estimated models, the percentage share of errors not exceeding 10% (in plus or in minus) was calculated. For OLS model, the share of that kind of appraisals was equal to 57.4%. For IRLS model, the share was slightly higher and equal to 60.9%. Generally, in 60% of the cases either model results in appraisals differing from the actual price by less than 10%.

PE and APE error distributions for the OLS and IRLS models are presented in Figs. 1 and 2. The overall conclusion is that appraisal errors were within a similar range for the analysed models. However, the percentage share of errors exceeding ±10% occurred to be slightly higher for the OLS model. In the case of this model, 42.6% of real properties were valued in such a way that a theoretical price differed

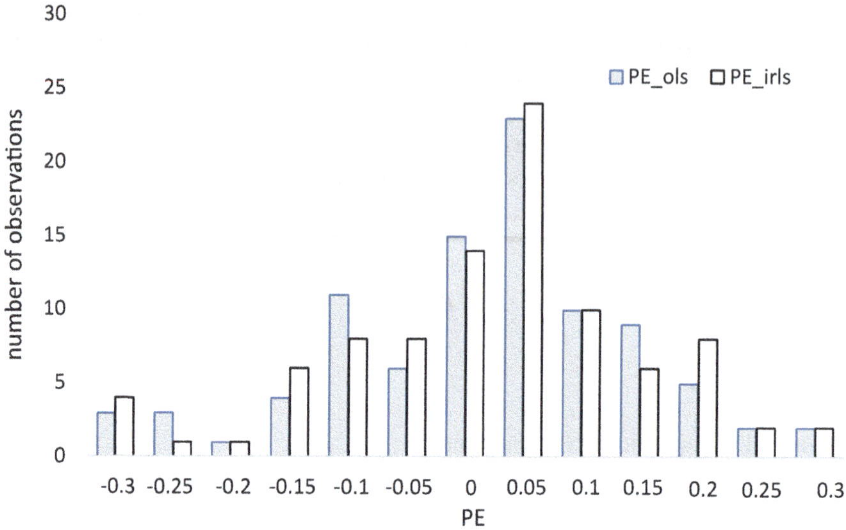

Fig. 1 PE error distribution determined for the OLS and IRLS models

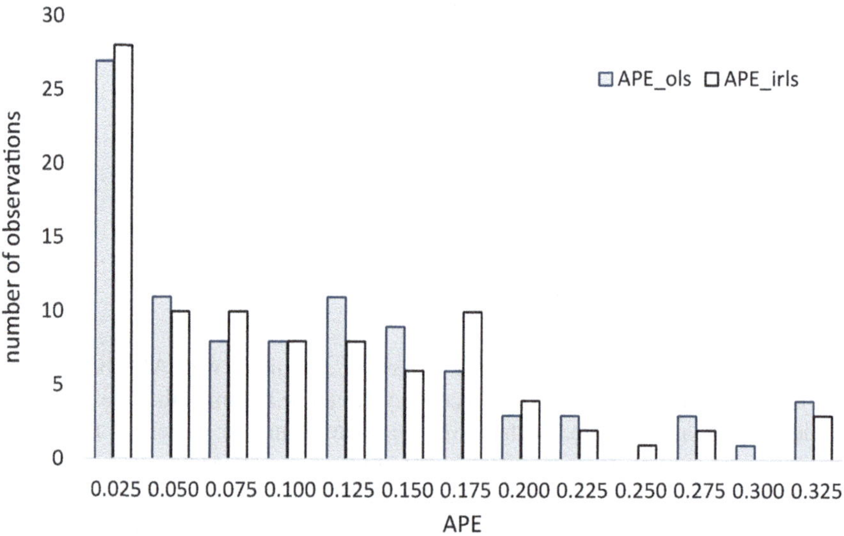

Fig. 2 APE error distribution determined for the OLS and IRLS models

from the actual price by more than 10%. In the case of the IRLS model, that percentage share was insignificantly lower and equal to 39.1%.

In Table 5, some statistics for PE and APE distributions are presented. In the context of appraisal, especially extreme values of PE and APE errors are interesting.

Table 5 Statistics for PE and APE error distributions

Error	Model	Min	First quartile	Median	Mean	Third quartile	Max
PE	OLS	−0.603	−0.091	0.000	−0.009	0.073	0.318
	IRLS	−0.622	−0.087	0.000	−0.009	0.070	0.320
APE	OLS	0.000	0.019	0.081	0.097	0.142	0.603
	IRLS	0.000	0.018	0.074	0.096	0.151	0.622

According to minimum and maximum of PE and APE, both OLS and IRLS models are similar, with slightly better results for OLS model. For OLS model, minimum was a little higher and maximum—a little lower.

Mean Percentage Errors (MPE) were almost the same for each model and very close to zero, so appraisals were unbiased. For PE and APE distributions, first and third quartiles as well as medians were very similar. Mean APE was equal to 0.097 for OLS model and 0.096 for IRLS model. It means that appraisals were different from transaction prices on average by 9.6–9.7%. This proves that appraisal accuracy based on presented models was rather good.

5 Conclusions

Recapitulating, in the case of econometric appraisal models the specificity of real estate and the real estate market frequently does not contribute to the satisfaction of the formal assumptions of a classic linear model. It applies to the majority of economic phenomena; hence, numerous proposals of solving this type of problems can be found.

For instance, it can be demonstrated that the randomness of explanatory variables leads to estimators that are asymptotically unbiased, i.e. in large samples, the lack of randomness of explanatory variables does not constitute a serious issue. The problem of the multicollinearity of explanatory variables may be resolved by applying ridge regression or, as in this article, by imposing restrictions on parameters in the form of inequalities. Spatial autocorrelation is taken into account in econometric spatial models. In turn, the generalized least squares method enables equalizing the negative effects of heteroscedasticity of an error term.

The verification of an econometric model should constitute a final evaluation of the model usefulness, with a particular consideration of the accuracy of appraisals generated by the model. The quality of the obtained results depends chiefly on the completeness, reliability and credibility of real estate databases. A frequently encountered problem is an insufficient number of observations, being the consequence of the fact that local real estate markets are poorly developed and transactions are rarely concluded on those markets. Real estate attributes are typically qualitative variables, therefore, when their states are determined, it is hard to avoid subjectivity. Moreover, the very quality of information on real estate is often poor, which further reduces data credibility.

The next questions related to the model development involve its proper specification and accurate attribute selection. In this study, the econometric model was specified on the basis of the so-called Szczecin algorithm of real estate mass appraisal, while the attributes were defined by a group of property surveyors and real estate analysts.

On the grounds of the estimated models, it is possible to conclude that the model with restrictions in the form of inequalities enables more accurate, from the standpoint of theoretical assumptions, determination of the impact of real estate attributes on their unitary price. In the case described in the article, the restriction in the form of inequalities contributed to removing the impact of those variables, for which the parameters did not differ from zero to a statistically significant degree. Nevertheless, for other databases the benefits of applying models with restrictions may be significantly greater. Above all, such types of restrictions enable limiting the negative effects of the multicollinearity of explanatory variables. The use of the a priori knowledge of the model parameters permits to improve the effectiveness of parameter estimators. However, it needs to be remembered that it occurs at the costs of their biasedness.

To sum up, the main contribution of the article is a construction of an econometric model that may support the process of real estate mass appraisal. The impact of a location was considered through dummy variables adopted for city surveying sections. All the variables were introduced into the model taking into account the proper measurement scales. Two types of restrictions (non-negativity of an attribute impact, monotonicity of an attribute impact) were imposed on the model parameters. The proposed model was verified with the use of a real transaction database from the Szczecin real estate market concerning undeveloped land properties designated for low-rise residential development. Because multicollinearity of explanatory variables is usually an important issue, IRLS models are recommended.

Acknowledgements The article was financed from the resources of the National Science Centre in Poland within the framework of project No. 2017/25/B/HS4/01813.

References

Benjamin JD, Randall S, Guttery RS, Sirmans CF (2004) Mass appraisal: an introduction to multiple regression analysis for real estate valuation. J Real Estate Pract Educ 7(1):65–77

Dell G (2017) Regression, critical thinking, and the valuation problem today. Appraisal J 85(3):217–230

Doszyń J, Hozer J (2017) Szczeciński algorytm masowej wyceny nieruchomości – podejście ekonometryczne (Szczecin's algorithm of real estate mass valuation—econometric approach). Studia i Prace WNEiZ 50(1):19–30

Fik TJ, Ling DC, Mulligan GF (2003) Modelling spatial variation in housing prices: a variable interaction approach. Real Estate Econ 31(4):623–646

Grömping U (2010) Inference with linear equality and inequality constraints using R: the package ic.infer. J Stat Softw 33(10):1–31
Hozer J, Foryś I, Zwolankowska M, Kokot S, Kuźmiński W (1999) Ekonometryczny model masowej wyceny nieruchomości gruntowych (Econometric algorithm of land properties mass valuation). Department of Econometrics and Statistics, University of Szczecin (Poland), Szczecin
Isakson HR (1998) The review of real estate appraisals using multiple regression analysis. J Real Estate Res 15(2):177–190
Jahanshiri E, Buyong T, Shariff ARM (2011) A review of property mass valuation models. Pertanika J Sci Technol 19:23–30
Judge GG, Takayama T (1966) Inequality restrictions in regression analysis. J Am Stat Assoc 61(313):166–181
Kauko T, D'amato M (eds) (2008) Mass appraisal methods: an international perspective for property valuers. Blackwell, London
Liew CK (1976) Inequality constrained least—squares estimation. J Am Stat Assoc 71(355):746–751
Lovell MC, Prescott E (1970) Multiple regression with inequality constraints: pretesting bias, hypothesis testing and efficiency. J Am Stat Assoc 65(330):913–925
McCluskey WJ, McCord M, Davis PT, Haran M, McIllhatton D (2013) Prediction accuracy in mass appraisal: a comparison of modern approaches. J Property Res 30(4):239–265
Pace RK, Gilley OW (1990) Estimation employing a priori information within mass appraisal and hedonic pricing models. J Real Estate Finan Econ 3:55–72
Pagourtzi E, Assimakopoulos V, Hatzichristos T, French N (2003) Real estate appraisal: a review of valuation methods. J Property Investment Finan 21(4):383–401
Pawłowski Z (1981) Zasady ekonometrii (Principles of econometrics). PWN, Warszawa
Wolak FA (1989) Testing inequality constraints in linear econometric models. J Econometrics 41(1989):205–235

Application of Hill Estimator to Assess Extreme Risks in the Metals Market

Dominik Krężołek

Abstract Rare phenomena are extremely important for risk assessment. The main reason is that their occurrence usually results in significant consequences. There are many methods for identifying such events. Some of these include extreme statistics, which are part of the extreme value theory. High-order quantiles allow to determine the level of risk for which the probability of a risky event occurring is negligible. In this paper, the issue of estimation of tail index of probability distribution using Hill estimator and its modification is presented. The results of comparative analysis for selected nonparametric and parametric models were compared. Empirical analysis was carried out on the example of assets from the base metals market: aluminium and copper.

Keywords Hill estimator · Extreme statistics · High-order quantiles · Value at risk · Metals market

1 Introduction

Analysing economic phenomena in recent years, one can observe multidirectional changes in the processes that define them. Often these changes are important as to the size and consequences they bring, and thus, these changes generate a risk. Taking into account the diversity and unpredictability of events that may affect market decisions, there is a fear that the activity may produce final results that significantly differ from expectations. Of course, the risk appears in many life situations, but in our paper we focus mainly on market risk, i.e. risk that is generated mainly by market factors.

There are many definitions of risk depending on the area in which it is identified. It can be said that risk arises when the results of our intentions differ from our expectations. Moreover, these results are unknown and uncertain. In decision-making theory, there are listed situations when the decision-maker, on the basis of the information available, has to choose the right decision option. The first one is when the decision

D. Krężołek (✉)
University of Economics in Katowice, Katowice, Poland
e-mail: dominik.krezolek@ue.katowice.pl

K. Jajuga et al. (eds.), *Classification and Data Analysis*, Studies in Classification, Data Analysis, and Knowledge Organization,
https://doi.org/10.1007/978-3-030-52348-0_7

is made under the condition of certainty. It means that each decision has specific and known consequences. If the decision is made under the condition of risk, each decision entails more than one consequence and both the set of possible consequences and the probability of their occurrence are known. But if we do not know the probability of occurrence of consequences of a given decision, such a decision is made under the condition of uncertainty.

Within risk, there are several approaches that define them. Mostly, there is a negative and neutral concept of risk. A negative concept means that the results of our actions differ from expectations and take values below the expected level. In a neutral situation, the final result differs from the expected one, but it is not important whether it is above or below that level. Regardless of the definition of risk, it can also be pointed out that is described by two dimensions. The first dimension is the probability of a risky event to be occurred, while the second dimension is the consequences of implementing such risky scenarios.

As mentioned at the beginning, there are two aspects of risk. One is volatility, and the other is uncertainty associated with the analysed process. Volatility is considered as a dispersion observed within the observation of the examined process. As far as it is concerned, there are two types of factors that determine it. The first one are internal factors that are specific and characteristic for a given process, while the second one are external factors which are not directly related to the process, but have a direct or indirect influence on it. The problem of uncertainty, on the other hand, is related to the situation when the decision-maker is forced to choose the right decision option. Also in this case, internal factors and external factors can be mentioned. Internal factors are related to the subjective approach of the decision-maker to the analysed problem and can usually be controlled. On the other hand, external factors are related to the environment created by the process and the way in which the decision-maker acts, and such factors do not necessarily need to be controlled. Thus, it can be said that there is a certain relationship between risk, volatility and uncertainty. Volatility reflects the mood of investors, and these emotions are observed in the process movements that are studied. In our case, these will be either price processes or return processes. On the other hand, uncertainty reflects a lack of knowledge about the direction of all these changes. Since prices and return are characterized by volatility and uncertainty, we can say that they generate risk, i.e. a result that will differ from the expected result.

2 Extreme Risk

Analysing various types of processes that we observe in the world around us, including the world of economy, it is possible to detect rare events that generate risk at a level which is very far from the expected one. We are talking about the so-called extreme risk, which is associated with events occurring with low probability, but if they do occur, they cause very important consequences, most often losses (Jajuga 2008). In the case of extreme risk analysis, the theory of extreme events and the so-called extreme statistics play a special role. Extreme statistics allow to estimate various parameters that can be used to define rare events, and these parameters are, e.g. quantiles of empirical distributions of examined phenomena or parameters defining periods, in which the analysed processes take some extreme values (Gumbel 2004). In terms of economics, there are many examples of events that have caused various types of drastic changes in the market, for example Black Thursday (24 October 1929), Black Monday (19 October 1987), World Trade Center (11 September 2001), recent financial crisis (2008–2009) or crisis on crude oil market (2014).

From a statistical point of view, two parameters can be distinguished, which refer to extreme events. The first one is a quantile, and the second one is an index of the thickness of the tail of the distribution, which indirectly determines the existence of the probability of the occurrence of a rare event. Turning to the methodology used in the paper at the beginning, the so-called extreme value distribution (EVD) should be defined. The EVD values can be described using the following density function (Gumbel 2004):

$$EVD_{\gamma}(x) = \begin{cases} \exp\left[-(1+\gamma x)^{-1/\gamma}\right], 1+\gamma x \geq 0 \text{ for } \gamma \neq 0 \\ \exp\left[-\exp(-x)\right], x \in \mathbb{R} \qquad \text{for } \gamma = 0 \end{cases} \tag{1}$$

where the parameter γ defines the extreme value index (EVI). It is the most important parameter in this distribution, which measures the thickness of the tail and thus the probability of an extreme event occurring. The tail index takes real values, and the heavier the tail, the higher the index value is. In case of a generalized extreme value distribution, the tail index parameter is the shape parameter, which is invariant of standardizing the distribution (Németh and Zempléni 2018). In our analysis, as a measure determining the quantile we use the value at risk, while as a measure determining the thickness of the tail of the distribution we use the Hill estimator.

3 Hill Estimator and Its Modification

For a heavy-tailed model, given a sample $\underline{\mathbf{X}}_n = (X_1, X_2, \ldots, X_n)$ and the associated sample of ascending order statistics $(X_{1:n} \leq X_{2:n} \leq \cdots \leq X_{n:n})$, the classical EVI estimator is the Hill estimator (Hill 1975) given by

$$H_{k,n} \equiv H_{k,n}(\underline{\mathbf{X}}_n) := \frac{1}{k}\sum_{i=1}^{k}\left[\ln\frac{X_{n-i+1:n}}{X_{n-k:n}}\right] \tag{2}$$

the average of $k-$ log excess over a random level $X_{n-k:n}$, which needs to be an intermediate order statistics satisfying $k = k_n \to \infty$ and $k/n \to 0$ as $n \to \infty$. If the value of k is chosen too high, the variance of the estimator increases. From the other hand, if k is too low, usually the bias of the estimator increases. The problem of the choice of k is also relevant to many other estimators (Fedotenkov 2018). The estimate of $\frac{1}{\gamma}$ varies with k quite substantially. This shows that the choice of k matters for obtaining the proper estimate (Danielson et al. 2016).

Hill estimator has some interesting properties that are discussed extensively in literature. Nevertheless, it can be said that it is scale-invariant, but not location-invariant. In addition, the Hill estimator usually exhibits a high asymptotic bias, but is asymptotically normal with an appropriate value of variance and a nonzero value of mean. Therefore, it is not robust against significant changes in the data range. For that reason, various modifications of the parameter are used in practice. The first one is so-called peak over random threshold estimator, denoted as PORT-Hill (Araújo Santos et al. 2006). The PORT-Hill estimators are functionals of a sample excesses over random level $X_{[nq]+1:n}$, i.e. functionals of $\underline{\mathbf{X}}_{k,n}^{(q)} := \left(X_{n:n} - X_{[nq]+1:n}, \ldots, X_{n-k:n} - X_{[nq]+1:n}\right)$ with $1 \leq k < n - [nq] - 1$. Therefore, the PORT-Hill estimator is given by

$$H_{k,n}^{(q)} := H_k\left(\underline{X}_{k,n}^{(q)}\right) = \frac{1}{k}\sum_{i=1}^{k}\left[\ln \frac{X_{n-i+1:n} - X_{[nq]+1:n}}{X_{n-k:n} - X_{[nq]+1:n}}\right] \tag{3}$$

for some tuning parameter q satisfies $0 \leq q < 1$. The PORT tail estimator is invariant for changes in location and scale, and the tuning parameter q makes it more flexible for abnormal data.

In terms of high quantiles, Gomes et al. (2010) have proposed another modification of classical Hill estimator called quasi-PORT-Hill estimator. Assume that the regularly varying function $L_U(t) = t^{-\gamma}U(t)$ tends to a finite non-null constant with $\gamma > 0, \rho < 0, \beta \neq 0$, where $U(t) = Ct^{\gamma}\left(1 + \frac{\gamma\beta t^{\rho}}{\rho} + o(t^{\rho})\right)$ as $t \to \infty$. Therefore, the quasi-PORT-Hill estimator is given by

$$\bar{H}_{k,n}^{(q)} \equiv \bar{H}_{k,n}^{(q)}\left(\widehat{\beta}, \widehat{\rho}\right) = H_{k,n}^{(q)}\left[1 - \frac{\widehat{\beta}\left(\frac{n}{k}\right)^{\widehat{\rho}}}{1 - \widehat{\rho}}\right] \tag{4}$$

where (β, ρ) are the second-order parameters. These parameters were extensively described by Caeiro et al. (2005). Some extension of the estimator given by (4) is proposed by Gomes et al. (2008) minimum-variance reduced-bias Hill estimator (MVRB-Hill), where they propose to reduce bias without increasing the asymptotic variance. This estimator can be expressed using the following formula:

$$\overline{H}_{k,n} \equiv \bar{H}_{k,n}\left(\widehat{\beta}, \widehat{\rho}\right) = H_{k,n}\left[1 - \frac{\widehat{\beta}\left(\frac{n}{k}\right)^{\widehat{\rho}}}{1 - \widehat{\rho}}\right] \tag{5}$$

This estimator, comparing to the previous once, is not location-invariant but approximately location-invariant, but reveals every asymptotical properties as those given by Formulas (3)–(4).

The Hill estimator allows us to estimate the thickness of the tail of the distribution, which is important for correct risk assessment which is measured by quantiles and, in the case of rare events, by high-order quantiles. In our case, we used a modification of classical Hill estimator to estimate VaR using the approach that it represents the quantile of any probability distribution. For the estimators given by (2)–(5), the formulas for calculating high-order quantiles using Hill, PORT-Hill, quasi-PORT-Hill and MVRB-Hill estimators are presented below:

$$Q_{p|H_{k,n}} := \widehat{\chi}_{p|H_{k,n}} := X_{n-k+1:n}\left(\frac{k}{np}\right)^{H_{k,n}} \tag{6}$$

$$Q_{p|H_{k,n}^{(q)}} := \widehat{\chi}_{p|H_{k,n}^{(q)}} := \left[X_{n-k:n} - X_{[nq]+1:n}\right]\left(\frac{k}{np}\right)^{H_{k,n}^{(q)}} + X_{[nq]+1:n} \tag{7}$$

$$Q_{p|\bar{H}_{k,n}^{(q)}} := \widehat{\chi}_{p|\bar{H}_{k,n}^{(q)}} := \left[X_{n-k:n} - X_{[nq]+1:n}\right]\left(\frac{k}{np}\right)^{\bar{H}_{k,n}^{(q)}} + X_{[nq]+1:n} \tag{8}$$

$$\ln Q_{p|\bar{H}_{k,n}} := \ln \widehat{\chi}_{p|\bar{H}_{k,n}} := \ln X_{n-k+1:n} + \bar{H}_{k,n}\left[\ln\left(\frac{k}{np}\right) + C_p\right] \tag{9}$$

where $H_{k,n}, H_{k,n}^{(q)}, \bar{H}_{k,n}^{(q)}, \bar{H}_{k,n}$ stand for Hill, PORT-Hill, quasi-PORT-Hill and MVRB-Hill estimators of EV, respectively, and $C_p = \widehat{\beta}\left(\frac{n}{k}\right)^{\widehat{\rho}} \frac{\left(\frac{k}{np}\right)^{\widehat{\rho}} - 1}{\widehat{\rho}}$ with the second-order parameters (β, ρ).

4 Empirical Study

The contemporary economy is an extremely complex structure, which is difficult to capture in one single, transparent model. Its form and directions of development depend on many areas and factors. Different market indicators, which correspond to the areas of economy, are barometers of economic development. The analysis concerns the financial, socio-economic, political and many other segments. In this study a specific market area has been chosen, namely the market of industrial (base) metals. Base metals are used in many industrial and commercial applications, including construction and manufacturing industry, military, aerospace, etc. In addition, their wide use in goods of everyday use makes them basic commodities in global markets. Two metals have been selected for the analysis: aluminium and copper. Their choice is not accidental. Both are very widely used in industry, so it is necessary to analyse these metals both in terms of supply and demand as well as their price volatility. The analysis of prices and returns of aluminium and copper indicates a very high level of volatility comparing with other metals traded on commodity markets. The research conducted by Krężołek (2015, 2016) and Krężołek and Trzpiot (2017) indicates that in various research periods significant changes were observed in both the values and directions of changes in prices and returns of these metals. Returns distributions are characterized by high level of kurtosis, skewness and the existence of outliers, which produces thick tails in the empirical distributions. The rapid and unpredictable changes of prices and returns significantly increase the level of extreme risk (see Sect. 2), which market participants want to avoid. One way to assess the risk of extreme changes in returns is to study the thickness of the tail of the returns distribution. In this study, the so-called Hill estimator and its modifications were proposed.

In the empirical part of the paper, extreme risk measures and methodology of extreme value theory, Hill's estimator and high-order quantiles were used to determine the risk measure related to the volatility observed on the aluminium and copper market. Given the high order of quantiles, it is necessary to have a sufficiently large dataset. In our case, we used a set of daily log returns for aluminium and copper spot prices from January 1989 to March 2019 (about 30 years), which gives us a total of 7742 daily observations. VaR was used as a risk measure to define high quantiles (0.001 and 0.0005). This value was estimated using Hill estimator and its modification. Additionally, some theoretical models were also used: GARCH and APARCH models with residuals following normal, t-Student and GED distributions.

Mathematically, any $GARCH(p, q)$ model can be described using equations as follows (Bollerslev 1986):

$$r_t - \mu = a_t = \sigma_t \varepsilon_t \tag{10}$$

$$\sigma_t^2 = \alpha_0 + \sum_{i=1}^{q} \alpha_i a_{t-i}^2 + \sum_{j=1}^{p} \beta_j \sigma_{t-j}^2 \tag{11}$$

where $\alpha_0 \geq 0$, $\alpha_i \geq 0$ for $i > 0$, $\beta_j \geq 0$, $\sum_{i=1}^{\max(p,q)} (\alpha_i + \beta_i) < 1$. Error term satisfies the assumptions of $\varepsilon_t \sim N(0, 1)$ and of iid. As the classical GARCH-type models do not describe asymmetry in the data (the impact of positive and negative information) and do not describe the leverage effect and long-memory effect, the class of APARCH models has been proposed (Ding et al. 1993):

$$\sigma_t^\delta = \alpha_0 + \sum_{i=1}^{q} \alpha_i (|a_{t-i}| - \gamma_i a_{t-i})^\delta + \sum_{j=1}^{p} \beta_j \sigma_{t-j}^\delta \tag{12}$$

where $\alpha_0 \geq 0$, $\alpha_i \geq 0$ for $i > 0$, $\beta_j \geq 0$, $\sum_{i=1}^{\max(p,q)} (\alpha_i + \beta_i) < 1$. The parameter δ plays the role of a Box–Cox transformation of the conditional standard deviation σ_t, while the parameters γ_i reflect the leverage effect. A positive (negative) value of the parameter γ_i means that past negative (positive) shocks have a deeper impact on current conditional volatility.

The conditional distribution of error term ε_t is described by normal, t-Student and GED distribution with the following density functions:

$$f_{Norm}\left(\varepsilon_t, \sigma_t^2; \theta\right) = \frac{1}{\sigma_t\sqrt{2\pi}} exp\left\{-\frac{\varepsilon_t^2}{2\sigma_t^2}\right\} \quad (13)$$

$$f_{t-Stud}\left(\varepsilon_t, \sigma_t^2; \theta\right) = \frac{\Gamma\left(\frac{v+1}{2}\right)}{\sigma_t\Gamma\left(\frac{v}{2}\right)\sqrt{\pi(v-2)}}\left(1+\frac{\varepsilon_t^2}{(v-2)\sigma_t^2}\right)^{\frac{v+1}{2}} \quad (14)$$

$$f_{GED}\left(\varepsilon_t, \sigma_t^2; \theta\right) = 2^{-\frac{v+1}{v}} \frac{v}{\sigma_t\sqrt{\frac{\Gamma(v^{-1})}{\Gamma(3v^{-1})}2^{-\frac{2}{v}}\Gamma(v^{-1})}} \exp\left\{-\frac{1}{2}\left|\frac{\varepsilon_t}{\sigma_t\sqrt{\frac{\Gamma(v^{-1})}{\Gamma(3v^{-1})}2^{-\frac{2}{v}}}}\right|^{v}\right\} \quad (15)$$

where $\{\varepsilon_t\}$ is the sequence of iid random variables, σ_t^2 is the conditional variance, θ is the vector of estimated parameters, v is the number of degrees of freedom, and $\Gamma(k) = \int_0^{+\infty} x^{k-1}e^{-1}dx$ is the gamma function with parameter k. The parameter v has to be estimated if t-Student and GED distribution are used.

Going into empirical results, the descriptive statistics for returns of the analysed metals are presented in Table 1.

The average return of copper is positive, whereas for aluminium—negative. The empirical distributions are leptokurtic and skewed to the left. These results show that we can expect some outliers (see boxplot at Fig. 1).

From Table 1, we can conclude also the high level of volatility, which can be verified graphically from Fig. 2.

We can easily find the periods with higher level of volatility from both metals. The phenomena of clustering variance are also observed.

At the final stage, the high-order quantiles (VaR) using Hill, PORT-Hill, quasi-PORT-Hill and MVRB-Hill estimators were calculated. The GARCH and APARCH models have been selected using AIC criterion. Additionally, the theoretical values

Table 1 Descriptive statistics

Statistics	Copper	Aluminium
Mean	0.000058	−0.000054
Standard error	0.000183	0.000152
Median	−0.000097	0.000000
Standard deviation	0.016142	0.013386
Coefficient of variation (%)	27632.04	−24895.45
Kurtosis	4.689797	3.628336
Skewness	−0.081365	−0.077063
Min	−0.118367	−0.098496
Max	0.156733	0.076409

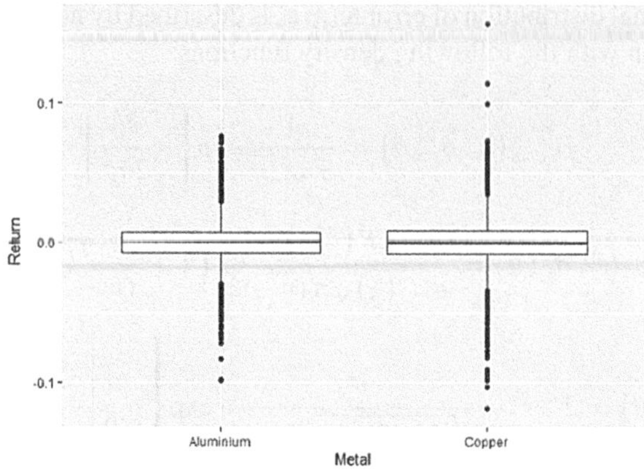

Fig. 1 Boxplots for aluminium and copper distribution

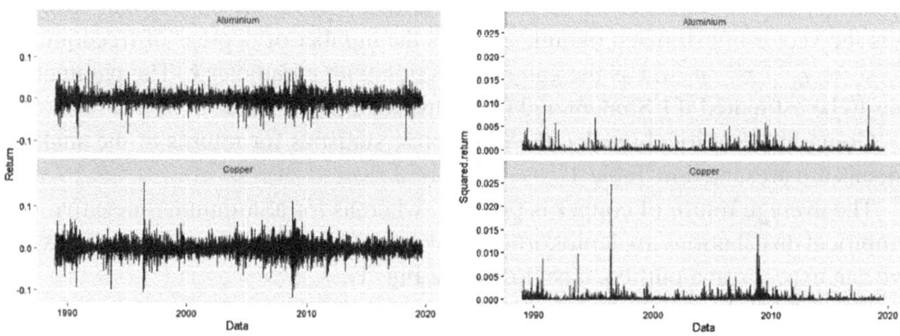

Fig. 2 Time series for aluminium and copper return (left) and squared return (right)

of normal and t-Student quantiles have been calculated. The results for copper and aluminium are presented in Tables 2, 3, 4 and 5. The values in bold refers to the lowest level of the RMSE.

Tables 2, 3, 4 and 5 summarize the results for copper and aluminium, respectively. In the last column, the root of mean-squared error has been estimated. This allowed us to identify the estimator that best approximates the empirical value of VaR.

As we can see, value at risk calculated under the normality assumption is rejected in terms of Kupiec test regardless the level of quantile. In other cases, the proportion of violation is acceptable. Comparing all estimators of VaR, the best results were obtained for parametric models GARCH (at 0.001) and APARCH (at 0.0005). If Hill-based estimators are compared, the results differ depending on metal and quantile level. For copper at 0.001, the best estimations were obtained for PORT-Hill estimator, whereas at 0.0005 for MVRB-Hill estimator. For aluminium, the best estimates of empirical VaR were obtained using quasi-PORT-Hill estimator. The main advantage of modifications of Hill estimator over the classical one is the existence of additional parameters such as tuning parameter q and second-order

Table 2 Hill-based estimators for VaR at the level 0.001—copper

Copper	VaR 0.001	No. of violations	Freq. of violations	Kupiec LR	Kupiec p-value	RMSE
Empirical	−0.08237	8	0.001033	0.008380	0.927064	–
Hill	−0.07964	11	0.001420	1.210863	0.271161	0.002735
PORT-Hill	−0.08129	9	0.001162	0.193716	0.659842	0.001083
Quasi-PORT-Hill ($p = 0.5$)	−0.08075	9	0.001162	0.193716	0.659842	0.001618
MVRB-Hill	−0.08119	9	0.001162	0.193716	0.659842	0.001183
GARCH-stud	**−0.08328**	**7**	**0.000904**	**0.073957**	**0.785661**	**0.000911**
APARCH-stud	−0.08333	7	0.000904	0.073957	0.785661	0.000956
Normal	−0.04982	50	0.006457	102.2297	0.000000	0.032549
t-Student	−0.08128	9	0.001162	0.193716	0.659842	0.001094

Table 3 Hill-based estimators for VaR at the level 0.0005—copper

Copper	VaR 0.0005	No. of violations	Freq. of violations	Kupiec LR	Kupiec p-value	RMSE
Empirical	−0.09678	4	0.000517	0.004188	0.948403	–
Hill	−0.09127	6	0.000775	1.000445	0.317203	0.005510
PORT-Hill	−0.09323	5	0.000646	0.300832	0.583361	0.003556
Quasi-PORT-Hill ($p = 0.5$)	−0.09418	5	0.000646	0.300832	0.583361	0.002601
MVRB-Hill	−0.09572	4	0.000517	0.004188	0.948403	0.001062
GARCH-stud	−0.09629	4	0.000517	0.004188	0.948403	0.000490
APARCH-stud	**−0.09692**	**3**	**0.000387**	**0.213145**	**0.644314**	**0.000137**
Normal	−0.05305	40	0.005165	114.7216	0.000000	0.043731
t-Student	−0.09697	3	0.000387	0.213145	0.644314	0.000185

Table 4 Hill-based estimators for VaR at the level 0.001—aluminium

Aluminium	VaR 0.001	No. of violations	Freq. of violations	Kupiec LR	Kupiec p-value	RMSE
Empirical	−0.06742	8	0.001033	0.008380	0.927064	–
Hill	−0.06276	11	0.001420	1.210863	0.271161	0.004652
PORT-Hill	−0.06420	10	0.001291	0.601993	0.437819	0.003220
Quasi-PORT-Hill ($p = 0.5$)	−0.06433	10	0.001291	0.601993	0.437819	0.003087
MVRB-Hill	−0.06388	11	0.001420	1.210863	0.271161	0.003539
GARCH-stud	**−0.06803**	**6**	**0.000775**	**0.426487**	**0.513718**	**0.000612**
APARCH-stud	−0.06927	4	0.000517	2.204820	0.137580	0.001858
Normal	−0.04142	50	0.006457	102.2297	0.000000	0.025998
t-Student	−0.06564	8	0.001033	0.008380	0.927064	0.001776

Table 5 Hill-based estimators for VaR at the level 0.0005—aluminium

Aluminium	VaR 0.0005	No. of violations	Freq. of violations	Kupiec LR	Kupiec *p*-value	RMSE
Empirical	−0.07176	3	0.000387	0.213145	0.644314	–
Hill	−0.06963	4	0.000517	0.004188	0.948403	0.002131
PORT-Hill	−0.06992	4	0.000517	0.004188	0.948403	0.001850
Quasi-PORT-Hill ($p = 0.5$)	−0.07065	4	0.000517	0.004188	0.948403	0.001116
MVRB-Hill	−0.06994	4	0.000517	0.004188	0.948403	0.001827
GARCH-stud	−0.07187	3	0.000387	0.213145	0.644314	0.000109
APARCH-stud	**−0.07181**	**3**	**0.000387**	**0.213145**	**0.644314**	**0.000043**
Normal	−0.04410	41	0.005294	119.4261	0.000000	0.027667
t-Student	−0.07773	3	0.000387	0.213145	0.644314	0.005964

parameters (β, ρ). It allows for better asymptotical properties if compared to the classical approach.

5 Conclusions

The paper presents the approach to estimate extreme risk using Hill estimator and its modification. As was examined, the processes observed on the base metals market are characterized by significant, unpredictable and multidirectional volatility. It was mentioned that extreme statistics play a particular role in risk analysis of rare events. Commenting the results, we pointed out that empirical distributions of returns for copper and aluminium were leptokurtic, skewed and thick-tailed in comparison with the normal distribution.

Comparing all models, we found that for parametrical approach, GARCH and APARCH models indicate the use of models with residuals described by the t-Student distribution (even if compared to classical normal and t-Student distribution). Normal models were rejected in terms of Kupiec test. VaR estimates for models based on the modification of Hill estimator are more accurate in terms of RMSE error. It was discovered that parameterization of Hill estimator improves estimates of VaR risk measure. Moreover, Hill-based estimators are much more stable due to location and scale and asymptotically normal if compared to the classical approach.

References

Araújo Santos P, Fraga Alves MI, Gomes MI (2006) Peaks over random threshold methodology for tail index and quantile estimation. Revstat Stat J 4(3):227–247

Bollerslev T (1986) Generalised autoregressive conditional heteroskedasticity. J Econometrics 31:307–327

Caeiro F, Gomes MI, Pestana D (2005) Direct reduction of bias of the classical Hill estimator. Revstat Stat. J 3(2):111–136

Danielson J, Ergun LM, de Haan L, de Vreis CG (2016) Tail index estimation: quantile driven threshold selection. Working paper, Jan 2016. Online at http://ssrn.com/abstract=2717478

Ding Z, Granger CWJ, Engle RF (1993) A long memory property of stock market returns and a new model. J Empir Finance 1:83–106

Fedotenkov I (2018) A review of more than one hundred Pareto-tail index estimators. MPRA paper No. 90072. Online at https://mpra.ub.uni-muenchen.de/90072/

Gomes MI, de Haan L, Henriques-Rodrigues L (2008) Tail index estimation for heavy-tailed models: accommodation of bias in weighted log-excesses. J R Stat Soc Ser B 70(1):31–52

Gomes MI, Figueiredo F, Henriques-Rodrigues L, Miranda MC (2010) A quasi-PORT methodology for VaR based on second-order reduced-bias estimation. Notas e Comunicações CEAUL

Gumbel EJ (2004) Statistics of extremes. Dover Publications Inc, Mineola, New York

Hill BM (1975) A simple general approach to inference about the tail of the distribution. Ann Stat 3(5):1163–1174

Jajuga K (2008) Zarządzanie ryzykiem. Polskie Wydawnictwo Naukowe PWN, Warszawa

Krężołek D (2015) Analiza ryzyka inwestycji na przykładzie wybranych dodatków stopowych, Studia Ekonomiczne. Zeszyty Naukowe Uniwersytetu Ekonomicznego w Katowicach 241:65–77

Krężołek D (2016) The GlueVaR risk measure and investor's attitudes to risk—an application to the non-ferrous metals market. Stat Trans New Ser 17(2):305–316

Krężołek D, Trzpiot G (2017) The application of the GlueVaR measure in risk assessment on the metal market. Arch Data Sci Ser A 2(1):1–17

Németh L, Zempléni A (2018) Regression estimator for the tail index. Online at https://arxiv.org/abs/1708.04815v3

Segmentation of Enterprises on the Basis of Their Duration Using Survival Trees—Results of an Analysis for Legal Persons and Organizational Entities Without Legal Personality in the Łódzkie Voivodship

Artur Mikulec and Małgorzata Misztal

Abstract The studies, carried out so far, on established and liquidated enterprises in the Łódzkie Voivodship show that in terms of legal persons and organizational entities without legal personality, firm's duration is significantly longer than in terms of natural persons conducting economic activity. These entities constitute a clearly different group of enterprises. The article presents the results of an analysis, whose aim was the segmentation of legal persons and organizational entities without legal personality by its duration. A total of 10,562 enterprises have been studied. Survival trees (CTree algorithm) have been used to define groups of enterprises similar in duration, while a specific legal form, firm's location (county—"powiat"), type of conducted activity, size (measured in terms of the number of employees) and type of ownership have been used as explanatory variables. The use of recursive partitioning method made it possible to divide sets of objects into homogenous subsets. Then, estimation of survival function has been made in each of the obtained subsets with the use of Kaplan–Meier method. Such an approach to analysis enables more precise estimate of firm's duration than the use of Kaplan–Meier function for the total data. Prediction error curves based on the bootstrap cross-validation estimates of the prediction error have been used to assess and compare predictions obtained from both models.

Keywords Enterprises · Legal persons and organizational entities without legal personality · Duration analysis · Kaplan–Meier survival curve · Survival trees

A. Mikulec (✉) · M. Misztal
Department of Statistical Methods, Faculty of Economics and Sociology, Institute of Statistics and Demography, University of Lodz, Łódź, Poland
e-mail: artur.mikulec@uni.lodz.pl

M. Misztal
e-mail: malgorzata.misztal@uni.lodz.pl

K. Jajuga et al. (eds.), *Classification and Data Analysis*, Studies in Classification, Data Analysis, and Knowledge Organization,
https://doi.org/10.1007/978-3-030-52348-0_8

1 Introduction

In the theory of economics, one of the three types of entities in the economy are enterprises—entities conducting economic activity. These create the market, generate income (national income), implement new technologies and invest, also their activities are a source of economic growth. Hence, the topic of their survival is essential, and establishment, duration and liquidation of enterprises are monitored in EUROSTAT, OECD statistics and EU member countries[1] (European Communities/OECD 2007). The aim of the European policy is to support entrepreneurship to facilitate the creation of enterprises and ensure favorable conditions for development of the existing ones, and one of the elements supporting this aim is an appropriate diagnosis of firm's survival.

The problem of the duration of enterprises and the factors determining their survival has been investigated by researchers from various countries, but not many works have dealt with the impact of special legal form on firm's survival. The results of survival analysis of enterprises, in which the specific legal form for legal persons and organizational entities without legal personality were considered, are presented, among others, by: Harhoff et al. (1998) for German enterprises; Mata and Portugal (2002) for firms in Portugal; Pérez et al. (2004) for firms in Spain; Szymański (2011), Śmiech (2011) and Ptak-Chmielewska (2013, 2016) for Polish enterprises.[2] One of the latest studies in this area is the analysis of enterprises from the so-called Visegrad Group (the Czech Republic, Hungary, Poland and Slovakia)—Baumöhl et al. (2017).

In the present work, the authors focused on legal persons and organizational entities without legal personality. The aim of the analyses was segmentation of enterprises by their duration and showing suitability of survival trees for the sake of analyses. In particular, a specific legal form, location (county—"powiat"), type of conducted activity, size (measured in terms of the number of employees) and type of ownership were considered as factors that could affect the firm's duration. Results of these analyses can be useful to support entrepreneurship locally—it can be crucial for creating favorable conditions for firm's establishment and channeling aid into different cohorts (groups) of enterprises, creating regional economic policy and preparation of regional development strategies.

The structure of entities of the national economy REGON in Poland in the years 2010–2015 shows that the share of legal persons and organizational entities without legal personality for all entities in the Łódzkie Voivodship (22.0–25.3%) was close to their average share by voivodships (24.0–27.8%).[3] However, the results of statistical research of enterprises show that in Poland in recent years (2010–2015) in terms of number of microenterprises, persons employed, average paid employment, gross wages and salaries, total revenues and costs, as well as the value of retail

[1] Since 2006 OECD—EUROSTAT—*Entrepreneurship Indicators Project* (*EIP*) and structural business statistics in respective countries has been developed.

[2] On the basis of data for the whole Poland, including microenterprises of the Małopolskie Voivodship.

[3] The authors' own calculations on the basis of the Local Data Bank, https://bdl.stat.gov.pl.

Table 1 Number of enterprises in the Łódzkie Voivodship according to the legal form

Legal form	Enterprises established		Enterprises liquidated	Enterprises censored
	Number	%		
Legal person	5990	5.02	89	5901
Organizational entity without legal personality	4572	3.83	556	4016
Natural person conducting economic activity	108,711	91.15	27,652	81,059
Total	119,273	100.00	28,297	90,976

sale and wholesale of microenterprises—the Łódzkie Voivodship was typical[4] when compared to the whole country (see, e.g., Statistics Poland 2017). Therefore, in the subject analysis the authors have used data on enterprises of the Łódzkie Voivodship assuming that it reflects well the changes taking place in the community of enterprises, on the voivodships level. It needs to be observed that the enterprises functioning in the Łódzkie Voivodship can be seen as a relatively uniform group which is influenced by similar external factors and conditions. This will allow giving a reliable assessment of the influence of the above-mentioned factors (specific legal form, location, type of conducted activity, size and type of ownership) on the survival of enterprises, without additional considerations on their connections with, e.g., geographical location, natural conditions or regional policy.

In the years 2010–2015 in the Łódzkie Voivodship, 119,273 companies were established. Among them, 28,297 entities went into liquidation until the end of 2015 (23.7%) and 90,976 (76.3%) were still functioning (censored data)—see Table 1.

The overwhelming majority of the analyzed enterprises were legal persons conducting economic activity (108,711; 91.15%). The studies, carried out so far for the Łódzkie Voivodship (Mikulec and Misztal 2018a), show that in terms of legal persons and organizational entities without legal personality, the probability of survival is significantly higher than in terms of natural persons conducting economic activity.[5] The Kaplan–Meier survival curve for natural persons conducting economic activity is significantly lower than survival curves for legal persons and organizational entities without legal personality (see Fig. 1). As one can see, probability of survival of 2190 days (6 years) for all companies is equal to 0.5847. A sharper decrease of the Kaplan–Meier survival curve for all firms can be observed after the first two years (730 days) of their activity. In case of natural persons conducting economic activity, probability of survival of 2190 days is 0.5587, and also a sharper decline of the survival curve, first after a year and then after 2 years of their activity can be noticed. For organizational entities without legal personality, probability of survival

[4]Values of the above-mentioned characteristics were closest to the average for all the voivodships.

[5]Detailed analysis for this group of entities was made in Mikulec and Misztal (2018b).

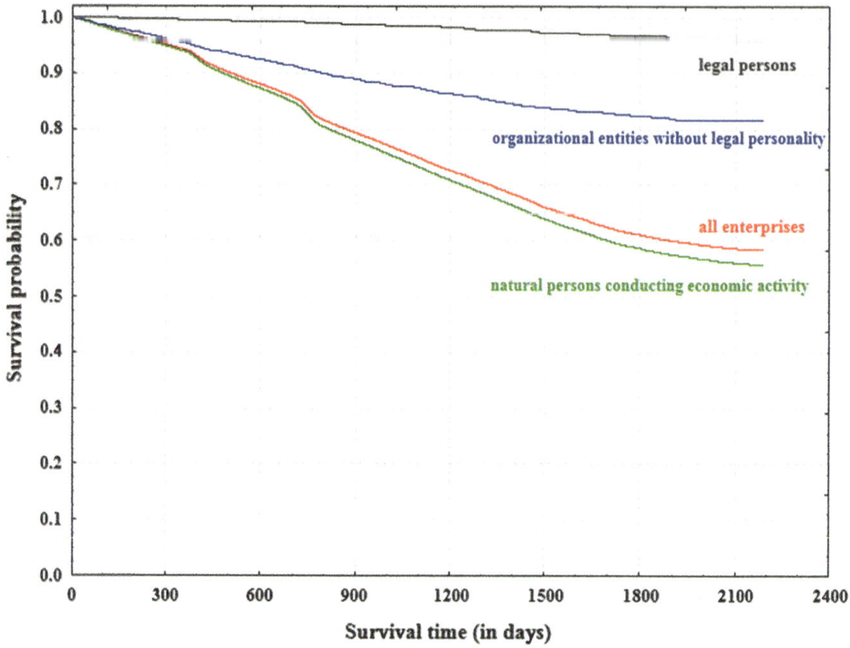

Fig. 1 Kaplan–Meier survival curves for all data and groups of enterprises according to the legal form

of 2190 days is 0.8159, but for legal persons 0.9586—these entities constitute a clearly different group of enterprises.

2 Data Characteristics and Methods

The individual data on 10,562 of enterprises—legal persons and organizational entities without legal personality established in the Łódzkie Voivodship in 2010–2015 were used in the work. Out of this number, 645 enterprises went into liquidation (6.1%) by the end of 2015, and 9917 (93.9%) continued their economic activity. These firms are treated as censored data. The structures of enterprises by the specific legal form are presented in Table 2.

However, in terms of size of the companies, microenterprises (0–9 employed persons) prevailed: 95.2% in a group of legal persons and 95.6% in a group of organizational entities without legal personality, respectively.

Taking into account location, a dominant position of companies conducting activities in the territory of Łódź can be observed—in terms of legal persons there were 61.8% companies, and in terms of organizational entities without legal personality—more than 51.5%. In the territory of two other counties ("powiat"), that is pabianicki

Table 2 Number of enterprises in the Łódzkie Voivodship according to the specific legal form

Specific legal form	Code	Legal persons		Organizational entities without legal personality	
		Number	%	Number	%
Civil law partnership[a]	019	x	x	3097	67.7
Other company[b]	023	16	0.3	x	x
Professional partnership	115	x	x	43	0.9
Joint-stock company	116	60	1.0	8	0.2
Limited liability company	117	5831	97.3	260	5.7
Unlimited company	118	x	x	449	9.8
Limited partnership	120	x	x	578	12.6
Joint-stock limited partnership	121	x	x	102	2.2
Main branches of foreign insurance[c]	136	x	x	22	0.5
Co-operative	140	83	1.4	x	x
Branches of foreign enterprises	179	x	x	10	0.2
Company without specific legal form	999	x	x	3	0.1
Total	x	5990	100.0	4572	100.0

[a]Partnership conducting its activities on the basis of the Civil Code
[b]Joint-stock companies foreseen in other provisions than the Commercial Companies Code and the Civil Code, or legal forms to which the provisions of the companies are applied
[c]Main branches of foreign branches of insurance companies (see Statistics Poland 2019)

and zgierski, economic activity was conducted by: 5.3% and 4.9%; and 5.0% and 5.6%, of enterprises, respectively.

When it comes to the type of ownership, both among legal persons and organizational entities without legal personality dominated (90.3% and 90.2%) enterprises in private sector ("pure" ownership).

For both legal persons and organizational entities without legal personality, the highest percentage was represented by trade companies (section G—*Trade; repair of motor vehicles*)—26.4% and 31.9%, respectively. A large group accounted for industrial companies (sections B+C+D+E)—15.4% and 12.7%, respectively, and also conducting professional, scientific or technical activity (section M—*Professional, scientific and technical activities*)—12.0% and 10.0%, respectively (see Table 3).

To identify groups of enterprises similar in terms of duration, survival trees were applied. The model was based on the CTree algorithm (*Conditional Inference Tree*) proposed by Hothorn et al. (2006). CTree is a nonparametric class of regression trees embedding recursive binary partitioning into theory of conditional inference procedures with stopping rule based on multiple test procedures. Estimation of survival function was made in the leaves of the tree with the use of Kaplan–Meier method.

Great flexibility, no assumptions on distributions of the survival times and the possibility to automatically detect interactions between covariates without the need to specify them beforehand are the advantages of the survival trees over the Cox

Table 3 Number of enterprises by Polish Classification of Activities (PKD 2007)

PKD 2007 section code	Enterprises established			
	Legal persons		Organizational entities without legal personality	
	Number	%	Number	%
A—Agriculture, forestry and fishing	11	0.2	8	0.2
B+C+D+E—Industry	925	15.4	582	12.7
F—Construction	505	8.4	383	8.4
G—Trade; repair of motor vehicles	1579	26.4	1458	31.9
H—Transportation and storage	278	4.6	183	4.0
I—Accommodation and food service activities	178	3.0	270	5.9
J—Information and communication	549	9.2	171	3.7
K—Financial and insurance activities	210	3.5	155	3.4
L—Real estate activities	334	5.6	214	4.7
M—Professional, scientific and technical activities	721	12.0	457	10.0
N—Administrative and support service activities	332	5.5	182	4.0
P—Education	98	1.6	92	2.0
Q—Human health and social work activities	55	0.9	70	1.5
R—Arts, entertainment and recreation	56	0.9	124	2.7
S—Other service activities	159	2.7	223	4.9
Total	5990	100.0	4572	100.0

proportional hazards model, being one of the most popular survival analysis methods. Survival tree can naturally group objects according to their survival behavior based on their covariates (Bou-Hamad et al. 2011).

Prediction error curves (i.e., time dependent estimates of the population average Brier score) based on the bootstrap cross-validation were used to assess the quality of estimates obtained by the CTree algorithm (Mogensen et al. 2012; Efron and Tibshirani 1997).

All the calculations were done with the use of the following packages: STATISTICA 13.3, SPSS 25 and R-project (packages: party, pec).

3 Results

The survival tree created with the use of the CTree algorithm has seven leaves, presented in Fig. 2. Four variables were used for splitting the nodes—a specific legal form, PKD section code, type of ownership and firm's location (county—"powiat").

Sample size (n) is given for each leaf. Additionally, survival curves for each terminal node are presented in Fig. 3. The probability of surviving 2190 days for enterprises separated in the leaves of the survival tree is presented in Table 4. It is not possible to determine the median duration (the time when half of the enterprises are expected to continue their economic activity) for any of the terminal nodes.

It can be observed that survival curve in the node no 3 is the highest. In this node, there are $n = 6253$ enterprises defined by a specific legal form: professional partnerships (code: 115), joint-stock companies (116), limited liability companies (117), main branches of foreign insurance companies (136), branches of foreign enterprises (179), other companies (023) and firms without specific legal form (999), where 97.4% entities in this node were limited liability companies (117). Regarding the type of ownership, 90.1% of all the companies in this node were from a private sector ("pure"). Additionally, 26.5% of companies conducted activity in section G (*Trade; repair of motor vehicles*), *Industry* (sections B+C+D+E) was also well represented—15.5%, and section M (*Professional, scientific and technical activities*)—12.3%. The probability of survival of 2190 days in this node is equal to 0.9592.

The second highest survival curve is observed in the node no 11. There are $n = 498$ enterprises defined by a specific legal form: civil law partnership (019)

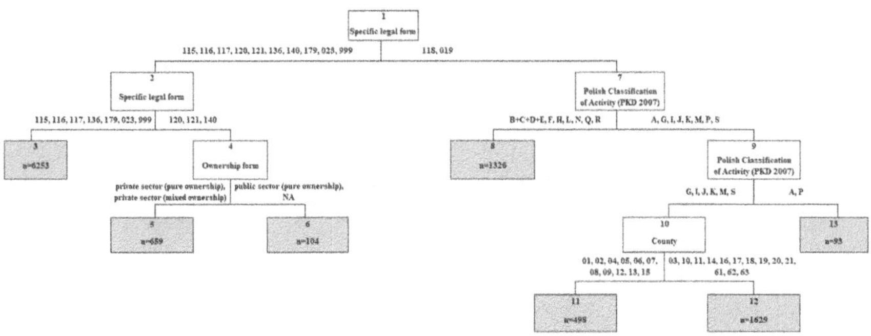

Fig. 2 Survival tree (CTree algorithm). Key: Specific legal form as described in Table 2; Polish Classification of Activity (PKD 2007) as described in Table 3; County ("powiat"): 01—bełchatowski; 02—kutnowski; 03—łaski; 04—łęczycki; 05—łowicki; 06—łódzki wschodni; 07—opoczyński; 08—pabianicki; 09—pajęczański; 10—piotrkowski; 11—poddębicki; 12—radomszczański; 13—rawski; 14—sieradzki; 15—skierniewicki; 16—tomaszowski; 17—wieluński; 18—wieruszowski; 19—zduńskowolski; 20—zgierski; 21—brzeziński; 61—Łódź city; 62—Piotrków Trybunalski city; 63—Skierniewice city; Ownership form: pure—the ownership of 100% of capital by one entity or more entities, provided that they represent the same type of ownership; mixed—the ownership of capital by two or more entities, provided that they represent at least two different types of ownership, NA—data not available

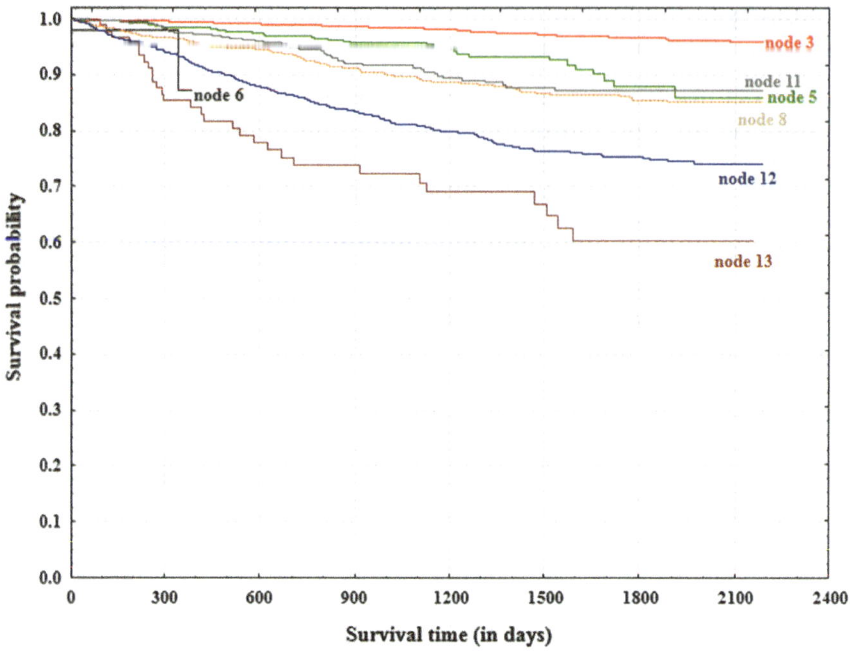

Fig. 3 Kaplan–Meier survival curves for seven groups of enterprises separated in the leaves of the CTree survival tree

Table 4 Probability of surviving 2190 days for enterprises grouped in the CTree leaves

Node number	Probability of survival of 2190 days	Number of observations in the node
3	0.9592	6253
5	0.8599	659
6	0.8720	104
8	0.8507	1326
11	0.8714	498
12	0.7407	1629
13	0.6025	93

and unlimited company (118)—organizational entities without legal personality, which accounted for 91.4% and 8.6%, respectively in this node. Among them, 62.7% conducted activity in section G (*Trade; repair of motor vehicles*); 14.7% in section M (*Professional, scientific and technical activities*) and 11.6% in section I (*Accommodation and food service activities*). In addition, above-mentioned companies were located in counties ("powiats") no: 01—bełchatowski, 02—kutnowski,

04—łęczycki, 05—łowicki, 06—łódzki wschodni, 07—opoczyński, 08—pabianicki, 09—pajęczański, 12—radomszczański, 13—rawski and 15—skierniewicki (see Fig. 4). The probability of survival of 2190 days in this node is 0.8714.

The third highest survival curve is obtained in the node no 5. In this node, there are $n = 659$ enterprises defined by a specific legal form: limited partnership (120), joint-stock limited partnership (121) and co-operative (140), where the largest share in this node (72.1%) had limited partnership. All these entities were from private sector ("pure" or "mixed" ownership). The probability of survival of 2190 days in this node is 0.8599.

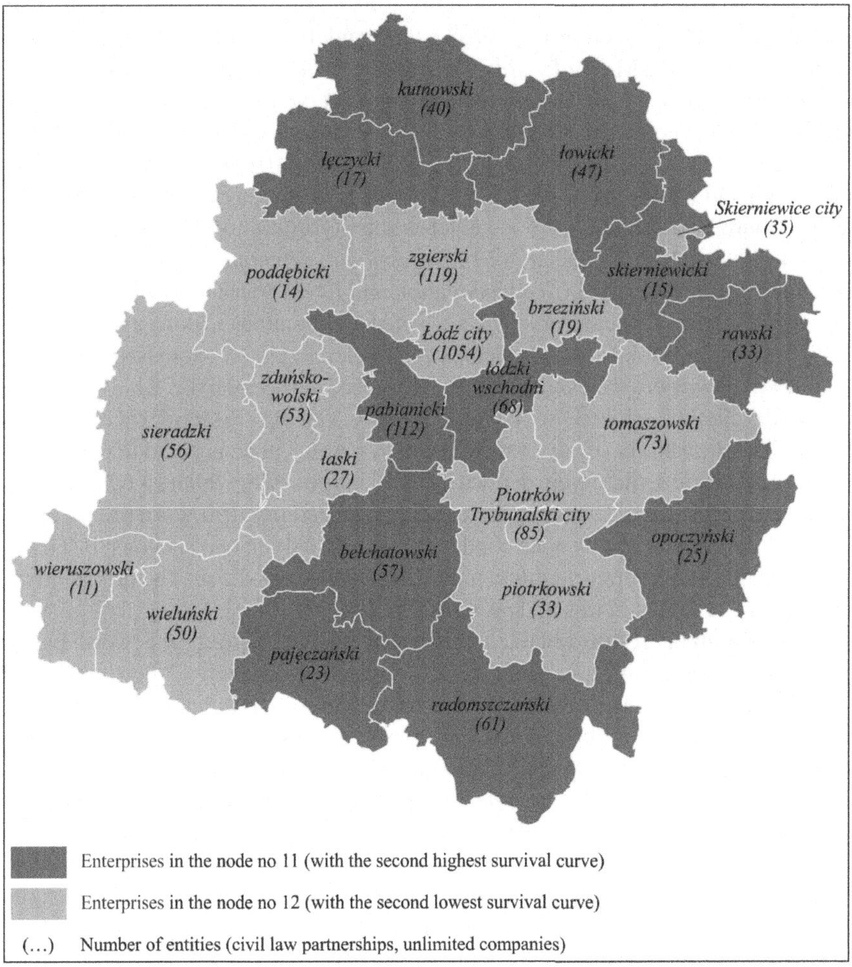

Fig. 4 Location of enterprises from the nodes no 11 and no 12 according to counties ("powiats") of the Łódzkie Voivodship

The node no 8, in which there are $n = 1326$ enterprises, mainly consists of civil law partnerships (019), which accounted for 83.3% and unlimited companies (118)—16.7%, which in 98.3% were privately owned ("pure"). More than a half of them conducted activity in *Industry* (sections B+C+D+E, 31.1%) or *Construction* (section F, 19.8%). The probability of survival of 2190 days in this node is 0.8507.

The node no 6, where $n = 104$ enterprises are located, including 103 (99.0%) limited partnerships (120), is specific. There is also lack of information on type of ownership for these companies. Their observation time is short (less than 1 year)—94 of them were established in 2015—hence, there are mainly censored observations in this node. When type of conducted activity is taken into account: 25.0% conducted activity in section G (*Trade; repair of motor vehicles*), 15.4% in sections B+C+D+E (*Industry*), 13.5% in section F (*Construction*) and 10.6% both in section L (*Real estate activities*) and M (*Professional, scientific and technical activities*).

Thus, the two lowest survival curves were observed for the nodes no: 12 and 13. In the node no 12, there are $n = 1629$ organizational entities without legal personality. These are mainly civil law partnerships (019)—89.0% and general partnerships (118)—11.0%. 98.2% of enterprises in this node belong to a private sector ("pure"). More than 54% conducted activity in section G (*Trade; repair of motor vehicles*); 15.5% in section M (*Professional, scientific and technical activities*) and 12.0% in section I (*Accommodation and food service activities*). Additionally, above-mentioned enterprises were located in counties ("powiats") no: 03—łaski, 10—piotrkowski, 11—poddębicki, 14—sieradzki, 16—tomaszowski, 17—wieluński, 18—wieruszowski, 19—zduńskowolski, 20—zgierski, 21—brzeziński, 61—Łódź city, 62—Piotrków Trybunalski city and 63—Skierniewice city (see Fig. 4). The probability of survival of 2190 days in this node is equal to 0.7407.

In the node no 13, there were defined $n = 93$ enterprises, of which 94.6% were civil law partnerships (019). Due to a type of conducted activity, 100% is a private sector ("pure"). Enterprises in this node conducted activity only in two sections: P (*Education*)—92.5% and—A (*Agriculture, forestry and fishing*)—7.5%. The probability of survival of 2190 days in this node is 0.6025.

Summing up the results obtained in the survival tree diagram, it is worth paying attention one more time to the location of enterprises (county—"powiat") which played a significant role in splitting the node no 10. Both terminal nodes (no 11 and no 12) comprised about 90% of civil partnerships mainly from sections: G (*Trade; repair of motor vehicles*), M (*Professional, scientific and technical activities*) and I (*Accommodation and food service activities*)—the structure according to the section was also similar. However, the place of conducting economic activity was the factor which determined whether the given partnerships were characterized by the second highest survival curve (the node no 11), or the second lowest survival curve (the node no 12). A slightly better situation was observed for entities (civil partnerships) conducting economic activity on the outskirts of the voivodship (northeast), and the counties ("powiats") located south of Łódź, excluding all cities with the rights of the county—Łódź city, Piotrków Trybunalski city and Skierniewice city (see Fig. 4).

The log-rank test (Mantel–Cox) was applied for overall comparison of the Kaplan–Meier curves. *P*-value < 0.001 indicates that the null hypothesis (all survival curves are the same) should be rejected for the joint analysis. Taking into account all the pairwise comparisons, the null hypothesis should be rejected for the following Kaplan–Meier curves: the node no 3 versus the nodes no: 5, 6, 8, 11, 12, 13; the node no 5 versus the nodes no: 6, 8, 12, 13; the node no 6 versus the node no 11; the node no 8 versus the nodes no: 12, 13; the node no 11 versus the nodes no: 12, 13; and the node no 12 versus the node no 13.

To assess and compare predictions obtained by the CTree survival tree and the Kaplan–Meier estimation of the survival function, prediction error curves were determined based on the bootstrap.632+ cross-validation estimates of the prediction error (see Fig. 5).

Both the curves (Kaplan–Meier ignoring covariates and the CTree) are practically the same during the first 300 days of observation, and then one can see the advantage of the CTree survival tree over the Kaplan–Meier estimation of the survival function.

The Integrated Brier Score between 0 and 2190 days for the bootstrap.632+ estimates of the prediction error is slightly better for the CTree survival tree (IBS = 0.056 vs. IBS = 0.057 for the Kaplan–Meier estimate).

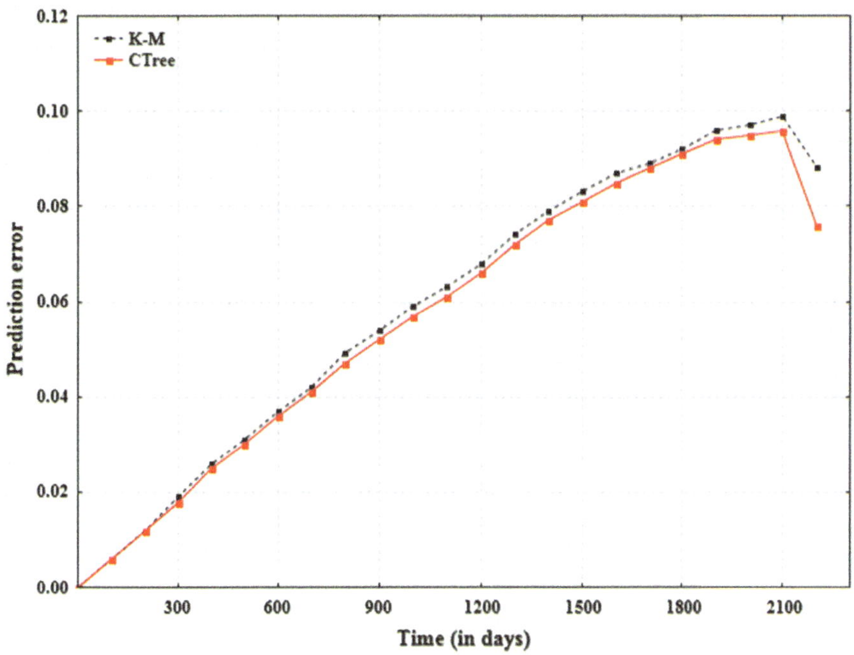

Fig. 5 Comparison of prediction error curves (the bootstrap.632+ estimates)

4 Conclusions

Four variables have been used for splitting the tree nodes—a specific legal form, PKD section code, type of ownership and location of the enterprise ("powiat"). The analysis itself using survival trees indicates extension of the existing approaches, since the Kaplan–Meier survival analysis or the Cox proportional hazards model is usually used in Polish or foreign literature. It turns out, however, that the use of a recursive partitioning method in combination with estimation of survival function in the nodes enables a more precise estimate of firm's duration than the use of Kaplan–Meier estimator for the total data.

The results of the firm's duration analysis, i.e., legal persons and organizational units without legal personality, established in the Łódzkie Voivodship in the years 2010–2015 and observed until the end of 2015, presented in this work are not only in line with outcomes of other analyses of this type, but they also provide additional and more specific information to reflect.

For instance, in Harhoff et al. (1998), concerning German business activity, first conclusions regarding enterprises going into liquidation (total, divided into bankruptcy and voluntary liquidations) by legal status and sector shall be found. The authors state that, among others, the highest indicator of voluntary liquidations is for "natural persons conducting economic activity" (sole proprietorship and artisan activity), next "partnership" groups (civil law partnership, unlimited companies, professional partnership, limited partnerships), then "limited liabilities companies" (as Ltd. and others) and finally "stock companies" (joint-stock companies and joint-stock limited partnerships). Taking into consideration sectors, the authors showed that indicators of liquidation are particularly high for trade and service, further in industry and construction. Confirmation of the above findings (taking into account the positioning of survival curves in the nodes) shall be found in the previous article (Mikulec and Misztal 2018a) and on the survival tree diagram (see Fig. 2), if we concentrate on splitting the nodes no. 1 and no. 2 of the survival tree (legal form) and the node no. 7 of the survival tree (PKD sections).

The results discussed in this work—compare splitting the node no. 1 and no. 2 of the survival tree—correspond to the results for Portuguese companies in Mata, Portugal (2002), claiming that "limited liability" companies are characterized by lower probability of exit from the market in comparison with other legal structure companies. They are also consistent with the results for Spanish firms, presented in Pérez et al. (2004), according to the null hypothesis of the equality of survival curves for analyzed enterprises, among others, "groups of the whole of limited liability companies" versus "groups of the whole of other special legal form" shall be rejected.

Hence, in Baumöhl et al. (2017), firm's survival in the Czech Republic, Hungary, Poland and Slovakia was analyzed, and it was stated, among others, that in Poland and Hungary, activity in the form of a joint-stock company is a significant factor which reduces the probability of the company's exit from the market. Additionally, in Poland also activity in the form of co-operatives and associations reduces the probability of liquidation. The above-mentioned dependencies can be found in the nodes no 3 and no 4 of the obtained survival tree.

First of all, results presented in this work, concerning enterprises segmentation of the Łódzkie Voivodship according to their duration and impact of, e.g., specific legal forms on their duration, are confirmed in Polish literature of the subject. In the works given below, one can find several conclusions which are reflected in the survival tree diagram and the survival curves for the leaves:

- The more complicated legal form of the company (unlimited companies, professional partnerships, limited liabilities companies, limited partnerships and joint-stock companies), the greater the probability of survival (higher positioning of survival curve)—it is connected in particular with an easier access to credit from financial institutions. In addition, such companies are usually bigger and hold more capital. On the other hand, chances for the earlier exit from the market are greater for entities with less complicated legal form (natural persons conducting economic activity or civil law partnerships)—the analysis for microenterprises in Poland in years 2002–2007—Szymański (2011).[6]
- There are significant differences in Kaplan–Meier survival curves among limited liability companies (the biggest chances of survival) and civil law partnerships—the analysis of companies from the Małopolskie Voivodship (2002–2007)—Śmiech (2011).
- Registered enterprises and conducting activity as partnership (legal persons), including companies with no specific legal form, are less exposed to the exit from the market by liquidation (Cox regression) than sole proprietorships—the analysis for microenterprises in the Małopolskie Voivodship in years 2006–2011—Ptak-Chmielewska (2013).
- Significant statistical differences occur between survival functions for the group of companies in the form of a partnership (with legal personality and without legal personality), and natural persons conducting activity (see Fig. 1)—data from panel research of enterprises established in 2004—Ptak-Chmielewska (2016).

References

Baumöhl E, Iwasaki I, Kočenda E (2017) Firms' survival in the new EU member state. Centre of Economic Institutions working paper series No. 2017-5. http://hermes-ir.lib.hit-u.ac.jp/rs/bitstream/10086/28883/1/wp2017-5.pdf. Accessed 30 Aug 2019

Bou-Hamad I, Larocque D, Ben-Ameur H (2011) A review of survival trees. Stat Surv 5:44–71

Efron B, Tibshirani R (1997) Improvements on cross-validation: the .632+ bootstrap method. J Am Stat Assoc 92(438):548–560

European Communities/OECD (2007) Eurostat-OECD manual on business demography statistics, 2007th edn. Office for Official Publications of the European Communities, Luxembourg

[6]Additionally, on the basis of conducted analyses the author states that the influence of legal form on firm's duration is relevant only in the first phase of its activity cycle. After the period of 2.5 years of activity, this variable does not differentiate the length of companies stay in business.

Harhoff D, Stahl K, Woywode M (1998) Legal form, growth and exit of West German firms—empirical results for manufacturing, construction, trade and service industries. J Ind Econ 46(4):453–488
Hothorn T, Hornik K, Zeileis A (2006) Unbiased recursive partitioning: a conditional inference framework. J Comput Graph Stat 15(3):651–674
Mata J, Portugal P (2002) The survival of new domestic and foreign-owned firms. Strat Manage J 23(4):323–343
Mikulec A, Misztal M (2018a) Zastosowanie metody rekurencyjnego podziału w analizie trwania przedsiębiorstw województwa łódzkiego. Pr Nauk Uniw Ekon Wroc 507:179–190
Mikulec A, Misztal M (2018b) Does the type of business activity and the enterprise location affect a firm's survival? Results of an analysis for natural persons conducting economic activity in the Łódzkie Voivodship. Econometrics. Ekonometria. Adv Appl Data Anal 22(3):23–40
Mogensen UB, Ishwaran H, Gerds TA (2012) Evaluating random forest for survival analysis using prediction error curves. J Stat Softw 50(11):1–23
Pérez SE, Llopis AS, Llopis JAS (2004) The determinants of survival of Spanish manufacturing firms. Rev Ind Organ 25(3):251–273
Ptak-Chmielewska A (2013) Semiparametric Cox regression model in estimation of small and micro enterprises' survival in the Malopolska Voivodeship. Quant Methods Econ XIV(2):169–180
Ptak-Chmielewska A (2016) Determinanty przeżywalności mikro i małych przedsiębiorstw w Polsce. Oficyna Wydawnicza SGH, Warszawa
Śmiech S (2011) Analiza przeżycia podmiotów gospodarczych w województwie małopolskim w latach 2002-2008. Zeszyty Naukowe Uniwersytetu Ekonomicznego w Krakowie 876:121–132
Statistics Poland (2017) Activity of enterprises with up to 9 persons employed in 2015. Statistical Publishing Establishment, Warsaw
Statistics Poland (2019) Structural changes of groups of the national economy entities in the REGON register, 2018. Statistical Publishing Establishment, Warsaw
Szymański D (2011) Badanie żywotności nowo powstałych mikroprzedsiębiorstw w Polsce w latach 2002–2007. Rozprawa doktorska. http://depotuw.ceon.pl/handle/item/177. Accessed 30 Aug 2019

Corporate Bankruptcy Prediction with the Use of the Logit Leaf Model

Barbara Pawełek and Józef Pociecha

Abstract Various data classification methods are used for bankruptcy prediction. Among them is the logit leaf model as a hybrid classification algorithm that enhances logistic regression and decision tree. The logit leaf model consists of two stages. In the first stage, company sets are identified using decision tree, and in the second stage a logit model is created for every leaf of this tree. The purpose of the paper is to present the results of the research on the usefulness of the logit leaf model for corporate bankruptcy prediction. A value added of the paper is the application of the logit leaf model to the prediction firms' bankruptcy. The research was carried out with the use of 61 financial ratios regarding enterprises from the manufacturing sector in Poland. The CART classification tree, the logit model and the logit leaf model were applied. Models were constructed for balanced and non-balanced data sets. The bankruptcy prediction was made for a year in advance. The following measures of prediction effectiveness of the analysed methods were used: sensitivity, specificity, precision, F1, G-mean and AUC. The results of the conducted research did not confirm an advantage of the hybrid approach over the use of individual classifiers. Calculations were performed in R program.

Keywords Bankruptcy prediction classification tree · Logit leaf model · Logit model · Prediction effectiveness

1 Introduction

There are many data classification methods that are applied to the bankruptcy prediction (Baesens et al. 2003; Lessmann et al. 2015). The main criterion of usefulness bankruptcy prediction models is their predictivity effectiveness. The multimodel

---

B. Pawełek (✉) · J. Pociecha
Department of Statistics, Cracow University of Economics, Kraków, Poland
e-mail: barbara.pawelek@uek.krakow.pl

J. Pociecha
e-mail: jozef.pociecha@uek.krakow.pl

K. Jajuga et al. (eds.), *Classification and Data Analysis*, Studies in Classification, Data Analysis, and Knowledge Organization,
https://doi.org/10.1007/978-3-030-52348-0_9

approach and the hybrid approach are often used for the purpose of bankruptcy prediction (Garcia et al. 2019; du Jardin 2018; Hastie et al. 2009). Many results of comparative research regarding the usefulness of different methods for bankruptcy prediction could be found in the literature (Brown and Mues 2012; Garcia et al. 2019). One of the sources of erroneous classification of firms into the bankrupts and non-bankrupts groups may be the non-homogeneity of the data set.

Earlier, authors' attempts to solve the problem of non-homogeneity of the data set in the bankruptcy prediction consisted in:

- Building prediction models for sets of firms that included bankrupts from a given year (Pawełek 2017).
- Building prediction models for groups of firms, which are similar in 'size' (measured, for example, by the total value of assets). Then the ABC method was used for division of the enterprises into groups (Pawełek et al. 2017; Ultsch and Lötsch 2015).

De Caigny et al. (2018a) recommended a hybrid classification algorithm—logit leaf model (LLM). The LLM method builds logit models for groups of firms received with the use of a classification tree. The paper has already been quoted 32 times (according to Google Scholar, as of 14/09/2019). However, none of the publications referring to the article (De Caigny et al. 2018a) presents the results of the empirical investigations carried out with the use of the LLM.

The algorithm presented by De Caigny et al. (2018a) was perceived as a new proposal of solving the non-homogeneity problem of data set in corporate bankruptcy prediction.

The purpose of the paper is to present the results of the empirical investigations on the usefulness of the logit leaf model for prediction of bankruptcy of enterprises. A value added of the paper is the application of the logit leaf model for corporate bankruptcy prediction. In compliance with the authors' best knowledge, this research is the first research described in the literature that uses the logit leaf model for bankruptcy prediction.

The results of the pilot research were presented during the 6th European Conference on Data Analysis (ECDA 2019) held in Bayreuth, Germany (18–20 March 2019) (Pawełek et al. 2019) and at the 28th Conference of the Section on Classification and Data Analysis of the Polish Statistical Association entitled Data Classification and Analysis—Theory and Applications (SKAD 2019) held in Szczecin, Poland (18–20 September 2019) (Pawełek and Pociecha 2019).

The study is divided as follows: Sect. 2 provides a description of the relevant database and the research procedure; the results of the empirical research are presented and discussed in Sects. 3 and 4; and the main findings of the study are summarised in Sect. 5.

2 Data and Research Procedure

The research makes use of the database available on the website: https://archive.ics.uci.edu/ml/datasets/Polish+companies+bankruptcy+data. The data were taken from the Emerging Markets Information Service (EMIS).

The objects of the research are enterprises operating in the manufacturing sector in Poland. The financial data regards the years 2000–2013. The bankruptcy of enterprises was predicted a year in advance.

The initial database contained 64 financial ratios that characterised 5910 enterprises. The objects included 410 bankrupts (B), which represented 6.9% of all objects, and 5500 non-bankrupts (NB), which represented 93.1% of all objects.

The database had certain missing data. For the purpose of obtaining a complete set of data, three financial ratios were removed from the database, since they had more than 20% of the missing data in the groups of bankrupts or non-bankrupts. Next, enterprises were removed for which data were missing. As a result of the removal of the missing data, the number of financial ratios which were left in the database was 61. The set of objects included 5329 enterprises (ca. 9.83% of objects were removed). Finally, the database had 360 bankrupts (as a result of removal of ca. 12.20% of bankrupts), which represented ca. 6.8% of all objects, and 4969 non-bankrupts (after removal of ca. 9.65% of non-bankrupts), which represented ca. 93.2% of all objects. Thus, the structure of the database with respect to bankrupts and non-bankrupts was not changed significantly.

The three binary classifiers were subject to the research, namely logit leaf model (LLM), logit model (LOGIT) and CART classification tree (TREE).

The prediction effectiveness of binary classifiers is assessed using, among others, confusion matrix (Table 1) (Bramer 2016; Ohsaki et al. 2017).

The research made use of the six measures of prediction effectiveness which are based on the confusion matrix:

- Sensitivity measure, which informs of the ratio of $\frac{TP}{TP+FN}$ (Altman and Bland 1994a)—the proportion of positive instances that are correctly classified as positive.

Table 1 Confusion matrix for a two-class problem

		Predicted		
		Negative	Positive	Total
Actual	Negative	TN	FP	AN
	Positive	FN	TP	AP
	Total	PN	PP	

Notes: AN—the number of actual negatives, AP—the number of actual positives, PN—the number of predicted negatives, PP—the number of predicted positives, TN—True Negatives, FN—False Negatives, FP—False Positives, TP—True Positives

- Specificity measure, which informs of the ratio of $\frac{TN}{TN+FP}$ (Altman and Bland 1994a)—the proportion of negative instances that are correctly classified as negative.
- Precision measure, which informs of the ratio of $\frac{TP}{TP+FP}$ (Altman and Bland 1994b)—proportion of instances classified as positive that are really positive.
- F1 measure, which informs of the ratio of $\frac{2 \cdot \text{Sensitivity} \cdot \text{Precision}}{\text{Sensitivity}+\text{Precision}}$ (Lewis and Gale 1994)—a measure that combines sensitivity and precision measures.
- G-mean measure, which informs of the geometric mean of the accuracies of the two classes $\sqrt{\text{Sensitivity} \cdot \text{Specificity}}$ (Kubat et al. 1998)—a measure that combines sensitivity and specificity measures.
- AUC measure, which informs of the area under the ROC curve, where ROC is the receiver operating characteristic (Altman and Bland 1994c; Zweig and Campbell 1993).

The research was divided into the five stages:

1. Construction of sets:

    - the balanced set X_{balanced}: 360 bankrupts + 360 non-bankrupts = 720 objects (i.e. 50% B + 50% NB; random selection of non-bankrupts was repeated five times),
    - the non-balanced set $X_{\text{non-balanced}}$: 360 bankrupts + 1440 non-bankrupts = 1800 objects (i.e. 20% B + 80% NB; random selection of non-bankrupts was repeated five times);

2. Random division of set X^v, where $X = X_{\text{balanced}}$ or $X = X_{\text{non-balanced}}$ and $v = 1, 2, 3, 4, 5$, into the training part U^v and the test part T^v ($X^v = U^v \cup T^v$, save that $\overline{\overline{U^v}} = \frac{2}{3}\overline{\overline{X^v}}$ and $\overline{\overline{T^v}} = \frac{1}{3}\overline{\overline{X^v}}$) in compliance with the structure with respect to bankrupts (B) and non-bankrupts (NB). The division was repeated 30 times, as a result of which the following sets were obtained:

    - $\{U_r^v\}_{r=1,\dots,30}^{v=1,\dots,5}$,
    - $\{T_r^v\}_{r=1,\dots,30}^{v=1,\dots,5}$;

3. Building models on the basis of each training sample U_r^v ($r = 1, \dots, 30$; $v = 1, \dots, 5$), created for balanced and non-balanced sets;
4. Assessment of prediction effectiveness of the built models on the basis of test samples T_r^v ($r = 1, \dots, 30$; $v = 1, \dots, 5$) with the use of the following measures: sensitivity, specificity, precision, F1, G-mean and AUC;
5. Verification of the hypothesis that the values of the given prediction effectiveness measures (i.e. sensitivity, specificity, precision, F1, G-mean and AUC) calculated for the analysed models come from the sets with the same average values. The following tests were applied: Kruskal–Wallis test and post hoc Dunn's test (with Bonferroni correction).

Calculations and graphs were made in R program, with the use of, in particular, the LLM package in version 1.0.0 entitled 'Logit Leaf Model Classifier for Binary Classification' (De Caigny et al. 2018b).

3 Empirical Results for the Balanced Sets

The empirical distributions of the selected prediction effectiveness measures (i.e. sensitivity, specificity, precision, F1, G-mean and AUC) calculated for the logit leaf model, the logit model and the classification tree on the basis of the test parts of the five balanced sets are presented in Figs. 1, 2, 3, 4, 5 and 6, and the empirical distributions obtained with respect to the five non-balanced sets were presented in Figs. 7, 8, 9, 10, 11 and 12. The average values of the analysed prediction effectiveness measures together with the results of the post hoc Dunn's test (with Bonferroni correction) are presented in Tables 2, 3, 4, 5, 6 and 7 with respect to the balanced sets and in Tables 8, 9, 10, 11, 12 and 13 with respect to the non-balanced sets.

The presentation of the empirical investigations was divided into the two parts. The first part (Sect. 3) contains a description of the results carried out on the basis of the balanced sets. The second part (Sect. 4) presents the results of the analysis carried out for the non-balanced sets.

On the basis of the results presented in Table 2, the conclusion may be made that for most analysed balanced sets, on the significance level of 0.10, there are no grounds for considering one of the analysed methods as more effective as regards prediction of bankruptcy a year in advance in the group of bankrupts.

The analysis of the boxplot charts presented in Fig. 1 leads to the conclusion that in most analysed balanced sets the greatest range of variability of the sensitivity measure (i.e. the greatest instability of results) was obtained for the logit leaf model. In turn, the logit model was characterised by the lowest variability of the sensitivity measure (i.e. the greatest stability of results).

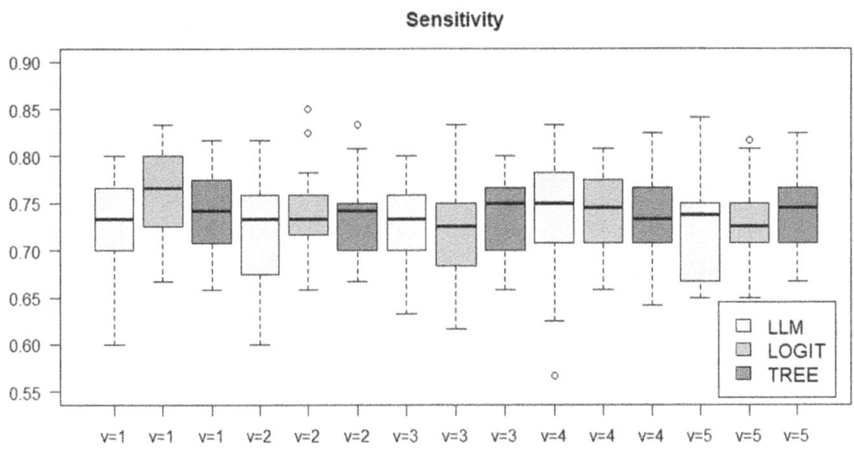

Fig. 1 Empirical distributions of the sensitivity measure calculated for the logit leaf model (LLM), the logit model (LOGIT) and the classification tree (TREE) on the basis of test parts of the five balanced sets ($v = 1, \ldots, 5$)

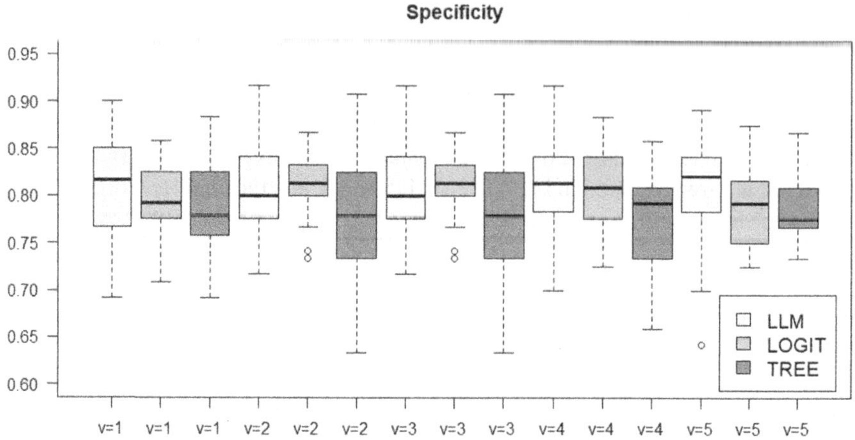

Fig. 2 Empirical distributions of the specificity measure calculated for the logit leaf model (LLM), the logit model (LOGIT) and the classification tree (TREE) on the basis of test parts of the five balanced sets ($v = 1, \ldots, 5$)

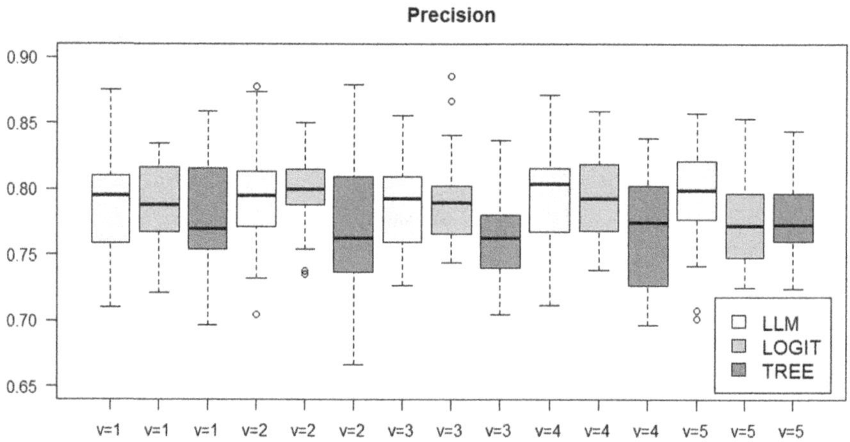

Fig. 3 Empirical distributions of the precision measure calculated for the logit leaf model (LLM), the logit model (LOGIT) and the classification tree (TREE) on the basis of test parts of the five balanced sets ($v = 1, \ldots, 5$)

In the event of non-threat of bankruptcy, carried out a year in advance, in the group of firms that continue the economic activity in the following year (i.e. non-bankrupts), one may distinguish the logit leaf model and the logit model (Table 3).

The graphs presented in Fig. 2 confirm the lower prediction effectiveness of the classification tree in the group of non-bankrupts as compared with the logit leaf

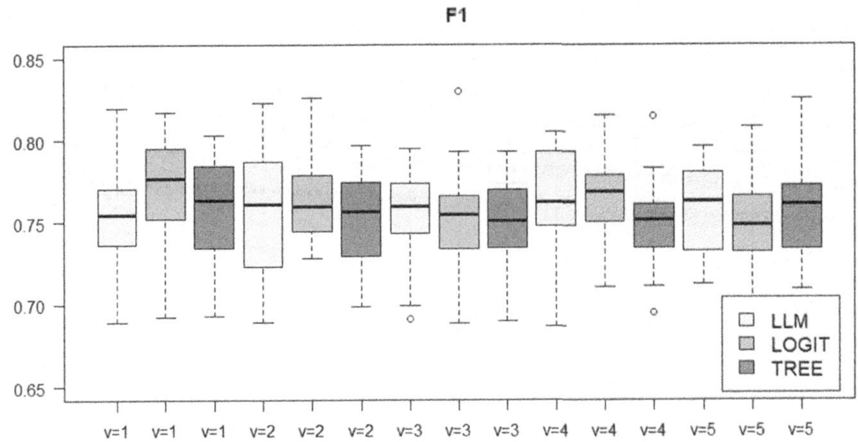

Fig. 4 Empirical distributions of the F1 measure calculated for the logit leaf model (LLM), the logit model (LOGIT) and the classification tree (TREE) on the basis of test parts of the five balanced sets ($v = 1, \ldots, 5$)

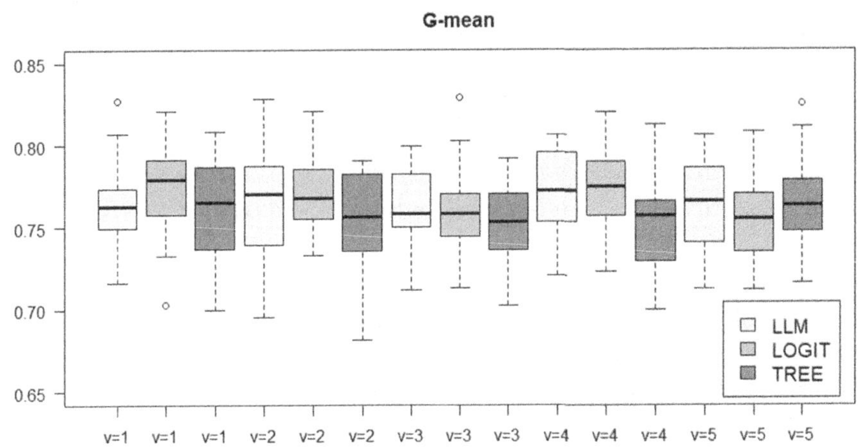

Fig. 5 Empirical distributions of the G-mean measure calculated for the logit leaf model (LLM), the logit model (LOGIT) and the classification tree (TREE) on the basis of test parts of the five balanced sets ($v = 1, \ldots, 5$)

model and the logit model. In most analysed balanced sets, the smallest range of variability specificity measure was obtained for the logit model.

The models characterised by the greatest precision of bankruptcy prediction a year in advance were the logit leaf model and the logit model (Table 4).

The graphs presented in Fig. 3 confirm the lower prediction precision of the classification tree as compared with the logit leaf model and the logit model. In most

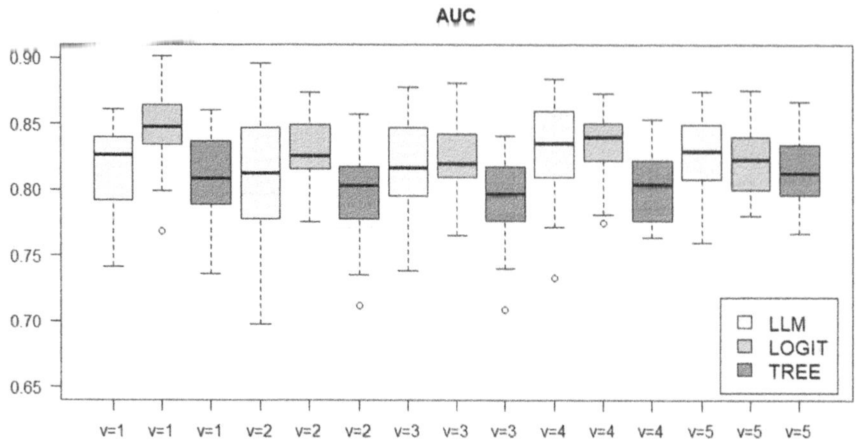

Fig. 6 Empirical distributions of the AUC measure calculated for the logit leaf model (LLM), the logit model (LOGIT) and the classification tree (TREE) on the basis of test parts of the five balanced sets ($v = 1, \ldots, 5$)

Table 2 Sensitivity measure—average value and *p*-value in Dunn's test for balanced sets

Average value					
Method	$v = 1$	$v = 2$	$v = 3$	$v = 4$	$v = 5$
LLM	0.7256	0.7236	0.7281	0.7431	0.7267
LOGIT	0.7606	0.7339	0.7197	0.7419	0.7289
TREE	0.7433	0.7356	0.7403	0.7342	0.7422
p-value					
Hypothesis	$v = 1$	$v = 2$	$v = 3$	$v = 4$	$v = 5$
LLM = LOGIT	0.0301	1.0000	1.0000	1.0000	1.0000
LLM = TREE	0.7972	1.0000	0.8514	0.7872	0.5711
LOGIT = TREE	0.4315	1.0000	0.2327	1.0000	0.5915

Notes: LLM—the logit leaf model, LOGIT—the logit model, TREE—the classification tree

analysed balanced sets, the smallest range of variability of the precision measure was obtained for the logit model.

On the basis of the results shown in Table 5, it can be concluded that for most of the analysed sets there is no reason for considering one of the methods as the most predictively effective.

Also, due to the stability of the results obtained for the F1 measure, it is not possible to indicate a better method than the others (Fig. 4).

For most of the analysed sets, at the level of significance of 0.10, there is no basis for recognising one of the considered methods as the most predictively effective a year in advance (Table 6).

Table 3 Specificity measure—average value and *p*-value in Dunn's test for balanced sets

Average value

Method	$v = 1$	$v = 2$	$v = 3$	$v = 4$	$v = 5$
LLM	0.8081	0.8097	0.8019	0.8036	0.8100
LOGIT	0.7950	0.8133	0.8078	0.8069	0.7875
TREE	0.7861	0.7775	0.7656	0.7736	0.7847

p-value

Hypothesis	$v = 1$	$v = 2$	$v = 3$	$v = 4$	$v = 5$
LLM = LOGIT	1.0000	1.0000	1.0000	1.0000	0.0333
LLM = TREE	0.1984	0.1960	0.0034	0.0736	0.0196
LOGIT = TREE	1.0000	0.0233	0.0012	0.0439	1.0000

Notes: LLM—the logit leaf model, LOGIT—the logit model, TREE—the classification tree

Table 4 Precision measure—average value and *p*-value in Dunn's test for balanced sets

Average value

Method	$v = 1$	$v = 2$	$v = 3$	$v = 4$	$v = 5$
LLM	0.7940	0.7941	0.7879	0.7937	0.7957
LOGIT	0.7886	0.7979	0.7908	0.7949	0.7756
TREE	0.7783	0.7709	0.7607	0.7666	0.7763

p-value

Hypothesis	$v = 1$	$v = 2$	$v = 3$	$v = 4$	$v = 5$
LLM = LOGIT	1.0000	1.0000	1.0000	1.0000	0.0483
LLM = TREE	0.2995	0.3413	0.0053	0.0228	0.0643
LOGIT = TREE	0.5990	0.0626	0.0044	0.0179	1.0000

Notes: LLM—the logit leaf model, LOGIT—the logit model, TREE—the classification tree

Table 5 F1 measure—average value and *p*-value in Dunn's test for balanced sets

Average value

Method	$v = 1$	$v = 2$	$v = 3$	$v = 4$	$v = 5$
LLM	0.7563	0.7555	0.7556	0.7655	0.7578
LOGIT	0.7735	0.7639	0.7524	0.7665	0.7506
TREE	0.7593	0.7514	0.7495	0.7487	0.7580

p-value

Hypothesis	$v = 1$	$v = 2$	$v = 3$	$v = 4$	$v = 5$
LLM = LOGIT	0.0483	1.0000	1.0000	1.0000	0.9437
LLM = TREE	1.0000	1.0000	0.8114	0.0541	1.0000
LOGIT = TREE	0.2138	0.6043	1.0000	0.0337	1.0000

Notes: LLM—the logit leaf model, LOGIT—the logit model, TREE—the classification tree

Table 6 G-mean measure—average value and *p*-value in Dunn's test for balanced sets

Average value

Method	$v = 1$	$v = 2$	$v = 3$	$v = 4$	$v = 5$
LLM	0.7643	0.7643	0.7632	0.7713	0.7657
LOGIT	0.7769	0.7721	0.7616	0.7730	0.7569
TREE	0.7636	0.7551	0.7521	0.7525	0.7625

p-value

Hypothesis	$v = 1$	$v = 2$	$v = 3$	$v = 4$	$v = 5$
LLM = LOGIT	0.1138	1.0000	1.0000	1.0000	0.4010
LLM = TREE	1.0000	0.9225	0.2245	0.0264	1.0000
LOGIT = TREE	0.2047	0.2103	0.6923	0.0111	1.0000

Notes: LLM—the logit leaf model, LOGIT—the logit model, TREE—the classification tree

Table 7 AUC measure—average value and *p*-value in Dunn's test for balanced sets

Average value

Method	$v = 1$	$v = 2$	$v = 3$	$v = 4$	$v = 5$
LLM	0.8187	0.8127	0.8186	0.8322	0.8279
LOGIT	0.8467	0.8302	0.8223	0.8333	0.8219
TREE	0.8086	0.7962	0.7952	0.8021	0.8146

p-value

Hypothesis	$v = 1$	$v = 2$	$v = 3$	$v = 4$	$v = 5$
LLM = LOGIT	0.0017	0.2092	1.0000	1.0000	1.0000
LLM = TREE	0.7335	0.1628	0.0239	0.0005	0.2185
LOGIT = TREE	0.0000	0.0006	0.0039	0.0005	1.0000

Notes: LLM—the logit leaf model, LOGIT—the logit model, TREE—the classification tree

The analysis of the boxplot graphs presented in Fig. 5 leads to the conclusion that in most of the analysed balanced sets, the ranges of variability of the G-mean measure obtained for the logit leaf model and the logit model are at a similar level. However, the variation of the G-mean value in the classification tree is generally similar or greater compared to the results obtained for the logit leaf model and the logit model.

The results of the post hoc test of significance of the differences between the average values obtained for the AUC measure allow to perceive the logit model as the most effective one of the analysed models regards to bankruptcy prediction a year in advance (Table 7). The second position in this ranking is taken by the logit leaf model.

The graphs presented in Fig. 6 confirm the greater prediction effectiveness of the logit leaf model and the logit model than the classification tree. As regards the AUC measure, the logit leaf model was characterised by the greatest instability of results, whereas the logit model demonstrated the greatest stability of results.

4 Empirical Results for the Non-balanced Sets

This part of the paper contains presentation of the results of the analysis conducted on the basis of the non-balanced sets.

The results of the analysis carried out for the non-balanced sets gave no basis for distinguishing one of the analysed methods as regards the prediction effectiveness in the group of bankrupts (Table 8).

In most analysed non-balanced sets, the smallest range of variability of the sensitivity measure was obtained for the logit model (Fig. 7).

The results of the post hoc test of significance of the differences between the average values obtained for the specificity measure allow to perceive the logit model as the most effective one of the analysed models as regards prediction of bankruptcy a year in advance in the group of non-bankrupts (Table 9). The second position in this ranking is taken by the logit leaf model.

The graphs presented in Fig. 8 confirm the greater predictive effectiveness of the logit leaf model and the logit model in the group of non-bankrupts than the classification tree. In most analysed non-balanced sets, the greatest range of variability of the specificity measure was obtained for the classification tree (Fig. 8).

The models characterised by the greatest precision of bankruptcy prediction a year in advance were the logit leaf model and the logit model (Table 10).

The graphs presented in Fig. 9 confirm the lower predictive precision of the classification tree as compared with the logit leaf model and the logit model in the case of the non-balanced sets. In most analysed non-balanced sets, the greatest range of variability of the precision measure was obtained for the classification tree.

Table 8 Sensitivity measure—average value and *p*-value in Dunn's test for non-balanced sets

Average value

Method	$v = 1$	$v = 2$	$v = 3$	$v = 4$	$v = 5$
LLM	0.4928	0.4928	0.4878	0.5072	0.5031
LOGIT	0.4831	0.4964	0.4839	0.4897	0.4992
TREE	0.4944	0.5139	0.4900	0.5008	0.5228

p-value

Hypothesis	$v = 1$	$v = 2$	$v = 3$	$v = 4$	$v = 5$
LLM = LOGIT	1.0000	1.0000	1.0000	0.7169	1.0000
LLM = TREE	1.0000	0.1558	1.0000	1.0000	0.3000
LOGIT = TREE	1.0000	0.2942	1.0000	0.9672	0.3998

Notes: LLM—the logit leaf model, LOGIT—the logit model, TREE—the classification tree

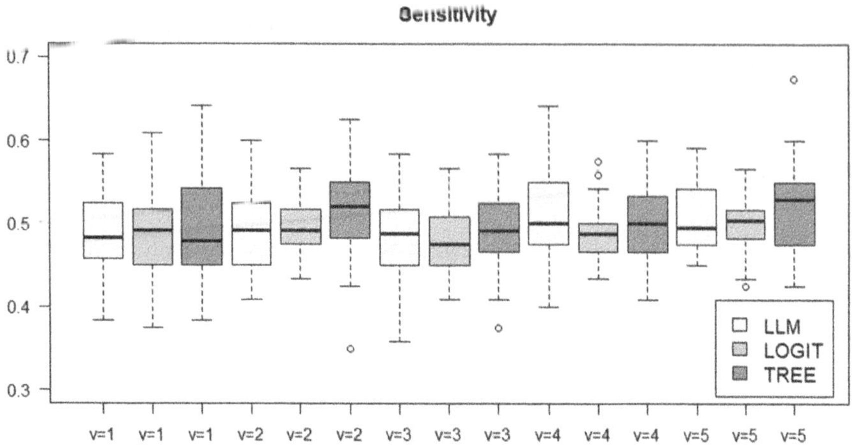

Fig. 7 Empirical distributions of the sensitivity measure calculated for the logit leaf model (LLM), the logit model (LOGIT) and the classification tree (TREE) on the basis of test parts of the five non-balanced sets ($v = 1, \ldots, 5$)

Table 9 Specificity measure—average value and p-value in Dunn's test for non-balanced sets

Average value

Method	$v = 1$	$v = 2$	$v = 3$	$v = 4$	$v = 5$
LLM	0.9541	0.9501	0.9510	0.9440	0.9531
LOGIT	0.9534	0.9507	0.9525	0.9529	0.9506
TREE	0.9430	0.9419	0.9441	0.9362	0.9394

p-value

Hypothesis	$v = 1$	$v = 2$	$v = 3$	$v = 4$	$v = 5$
LLM = LOGIT	1.0000	1.0000	1.0000	0.0234	1.0000
LLM = TREE	0.0181	0.2859	0.1680	0.1595	0.0003
LOGIT = TREE	0.0307	0.1893	0.0604	0.0000	0.0056

Notes: LLM—the logit leaf model, LOGIT—the logit model, TREE—the classification tree

Based on the analysis of the results obtained for non-balanced sets, similar conclusions were formulated as for the balanced sets. Also in this case, at 0.10 significance level, there was no reason to reject the hypothesis, which proclaimed a similar average prognostic effectiveness of the methods under consideration (Table 11).

For most non-balanced analysed sets, the most stable value of the F1 measure was characterised by a logit model (Fig. 10).

The results of the analysis carried out for non-balanced sets did not give the basis to the distinction of one of the methods the most effective in terms of predictive performance (measured by the G-mean measure) for bankruptcy prediction a year in advance (Table 12).

Table 10 Precision measure—average value and p-value in Dunn's test for non-balanced sets

Average value

Method	$v = 1$	$v = 2$	$v = 3$	$v = 4$	$v = 5$
LLM	0.7307	0.7140	0.7162	0.6957	0.7292
LOGIT	0.7240	0.7172	0.7202	0.7263	0.7178
TREE	0.6894	0.6944	0.6899	0.6643	0.6864

p-value

Hypothesis	$v = 1$	$v = 2$	$v = 3$	$v = 4$	$v = 5$
LLM = LOGIT	1.0000	1.0000	1.0000	0.1476	1.0000
LLM = TREE	0.0049	0.4205	0.2445	0.0695	0.0026
LOGIT = TREE	0.0626	0.2730	0.0634	0.0001	0.0323

Notes: LLM—the logit leaf model, LOGIT—the logit model, TREE—the classification tree

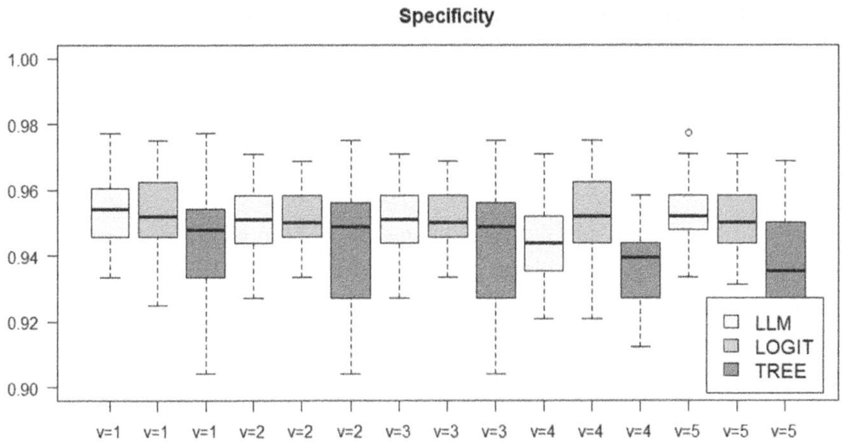

Fig. 8 Empirical distributions of the specificity measure calculated for the logit leaf model (LLM), the logit model (LOGIT) and the classification tree (TREE) on the basis of test parts of the five non-balanced sets ($v = 1, \ldots, 5$)

In most non-balanced sets analysed, the smallest variation range of G-mean values was obtained for the logit model (Fig. 11).

The results of the post hoc test of significance of the differences between the average values obtained for the AUC measure allow to state that the logit model is the most effective one of the analysed models regards to bankruptcy prediction a year in advance (Table 13). The second position in this ranking is taken by the logit leaf model.

The graphs presented in Fig. 12 confirm the greater prediction effectiveness of the logit leaf model and the logit model than the classification tree. In most analysed

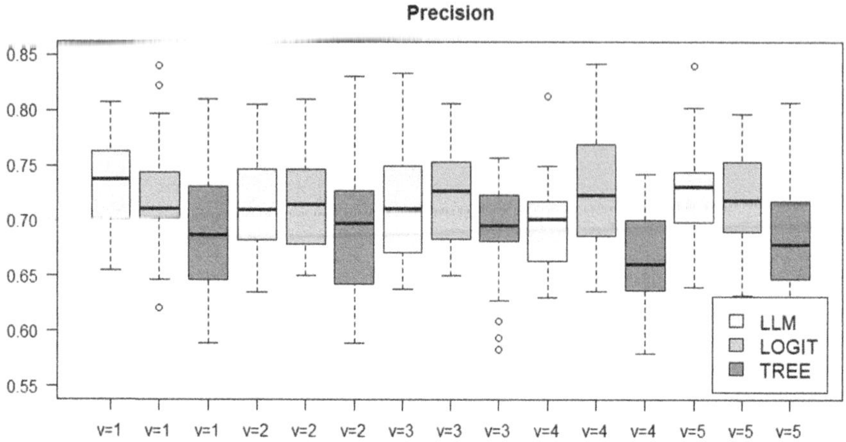

Fig. 9 Empirical distributions of the precision measure calculated for the logit leaf model (LLM), the logit model (LOGIT) and the classification tree (TREE) on the basis of test parts of the five non-balanced sets ($v = 1, \ldots, 5$)

Table 11 F1 measure—average value and p-value in Dunn's test for non-balanced sets

Average value					
Method	$v = 1$	$v = 2$	$v = 3$	$v = 4$	$v = 5$
LLM	0.5867	0.5811	0.5784	0.5843	0.5944
LOGIT	0.5776	0.5860	0.5775	0.5835	0.5881
TREE	0.5726	0.5876	0.5710	0.5696	0.5914

p-value					
Hypothesis	$v = 1$	$v = 2$	$v = 3$	$v = 4$	$v = 5$
LLM = LOGIT	1.0000	1.0000	1.0000	1.0000	1.0000
LLM = TREE	0.4858	0.7859	1.0000	0.5635	1.0000
LOGIT = TREE	1.0000	1.0000	1.0000	0.4429	1.0000

Notes: LLM—the logit leaf model, LOGIT—the logit model, TREE—the classification tree

non-balanced sets, the smallest range of variability of the AUC measure was obtained for the logit model.

Summing up the deliberations contained in Sects. 3 and 4, one may conclude that the results obtained on the basis of the balanced or non-balanced sets are very similar.

5 Conclusions

For the most analysed balanced and non-balanced sets, on the significance level of 0.10, there were no grounds for rejection of the zero hypothesis, according to which

Table 12 G-mean measure—average value and p-value in Dunn's test for non balanced sets

Average value					
Method	$v = 1$	$v = 2$	$v = 3$	$v = 4$	$v = 5$
LLM	0.6847	0.6832	0.6801	0.6906	0.6918
LOGIT	0.6777	0.6866	0.6783	0.6826	0.6884
TREE	0.6814	0.6945	0.6792	0.6839	0.6997
p-value					
Hypothesis	$v = 1$	$v = 2$	$v = 3$	$v = 4$	$v = 5$
LLM = LOGIT	1.0000	1.0000	1.0000	1.0000	1.0000
LLM = TREE	1.0000	0.2788	1.0000	1.0000	0.8374
LOGIT = TREE	1.0000	0.5367	1.0000	1.0000	0.7426

Notes: LLM—the logit leaf model, LOGIT—the logit model, TREE—the classification tree

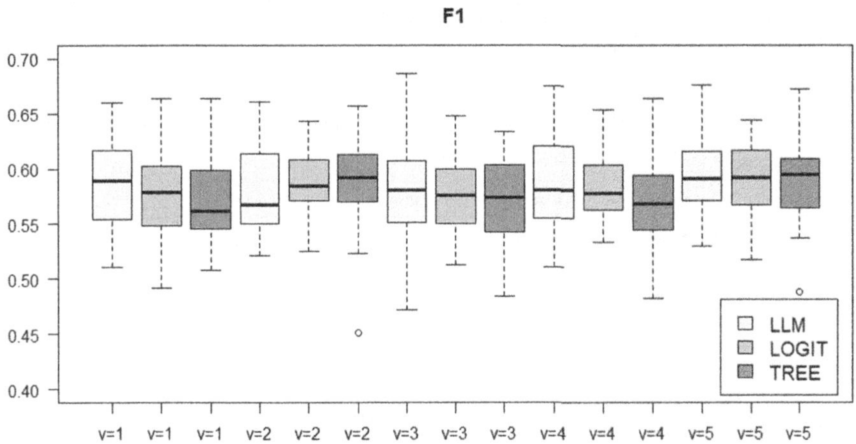

Fig. 10 Empirical distributions of the F1 measure calculated for the logit leaf model (LLM), the logit model (LOGIT) and the classification tree (TREE) on the basis of test parts of the five non-balanced sets ($v = 1, \ldots, 5$)

prediction effectiveness (measured with the use of the sensitivity, F1 and G-mean measures) of the analysed methods (i.e. the logit leaf model, the logit model, the classification tree) is, on average, on the same level.

As regards the other analysed measures (i.e. specificity, precision and AUC), either the logit model had the greatest prediction effectiveness, with the second position in the ranking taken by the logit leaf model, or the logit model and the logit leaf model had a greater prediction effectiveness than the classification tree.

In most analysed sets (both balanced and non-balanced), the logit model demonstrated the greatest stability (measured with the use of the range of variability) of the results obtained for particular prediction effectiveness measures.

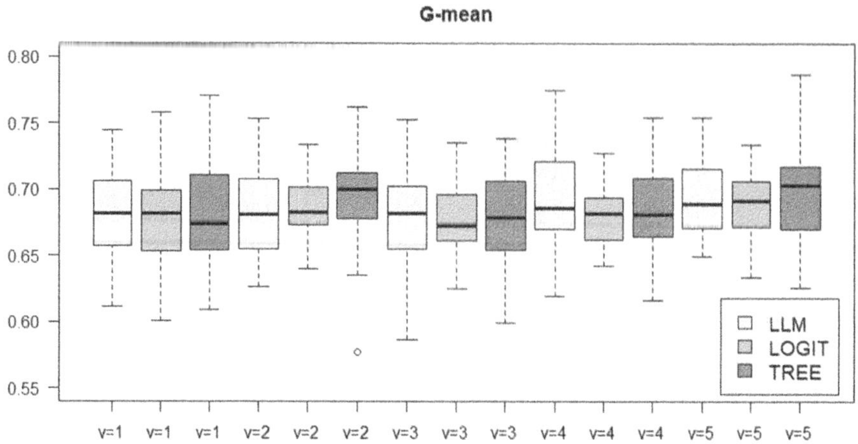

Fig. 11 Empirical distributions of the G-mean measure calculated for the logit leaf model (LLM), the logit model (LOGIT) and the classification tree (TREE) on the basis of test parts of the five non-balanced sets ($v = 1, \ldots, 5$)

Table 13 AUC measure—average value and p-value in Dunn's test for non-balanced sets

Average value

Method	$v = 1$	$v = 2$	$v = 3$	$v = 4$	$v = 5$
LLM	0.8064	0.8207	0.8133	0.8318	0.8149
LOGIT	0.8387	0.8521	0.8441	0.8507	0.8506
TREE	0.7969	0.7875	0.7768	0.7848	0.7921

p-value

Hypothesis	$v = 1$	$v = 2$	$v = 3$	$v = 4$	$v = 5$
LLM = LOGIT	0.0011	0.0026	0.0039	0.0785	0.0001
LLM = TREE	0.3088	0.0029	0.0008	0.0000	0.0464
LOGIT = TREE	0.0000	0.0000	0.0000	0.0000	0.0000

Notes: LLM—the logit leaf model, LOGIT—the logit model, TREE—the classification tree

Thus, on the basis of obtained results, the conclusion can be drawn that in bankruptcy prediction a year in advance of enterprises operating in the manufacturing sector in Poland, the hybrid model logit leaf model has, in most cases, prediction effectiveness that is not greater than the prediction effectiveness of the logit model, and not lower than the prediction effectiveness of the classification tree.

Summing up, it has to be stated that the results of empirical research did not confirm an advantage of the hybrid approach over the use of individual classifiers.

A value added of application of the logit leaf model in bankruptcy prediction is obtaining logit models for the groups of enterprises that have a similar financial

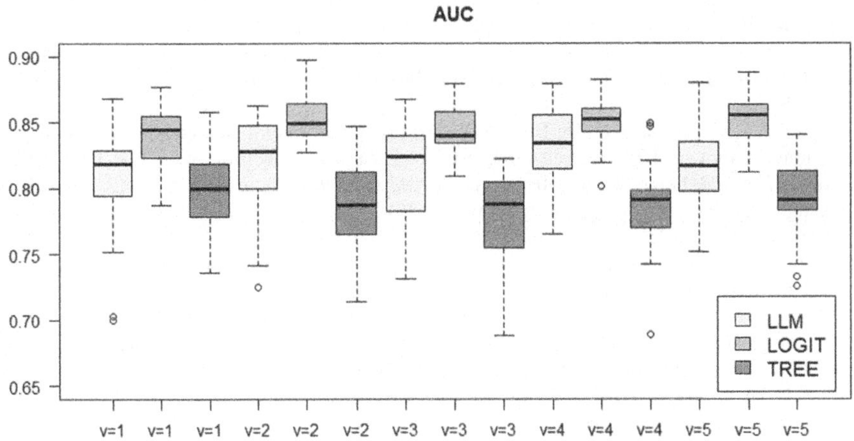

Fig. 12 Empirical distributions of the AUC measure calculated for the logit leaf model (LLM), the logit model (LOGIT) and the classification tree (TREE) on the basis of test parts of the five non-balanced sets ($v = 1, \ldots, 5$)

situation. Such groups of similar enterprises are obtained with the use of decision-making rules (the classification tree); thus, one may consider it as enriching the interpretation aspect of investigation.

In further research, the authors intend to (1) conduct an analysis of the usefulness of the considered financial ratios for prediction, a year in advance, bankruptcy of enterprises operating in the manufacturing sector in Poland, (2) expand the research with the bankruptcy prediction two and three years in advance and (3) analyse the impact of cleaning data from the outliers on the prediction effectiveness of the hybrid model logit leaf model and individual classifiers (the logit model and the classification tree).

Acknowledgements The research behind this publication was financed from the funds granted to the Cracow University of Economics.

References

Altman DG, Bland JM (1994a) Diagnostic tests 1: sensitivity and specificity. BMJ 308(6943):1552. https://doi.org/10.1136/bmj.308.6943.1552

Altman DG, Bland JM (1994b) Diagnostic tests 2: predictive values. BMJ 309(6947):102. https://doi.org/10.1136/bmj.309.6947.102

Altman DG, Bland JM (1994c) Diagnostic tests 3: receiver operating characteristic plots. BMJ 309(6948):188. https://doi.org/10.1136/bmj.309.6948.188

Baesens B, Van Gestel T, Viaene S, Stepanova M, Suykens J, Vanthie J (2003) Benchmarking state-of-the-art classification algorithms for credit scoring. J Oper Res Soc 54:627–635. https://doi.org/10.1057/palgrave.jors.2601545

Bramer M (2016) Principles of data mining, 3rd edn. Springer, London
Brown I, Mues Ch (2012) An experimental comparison of classification algorithms for imbalanced credit scoring data sets. Expert Syst Appl 39(3):3446–3453. https://doi.org/10.1016/j.eswa.2011.09.033
De Caigny A, Coussement K, De Bock KW (2018a) A new hybrid classification algorithm for customer churn prediction based on logistic regression and decision trees. Eur J Oper Res 269(2):760–772. https://doi.org/10.1016/j.ejor.2018.02.009
De Caigny A, Coussement K, De Bock KW (2018b) LLM: logit leaf model classifier for binary classification. R package version 1.0.0. https://CRAN.R-project.org/package=LLM
du Jardin P (2018) Failure pattern-based ensembles applied to bankruptcy forecasting. Decis Support Syst 107:64–77. https://doi.org/10.1016/j.dss.2018.01.003
Garcia V, Marques AI, Sanchez JS (2019) Exploring the synergetic effects of sample types on the performance of ensembles for credit risk and corporate bankruptcy prediction. Inf Fusion 47:88–101. https://doi.org/10.1016/j.inffus.2018.07.004
Hastie T, Tibshirani R, Friedman J (2009) The elements of statistical learning. Data mining, inference, and prediction, 2nd edn. Springer, New York
Kubat M, Holte RC, Matwin S (1998) Machine learning for the detection of oil spills in satellite radar images. Mach Learn 30:195–215
Lessmann S, Baesens B, Seow HV, Thomas LC (2015) Benchmarking state-of-the-art classification algorithms for credit scoring: an update of research. Eur J Oper Res 247(1):124–136. https://doi.org/10.1016/j.ejor.2015.05.030
Lewis DD, Gale WA (1994) A sequential algorithm for training text classifiers. In: Proceedings of the seventeenth annual international ACM SIGIR conference on research and development in information retrieval, Springer, pp 3–12
Ohsaki M, Wang P, Matsuda K, Katagiri S, Watanabe H, Ralescu A (2017) Confusion-matrix-based kernel logistic regression for imbalanced data classification. IEEE Trans Knowl Data Eng 29(9):1806–1819. https://doi.org/10.1109/TKDE.2017.2682249
Pawełek B (2017) Prediction of company Bankruptcy in the context of changes in the economic situation. In: Papież M, Śmiech S (eds) The 10th Professor Aleksander Zeliaś international conference on modelling and forecasting of socio-economic phenomena. Conference proceedings. Foundation of the Cracow University of Economics, Cracow, pp 290–299
Pawełek B, Pociecha J (2019) The problem of outliers in the prediction of corporate bankruptcy using the Logit leaf model. Paper presented at the 28th conference of the section on classification and data analysis of the polish statistical association entitled data classification and analysis—theory and applications (SKAD 2019), Szczecin, Poland, 18–20 Sept 2019
Pawełek B, Pociecha J, Baryła M (2017) Evaluation of the suitability financial indicators for corporate bankruptcy prediction depending on their size. Paper presented at the 4th conference on data analysis (ECDA 2017), Wroclaw, Poland, 27–29 Sept 2017
Pawełek B, Pociecha J, Grabarz S (2019) Logit leaf model in prediction of corporate bankruptcy. Paper presented at the 6th European conference on data analysis (ECDA 2019), Bayreuth, Germany, 18–20 Mar 2019
Ultsch A, Lötsch J (2015) Computed ABC analysis for rational selection of most informative variables in multivariate data. PLoS ONE 10(6):e0129767. https://doi.org/10.1371/journal.pone.0129767
Zweig MH, Campbell G (1993) Receiver-operating characteristic (ROC) plots: a fundamental evaluation tool in clinical medicine. Cunical Chem 39(4):561–577

The Impact of Longevity on a Valuation of Long-Term Investments Returns: The Case of Selected European Countries

Grażyna Trzpiot

Abstract The impact of longevity risk is a topic of growing importance in academic research and public debate over the past few years. From the individual's perspective, the need for long-term investment increases as the life expectancy increases. Improvements in longevity and changing structure of population impact economy and financial stability. In this paper, we consider some economic, financial and demographic variables in the context of their impact on longevity in terms of long-term investment. The principal component regression is used in order to construct long-term investment portfolios that are sensitive to risk factors according to the APT portfolio factor model for selected European countries. Three investment portfolios with different fixed risk profiles (low, medium and high) have been proposed as the final results of the main research. For selected European countries, PCA's longevity risk factors associated with longevity risk have a significant impact on return on long-term portfolio.

Keywords Longevity risk · Long-term portfolios · Portfolio return · Portfolio risk

1 Introduction

The recent analysis on considerably lower long-term investment returns expectations over the next twenty years than they were in the past three decades was the inspiration for this paper. Looking for economic and business conditions that are weakening or even reversing individuals would need to save more for retirement, retire later, or reduce consumption during retirement. The main aim of this paper is to look close to the problem: How global longevity trend can impact long-term investments returns? We attempt to identify risk factors that could have influence on the long-term investment return. An evaluation of the impact of each risk factor on portfolio return rates is presented, regardless of the portfolio risk level set in relation to the composition of the APT portfolio factor model.

G. Trzpiot (✉)
University of Economics in Katowice, Katowice, Poland
e-mail: grazyna.trzpiot@ue.katowice.pl

K. Jajuga et al. (eds.), *Classification and Data Analysis*, Studies in Classification, Data Analysis, and Knowledge Organization,
https://doi.org/10.1007/978-3-030-52348-0_10

Representative countries with different economy growth level and demographic situation are selected by cluster analysis. Next the principal component analysis is used to specify risk factors. Finally, the multifactor regression models (the principal component regression) were used to build portfolios that are sensitive to risk factors.

2 Longevity and Financial Markets

Currently in Europe and USA, at least one member of a 65-year-old couple now has a 90% chance of living to 80 or beyond, a nearly 50% chance of reaching 90 and a one-in-five chance of turning 95 or older. Living longer affects key retirement decisions. Changing in the demographic structure will have an impact on the asset returns. Declining investment in equities, population shifts can have a significant impact on investor behavior and equity values (Arnott and Chaves 2012). The older generation that generally invests the most will move increasingly into retirement. The post-war generations will convert their investments to cash, in order to consume more. The declining number of younger people, who tend to buy rather than save, will further reduce the demand for all kinds of investments.

2.1 Historical Trends and Projections Review

Lower returns are decreasing gradually since the financial crisis. Looking for past opportunity for long-run average returns, next making projection for the next 20 years we can observe decline on the level on the projected rate of return as a reaction to slowing economic growth in aging society. Long-run average returns in last 30 years were different in the USA and in Europe and essentially depend on the asset class. Average historical US bonds returns were equal 5% and in average historical European bonds returns were equal 5.9%. In the same period of time, average equities' returns in USA and in Europe were on the same level (about 7.9%).

In the literature, we can find two types of projection based on historical trends: slow-growth scenario and growth-recovery scenario (McKinsey report 2016) for economic markets. Slow-growth scenario means that demographic changes result in slow employment growth, and productivity growth remains on a par with the past 50 years; GDP growth will fall below the average of the past 50 years, interest rates will rise, but only slowly, and inflation remains lower, below the 2%, and competitive pressures result in declining margins. As a consequence of the assumptions made, as the results, it can be predicted that the average return on investment in the USA and European bonds will be on the very low level (0.1%), and in addition, it can be predicted that the average returns on equities in the USA will be 2.5%, and in Europe there will be no more than 1.5%.

Growth-recovery scenario assume: GDP growth will pick up as the result of a productivity surge, inflation rises rapidly, as do interest rates, and companies will be able to innovate and adapt to maintain their profit margins at today's levels. As a consequence of these assumptions, it can be predicted that the average return on investment in the US and European bonds will be on lower level (1.9%), and in addition it can be predicted that the average returns on equities in the USA will be 6.5%, and in Europe there will be no more than 4.5%.

For long-term portfolio returns we can also discuss historical return by different asset class. In J.P. Morgan report (2016) we can find some information about average historical return on selected investment in the USA during 1995–2014 period. Historical returns on a specific portfolio were used in this analysis. The financial instruments and indexes along with their rates of return used were as follows: NAREIT Equity REIT Index (11.5%), S&P 500 Index (9.9%), MSCI EAFE (5.4%), WTI Index (5.7%), Barclays US Aggregate Index (6.2%), median sale price of existing single-family homes (3.2%), Gold as USD/troy oz (5.9%), the US CPI (2.4), a balanced portfolio 60/40 with 60% invested in S&P 500 Index and 40% invested in high quality US fixed income, represented by the Barclays US Aggregate Index (8.7%) and similarly a balanced portfolio 40/60 (8.1%).

As a consequence of historical trends and presented projection individuals, elderly people would need to save more for retirement, retire later, or reduce consumption during retirement. Looking for alternative to the longevity risk, we try to consider possible investment path. This investment path is different in every country but you can also point out common factor, and you can indicate common features and possibilities.

In our research, we can construct the factor portfolio, long-term portfolios for elderly people with the main aim reducing the longevity risk. Starting point was information on the regular long-term portfolio returns and return by fixed asset class. Using this information, macroeconomic variables were selected for our research. Additionally, the balanced portfolio, with a known risk profile in the final part of our research, was modeled as 60/40 and 40/60 accordingly to this analysis.

2.2 *The Implication of Longevity and Aging on Investments Returns*

The impact of aging on the economy and financial markets is very important research area. We should confront with several interrelated issues: a decline in the working-age population, increased healthcare costs, unsustainable pension commitments and changing demand drivers within the economy.

Firstly, a rapidly aging population means there are fewer working-age people in the economy, which means lack of demographic dividend. This leads to a supply shortage of qualified workers, making it more difficult for businesses to fill up demand roles. For economy that cannot fill up work demand, we can expect: declining productivity, higher labor costs, lower tax revenue, etc.

Table 1 Examples of new allocation focus on drivers on return

Risk factors	Macroeconomic	Thematic
Equity	Growth	Aging
Liquidity	Stagnation	Knowledge economy
Term	Dislocation	Recourse scarcity
Credit	Inflation	Low-cost production

Source The Future of Long-term Investing, World Economic Forum USA Inc. (2011)

Secondly, demand for health care will rise with age, so countries with rapidly aging populations must allocate more money and resources to their healthcare systems. Demographic trends and economic consequences create challenges as well opportunities. The combination of lower tax revenue and higher spending commitments on health care, pension and other benefits is a major concern.

Taking into consideration the effect of longevity on investment returns, the main effects of aging are focused on the following three related issues: The balance between the supply of savings and the demand for savings determines the rate of return earned by investors,[1] investments demand, the demand for different classes of financial market assets. We have from the literature some evidence on demographic structure and asset returns (Arnott and Chaves 2012). A new approach to asset allocation has been observed, after the financial crisis, many investors are clearly changing the portfolio strategy. Many long-term investors have been transitioning to a simplified asset allocation (Table 1).

3 Methodology

As an objective in research part, we attempt to identify risk factors that could have influence on the long-term investment return. An evaluation of the impact of each risk factor on portfolio return rates is presented, regardless of the portfolio risk profile in relation to the fixed composition of the portfolio.

Three investment portfolios with different fixed risk profiles (low, medium and high) have been proposed as the final results of the main research as particularly likely investments and can be considered as scenarios for the future level of long-term return on investment for selected countries. PCA's longevity risk factors have a significant impact on the long-term return on investment portfolios.

The main contribution is proposed method for building portfolios that are sensitive to risk factors—based on APT portfolio methodology (Ross 1976).[2] We propose new

[1] Aging and the Macroeconomy: Long-Term Implications of an Older Population. Institute of Medicine (US) Committee on the Long-Run Macroeconomic Effects of the Aging U.S. Population, National Academies Press, 2012.

[2] In finance, arbitrage pricing theory (APT) is a general theory of asset pricing that holds that the expected return of a financial asset can be modeled as a linear function of various factors or

approach to constructing the risk factors (as indicators) that have impact on the long-term investment returns. Risk factor defined by PCA longevity risk factor has a significant impact on long-term rates of return of investment portfolios. At the end, we can make valuation our portfolios. Our research proceeds with the three main steps.

First step: selection of the European countries to the analysis. The cluster analysis is applied to choose representative countries from each cluster of countries due to the macroeconomic variables. Hierarchical method allows determining the best number of clusters as well as to see the hierarchical relations between obtained groups of countries. Steps 2 and 3 are conducted for each of the selected countries.

Second step: identification factors that could have influence on the long-term investment return. Dimension reduction by PCA is used for transformation of highly correlating variables into set of uncorrelated latent variables, and combination of several variables that characterize demographic changes and economic development into uncorrelated factors. Factors are associated with risks related to investments.

Third step: Simulation of three investment portfolios with different risk level (low, medium and high) as a particularly possibly investments. The level of the risk for long-term investment is determined by fixed percentage share of stocks and bonds. The investment rates of return were modeled through the PCR: risk factors—obtained in the step 2—were used as predictors in a regression model fitted using the least squares procedure. There are two main reasons for regressing the investment return on the risk factors rather than directly on the explanatory variables. Firstly, the explanatory variables are often highly correlated (multicollinearity) which may cause inaccurate estimations of the least squares regression coefficients. Secondly, the dimensionality of the regressors is reduced by taking only a subset of PCs for prediction. A method does not require uncorrelated variables or normal distribution of the residuals.

Two methods, PCR and PCA, are both good techniques for dimensionality reduction in modeling datasets. There are especially useful when the independent variables are highly multicollinear (Hotelling 1933; Jolliffe 1982, 2002).

The selection of variables was preceded by an analysis of literature in the field of research on determinants of macroeconomic and financial implications of aging. In the process of identification of risk factors, the following variables are taken into consideration:

1. Demographic old-age dependency ratio—traditionally seen as an indication of the level of support available to older persons (those aged 65 or over, i.e., age when they are generally economically inactive) by the working-age population (those aged between 15 and 64) [expressed per 100 persons of working age (15–64)].

---

theoretical market indices, where sensitivity to changes in each factor is represented by a factor-specific beta coefficient. The linear factor model structure of the APT is used as the basis for many of the commercial risk systems employed by asset managers.

2. Life expectancy at birth—the mean number of years that a newborn child can expect to live if subjected throughout his or her life to the current mortality conditions (age-specific probabilities of dying) [expressed in years].
3. Life expectancy at age 65—the mean number of years still to be lived by a man or a woman who has reached the age 65, if subjected throughout the rest of his or her life to the current mortality conditions (age-specific probabilities of dying) [expressed in years].
4. Consumer Price Index (CPI)—the change over time in the prices of consumer goods and services acquired, used or paid for by households [measured in an index, 2015 base year].
5. Real GDP per capita—the ratio of real GDP to the average population of a specific year; a measure of economic activity, used as a proxy for the development in a country's material living standards (a limited measure of economic welfare) [per capita, in current prices].
6. Unemployment rate—represents unemployed persons as a percentage of the labor force (the total number of people employed and unemployed) [% of active population].
7. Real effective exchange rates (REER)—aims to assess a country's price or cost competitiveness relative to its principal competitors in international markets; changes in cost and price competitiveness depend not only on exchange rate movements but also on cost and price trends [indices].
8. Gross saving—measures the portion of gross national disposable income that is not used for final consumption expenditure; gross national saving is the sum of the gross savings of the various institutional sectors [current prices].
9. Long-term government bond yields—refer to central government bond yields on the secondary market, gross of tax, with residual maturity of around 10 years; the bond or the bonds of the basket have to be replaced regularly to avoid any maturity drift [%].
10. Long-term care (health) expenditures—expenditures on a range of medical and personal care services that are consumed with the primary goal of alleviating pain and suffering and reducing or managing the deterioration in health status in patients with a degree of long-term dependency [share of current expenditures on health].
11. Currency exchange rates: EUR/USD, EUR/PLN.
12. Stock market a main index: DAX in Germany, IBEX35 in Spain, WIG20 in Poland.
13. Real Estate Funds and Equity/Dividend Funds: Unilmmo Deutchland and Allianz Vermögensbildung Deutschland (Germany), Seguffondo Inversion and Bankia Dividendo España FI (Spain), PZU UFK Investor Nieruchomości i Budownictwa and Investor FIO Subfundusz Akcji Spółek Dywidendowych (Poland).

Economic and demographic variables are derived from Eurostat database (variables 1–9) and OECD (variable 10), stock quotes—from stock exchange (Frankfurt, Madrid, Warsaw) and financial database (the Yahoo Finance) (variables 12 and 13).

Time series were obtained for the time period 2010–2016. It is not wide period of time; thus, some data were converted to monthly frequency (and then all variables were expressed as chain indices using a base previous observation in monthly aggregation), maintaining the strength and direction of correlation between variables. The period does not cover years from the financial crisis to avoid unusual observations from financial market.

Relations between the above-mentioned variables and longevity are analyzed in empirical studies. Some relations are clear, while others are still a subject of debate (in particular, the impact of longevity on inflation is unclear). Due to the complexity of these relations and their multidimensionality, it is worth mentioning a few confirmed consequences of longevity (e.g., Bloom et al. 2010; Rachel and Smith 2015; Acemoglu and Restrepo 2017): reducing investment return, reducing public saving, reducing growth rates, reducing real interest rates, affecting labor supply and returns, reallocation of saving from riskier to safe assets may lead to potential mispricing of risk, and running down assets may result in negative wealth effects.

4 Results

First step: selection of the European countries to the analysis. The cluster analysis was conducted according to the following variables: GDP growth rate (%), inflation rate (%), real productivity per hour worked, national savings, proportion of population aged 65 and over, old-age-dependency ratio (Ward linkage, Euclidean distance). As the result of cluster analysis (Gordon 1999), we obtain four groups of countries (Trzpiot and Majewska 2016). The conclusion on this part of the analysis Luxembourg, as the outlier, was excluded from the analysis (Fig. 1). We choose one representative country from each cluster: Germany, Spain and Poland. Each of these countries represents different level of economic growth and life expectancy.

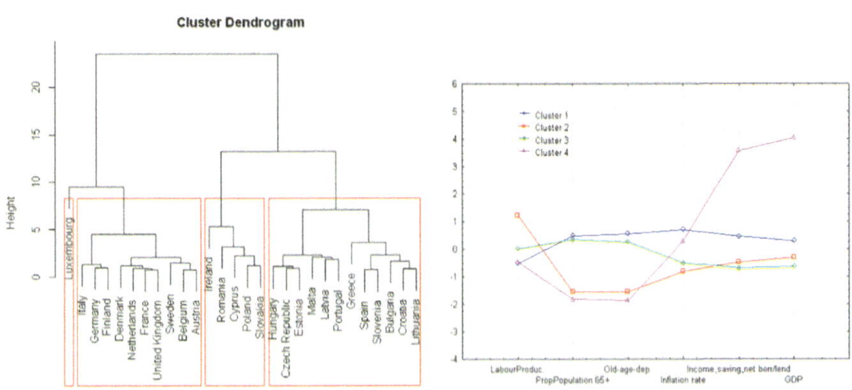

Fig. 1 Tree diagram (left) and plot of means of each cluster (right). *Source* Trzpiot and Majewska (2016)

Empirical investigation of relations between longevity phenomenon and selected macroeconomic and financial variables is made for selected European countries with different levels of economic growth and life expectancy, i.e., for Germany, Spain and Poland. From longevity perspective, life expectancy (at birth and at aged 65, for both sexes) in Poland is shorter than in Germany and Spain, while life expectancy is the highest in Spain. Spain is expected to become the world's second oldest country by 2050, behind Japan. According to HDI index Germany since 2010 has been in the group of five the most developed countries, Spain—in the second ten, and Poland—in the third ten the most developed countries in world (UNDP 2018).

Second step: identification factors that could have influence on the long-term investment return. It is not wide period of time; thus, some data were converted to monthly frequency (and then all variables were expressed as chain indices using a base previous observation—this is new approach in main research), with maintaining the strength and direction of correlation between variables. This analysis is new in two points: All variables were expressed as chain indices using a base previous observation and estimation of portfolio will depend on this new filters. In the previous paper (Trzpiot and Majewska 2016; Majewska and Trzpiot 2019), variables were expressed as indices using a base year of 2010.

For Germany (Table 2), the first principal component explains 27.2% of the variation, while all components—73.2%, we revived a set of linearly uncorrelated variables: The first component is identified as elderly needs risk especially because of

Table 2 Risk factor loads of principal components: Germany

	F1	F2	F3	F4	F5	F6
OLDAGERATIO	−0.78					
LEBIRTH	0.89					
LE65	0.91					
LT10					0.58	
EUR/USD		−0.60				
EUR/PLN		0.76				
EUR/GPB				0.61		
SAVING						
GDP	0.59					
UNEMP			0.62			
LONGCARE	−0.88					
DAX		−0.74				
FUNDRAELESTATE						
FUNDEQUITY						
INDUSTEFFEXCHRATE					0.60	
CPI						0.66

Source Own calculations

Table 3 Risk factor loads of principal components: Spain

	F1	F2	F3	F4	F5	F6
OLDAGERATIO					0.68	
LEBIRTH		−0.62				
LE65		−0.63				
LT10						0.54
EUR/USD		−0.56				
EUR/PLN			0.63			
EUR/GPB		−0.58				
SAVING				−0.61		
GDP	0.89					
UNEMP	−0.82					
LONGCARE	0.84					
IBEX35		−0.55				
FUNDRAELESTATE					−0.54	
FUNDEQUITY						0.68
INDUSTEFFEXCHRATE			0.49			
CPI	−0.86					

Source Own calculations

the high positive factor loadings life expectancy at birth and life expectancy at 65. The second component has been a set of variables that reflect financial risk. Next to component was identified as labor market and market risk. Advancing age due to increased life expectancy itself is a risk factor. The next is connected with long-term investment. The last component explains 7.6% of total variance, and it would associate with local economy risk.

For Spain (Table 3), the first principal component explains 22.2% of the variation, while all components—76.4%. The first component reflects economy risk. The second was call long-term standard of living (explains 16.2% of the variation). Next component has been clustered with currency exchange rates and real effective exchange rates so was call market risk, the fourth with gross savings—individual wealth risk. The last two components was explain 12.1% of total variance: old-dependency ratio—elderly needs risk and with long-term government bond yield, it would be associated with long-term investment risk.

For Poland (Table 4), the first component has been loaded with variables related with market risk and explains 23.3% of variance, and all components was explain—76.6% of the total variance. The second component has been identified as elderly needs risk. Next two components have been clustered with long-term standard of living risk and financial risk. The last two components explain 14.2% of total variance: long-term government bond yield and long-term care expenditures—we call them long-term investment risk—and a gross saving would be associated with individual wealth risk.

Table 4 Risk factor loads of principal components: Poland

	F1	F2	F3	F4	F5	F6
OLDAGERATIO		−0.88				
LEBIRTH			−0.88			
LE65			−0.85			
LT10					0.49	
EUR/USD	−0.63					
EUR/PLN	0.76					
EUR/GPB				0.68		
SAVING						0.89
GDP			0.48			
UNEMP		0.74				
LONGCARE					0.76	
WIG20						
FUNDRAELESTATE						
FUNDEQUITY						
INDUSTEFFEXCHRATE						
CPI						

Source Own calculations

Third step: Simulation of three investment portfolios with fixed risk profiles (low, medium and high) as a particularly possibly investments. Based on risk factors received by dimension reduction (by using PCA), we started to estimate portfolio according to APT theory (Ross 1976), where sensitivity to changes in each factor is represented by a factor-specific beta coefficient. All received factors are associated with risks related with investments (Table 5).

Construction of portfolios uses weighted return of stock and bonds, i.e., rate of return of main index on stock exchange and relative change of monthly return rate of long-term government bond yields 10 years. We construct different portfolios using PCR with fixed weights (proportion of stock and bond, respectively): 40/60 (low risk), 50/50 (medium risk) and 60/40 (high risk). *Rp* means portfolio rate of return.

Scenario #1: Portfolio return rate 40s/60b:

$$R_{pGERMANY} = -0.26F2 - 0.27F4 + 0.572F5 + 0.408F6 \quad R^2 = 0.641$$

The interpretation for this result for Germany is as follows: If risk represented by $F2$ increases by 1, then *Rp* will decrease by 0.26%, if risk represented by $F4$ increases by 1, then *Rp* will decrease by 0.27%, if risk represented by $F5$ increases by 1, then *Rp* will increase by 0.572%, and if risk represented by $F6$ increases by 1, then *Rp* will increase by 0.408%.

Table 5 Defined risk factors for selected European country 2010–2016 (chain indexes)

	Germany	Spain	Poland
Factor 1	**Elderly needs risk** Old-dependency ratio Life expectancy at birth, Life expectancy at 65 Real growth rate of GDP Long-term care expenditures	**Economy risk** Real growth rate of GDP Unemployment rate Long-term care expenditures CPI	**Market risk** Currency exchange rates EUR/USD, EUR/PL WIG20 return rates Fundequity Fund raelestate Industeffexch rate
Factor 2	**Financial risk** Currency exchange rates EUR/USD, EUR/PL DAX return rates	**Long-term standard of living risk** Life expectancy at birth, Life expectancy at 65 Currency exchange rates EUR/USD, EUR/GPB IBEX35 return rates	**Elderly needs risk** Old-dependency ratio Unemployment rate
Factor 3	**Labor market risk** Unemployment rate	**Market risk** Currency exchange rates EUR/PL Industeffexch rate	**Long-term standard of living risk** Life expectancy at birth, Life expectancy at 65 Real growth rate of GDP
Factor 4	**Market risk** Currency exchange rates EUR/GPB Fundequity	**Individual wealth risk** Gross savings	**Financial risk** Currency exchange rates EUR/GPB z
Factor 5	**Long-term investment risk** Long-term government bond yields Gross saving Fund raelestate Industeffexch rate	**Elderly needs risk** Old-dependency ratio Fund raelestate	**Long-term investment risk** Long-term government bond yield Long-term care expenditures
Factor 6	**Local economy risk** CPI	**Long-term investment risk** Long-term government bond yield Fund raelestate	**Individual wealth risk** Gross saving

Source Own calculations

$$R_{pSPAIN} = -0.19F4 + 0.537F6 \quad R^2 = 0.377$$

The interpretation of this equation for Spain is as follows: If risk represented by $F4$ increases by 1, then Rp will decrease by 0.19%, and if risk represented by $F6$ increases by 1, then R_p will increase by 0.537%.

$$R_{pPOLAND} = 0.504F5 \quad R^2 = 0.26$$

The interpretation of this equation for Poland is as follows: If risk represented by $F5$ increases by 1, then Rp will decrease and will increase by 0.504%.

Scenario #2: Portfolio return rate 50s/50b:

$$R_{pGERMANY} = -0.27F2 - 0.28F4 + 0.567F5 + 0.407F6 \quad R^2 = 0.644$$

The interpretation for this result for Germany is as follows: If risk represented by $F2$ increases by 1, then Rp will decrease by 0.27%, if risk represented by $F4$ increases by 1, then Rp will decrease by 0.28%, if risk represented by $F5$ increases by 1, then Rp will increase by 0.567%, and if risk represented by $F6$ increases by 1, then Rp will increase by 0.407%.

$$R_{pSPAIN} = 0.503F6 \quad R^2 = 0.3$$

The interpretation of this equation for Spain is as follows: If risk represented by $F6$ increases by 1, then Rp will increase by 0.503%.

$$R_{pPOLAND} = -0.29F1 + 0.464F5 \quad R^2 = 0.33$$

The interpretation of this equation for Poland is as follows: If risk represented by $F1$ increases by 1, then Rp will decrease by 0.29%, and if risk represented by $F5$ increases by 1, then R_p will increase by 0.464%.

Scenario #3: Portfolio return rate 60s/40b:

$$R_{pGERMANY} = -0.29F2 - 0.29F4 + 0.559F5 + 0.406F6 \quad R^2 = 0.649$$

The interpretation for this result for Germany is as follows: If risk represented by $F2$ increases by 1, then Rp will decrease by 0.29%, if risk represented by $F4$ increases by 1, then Rp will decrease by 0.29%, if risk represented by $F5$ increases by 1, then Rp will increase by 0.559%, and if risk represented by $F6$ increases by 1, then Rp will increase by 0.406%.

$$R_{pSPAIN} = 0.432F6 \quad R^2 = 0.26$$

The interpretation of this equation for Spain is as follows: If risk represented by $F6$ increases by 1, then Rp will increase by 0.432%.

$$R_{pPOLAND} = -0.49F1 + 0.391F5 \quad R^2 = 0.4$$

The interpretation of this equation for Poland is as follows: If risk represented by $F1$ increases by 1, then Rp will decrease by 0.49%, and if risk represented by $F5$ increases by 1, then R_p will increase by 0.391%.

5 Conclusions

There is statistically significant effect extracted by PCA risk factors on investment returns. In Germany, we can point out four factors, all are connected with financial risk and market risk, long-term investment risk and local economy risk, that impact statistically significant effect on each portfolio returns. In Spain, we have one main factor: long-term financial market risk which impact statistically significant effect on each portfolio returns. At the end, for Poland we receive two factors: financial market risk and long-term investment risk. Calibrated models are statistically significant, so we can use these models for prediction return of portfolio.

Our results based on reduction on number of variables are not in conflict with mentioned early empirical studies (e.g., Bloom et al. 2010; Rachel and Smith 2015; Fernald 2016; Acemoglu and Restrepo 2017). Rather we confirmed some projected consequences of longevity especially in valuation long-term investments returns. This analysis is varied from the previous one (Trzpiot and Majewska 2016; Majewska and Trzpiot 2019) in two points: All variables were expressed as chain indices using a base previous observation, and estimation of portfolio was based on this new filters. The previous paper was based on cumulative information—all variables were expressed as indices using a base year of 2010.

Longevity risk appears to be a very complex risk. From our research, we can claim that in chosen country each of the appointed factors include different levels of the impact of specific risk. Longevity analysis of a population or analysis an insured portfolio depends on the available data and their reliability. In particular the trend sensitivity on the modeling part of analysis. In addition, local variability of selected factors and associated long-term portfolio maturities were observed, as well as potential non-linear correlations with other sources of risk, both financial and non-financial. The European country moves through the "demographic transition," the slowdown in population growth and clear shift in age structure. Longevity risk should be perceived as very important on the macro-level, and we should looking for the impacts of longevity on the whole economy and the environment.

The effect of large long-term investors on both their investments and on the markets generally has prompted a key debate in academic literature. The presented results are important in this trend of research, at the time of population aging in European countries.

References

Acemoglu D, Restrepo P (2017) Secular stagnation? The effect of aging on economic growth in the age of automation. Am Econ Rev 107(5):174–179

Arnott RA, Chaves DB (2012) Demographic changes, financial markets, and the economy. Financ Anal J 68(1):23–46. https://doi.org/10.2469/faj.v68.n1.4

Bloom DE, Canning D, Fink G (2010) Implications of population ageing for economic growth. Oxford Rev Econ Policy 26(4):583–612

Fernald JG (2016) Reassessing longer-run U.S. growth: How low? Federal Reserve Bank of San Francisco working paper. http://www.frbsf.org/economic-research/publications/working-papers/wp2016-18.pdf. Accessed 18 Dec 2019

Gordon AD (1999) Classification. Chapman & Hall, London New York Washington

Hotelling H (1933) Analysis of a complex of statistical variables into principal components. J Educ Psychol 24:417–441

Jolliffe IT (1982) A note on the use of principal components in regression. J R Stat Soc Ser C (Appl Stat) 31(3):300–303

Jolliffe IT (2002) Principal component analysis, series: springer series in statistics, 2nd edn. Springer, New York

J.P. Morgan Asset Management (2016) Principles for successful long-term investing, UK

Majewska J, Trzpiot G (2019) Longevity risk factors: the perspectives of selected European countries. In: Papież M, Śmiech S (eds) The 13th Professor Aleksander Zelias international conference on modelling and forecasting of socio-economic phenomena. Conference proceedings. C.H. Beck, Warszawa, pp 132–140. ISBN: 978-83-8158-734-1 (pdf)

McKinsey Global Institute (2016) Diminishing returns: why investors may need to lower their expectations

Rachel L, Smith T (2015) Secular drivers of the global real interest rate. Bank of England Staff Working 571. J.P. Morgan Asset Management Multi-Asset Solutions

Ross SA (1976) The arbitrage theory of capital asset pricing. J Econ Theor 13:341–360

Trzpiot G, Majewska J (2016) The impact of longevity on long-term investments returns: scenarios for Europe. Available at https://www.cass.city.ac.uk/__data/assets/pdf_file/0020/334082/L12-32-TRZPIOT-and-MAJEWSKA.pdf

UNDP (2018) Human development indices and indicators 2018: statistical update. UN, New York

World Economic Forum USA Inc. (2011) The future of long-term investing

Applications in Economics

Sustainable Development and Green Economy in the European Union Countries—Statistical Analysis

Katarzyna Cheba and Iwona Bąk

Abstract The literature on the subject indicates that green growth is a direct result of implementing a sustainable development strategy. According to this assumption, countries that include sustainable development goals in their strategic documents should also achieve results in the area of green growth or the green economy. The monitoring of green growth should be based on indicators that make it possible to distinguish the green economy from the traditional one by taking into account, inter alia, indicators covering: green products, services, investments, and public procurement as well as green jobs in the green sectors of the economy. OECD proposes to use for this purpose indicators divided into four main groups: environmental and resource productivity, natural asset base, the environmental dimension of the quality of life and economic opportunities and policy responses. The aim of the work is to examine the relationships between sustainable development and green economy, especially in the area of their measurement and to determine the relationship between the results achieved by the EU countries in this area. The result of the research is the assessment of the results obtained by the EU countries in each of the analyzed areas using a taxonomic development measure based on the Weber median and the identification of relation between the results.

Keywords Sustainable development · Green economy · Relationships · Selection of indicators · Weber median

K. Cheba (✉) · I. Bąk
West Pomeranian University of Technology, Szczecin, Poland
e-mail: katarzyna.cheba@zut.edu.pl

I. Bąk
e-mail: iwona.bak@zut.edu.pl

K. Jajuga et al. (eds.), *Classification and Data Analysis*, Studies in Classification, Data Analysis, and Knowledge Organization,
https://doi.org/10.1007/978-3-030-52348-0_11

1 Introduction

The 2030 Agenda for Sustainable Development (Agenda 2030) signed in 2015 by the representatives of 193 UN member countries has contributed to the renewed interest in the concept of sustainable development. A more comprehensive understanding of the concept of sustainable development in light of its dynamic evolution also requires looking at the connections of this idea with other, often complementary directions of development in the countries of the world. In the contemporary literature on the subject, there are more and more attempts to integrate sustainable development into other areas important for the development of most countries in the world. Hence, one can observe an increasing popularity of terms such as: sustainable agriculture (Altieri 2018), sustainable finances (Fullwiler 2015; Zioło et al. 2019), sustainable innovations (Burget et al. 2016), responsible innovations (Cheba and Szopik-Depczyńska 2019), sustainable logistics, sustainable transport (Kiba-Janiak 2015), sustainable logistics infrastructure (Dembińska 2018) and sustainable competitiveness (Aiginger et al. 2013; Schwab and Sala-i-Martin 2012; The Global Sustainable Competitiveness Index 2012; Cheba 2019). One can basically talk about the spreading of this concept to other fields of science and business practice (Borys and Czaja 2009). There are also supporting or complementing ideas for the basic concept of sustainable development. Such supplementation can be discussed, e.g., in the context of so-called green growth or green economy, which are considered to be the direct effects of the implementation of sustainable development strategy. According to this assumption, countries that include sustainable development goals in their strategic documents should also achieve results in the area of green growth or green economy. The study of the relationships between the results of ordering obtained by EU countries in these two development concepts, in particular in the area of measuring the current state of their implementation, is the main purpose of the work. The result of the research in this area will be, among others, the assessment and comparative analysis of the results achieved by the European Union countries preformed with the use of selected methods of multidimensional statistical analysis and the identification of relationships between the obtained results.

In the paper, the following research questions were formulated:

1. Is there compliance between the results of ordering EU countries in the case of individual sustainable development orders and areas of the green economy (green growth), considered separately?
2. Is there a consistency of the results of ordering EU countries between individual sustainable development orders and green areas of the economy?
3. What is the impact of the selection of diagnostic features adopted for the study (two variants of sets of diagnostic features are considered) on the results of ordering EU countries in both the area of sustainable development and the green economy?

The work was divided into five thematically related parts. The introduction presents the main purpose of the work. Therefrom, the second part contains a review of the literature on the subject and the research assumptions resulting. The third part discusses the methodological basis of the study, including the description of statistical data and applied research methods. The results of the study and the discussion are contained in part four. The work ends with a summary of information about the possibilities of implementation of the results presented in the study and about the subsequent directions of analysis in this regard. The added value of the paper is a profound analysis of the results of EU countries obtained in two important areas of strategic development of world: sustainable development and green growth or green economy. It is a new proposal of analysis in two areas concentrated on the study of relationship between these two important concepts of world countries' development. The authors have also made attempt to verify the impact of selection of diagnostic variables on the results of ordering of EU countries. According to the authors in the case of analyses based on indicators elaborated for the strategic plans, the use of the formal and statistic verification can change significantly the obtained results.

2 Literature Review

The implementation of the main goal of the research concerning the examination of the relationship between sustainable development and green economy or green growth first of all requires defining of the terms. Although the concept of sustainable development in political debate and scientific considerations has been present for many years, new proposals for explaining this term are still appearing in the literature. The best-known definition of this concept as a new global development path was formulated in the Gro Harlem Brundtland Report published in 1987 as: "development that meets current needs without depriving future generations of the ability to meet their own needs" (WCED (World Commission on Environment and Development) 1987). However, in the new definitions of sustainable development appearing today, it is worth paying attention to terms such as (Cheba 2019):

- development (Ciegis et al. 2009; Flint 2012; Piontek 2016), which is usually one of the first terms referred to in constructed definitions,
- integration (Borys 2011; Stafford-Smith et al. 2017) also understood as an integration process and balancing most often related to three main orders: social, economic and environmental ones,
- the quality of life (Pearce et al. 1989; Ostasiewicz 2004) indicated as the overarching goal of sustainable development.

The simultaneous consideration of these distinguishing features in the proposed definitions allows a closer understanding of the concept of sustainable development. If we assume that sustainable development is a process in which we strive for the balance between the most important economic, social and environmental orders, it will be obvious that achieving a full equilibrium, in particular in the case of economic and environmental governance, is impossible. At the current level of technology development and the acceptance of world governments for restrictions resulting from the need to protect the environment, it will not be possible to achieve a state in which economic development will occur without harming the natural environment. Understanding the relationships between the orders of sustainable development changes the current, quite frequent measurement practices consisting, e.g., in determining one synthetic measure that takes into account the results obtained in all the analyzed orders. In the case of measuring a real progress in implementing the idea of sustainable development, it is more important to study the interrelationships between the results achieved within individual governance. This is also in line with so-called strong principle of sustainability, according to which the replacement of a defect of a resource can only take place within the resources qualified for the same order. Therefore, e.g., the negative effects related to environmental pollution cannot be replaced by the development of environmentally friendly technologies. Hence, the results in the field of sustainable development are considered separately for each analyzed order. It is also necessary to pay attention to the problem of the importance of individual orders. In the literature on the subject (Cox and Cox 2000; Walesiak 2002; Wagenmakers and Farrell 2004), often very complicated weight systems are used to determine the importance of various test areas. In the case of research on sustainable development and the analyses conducted separately for each distinguished order, the authors assumed that each of them is equally important and differentiating their importance would not be in line with the idea of sustainable development, which assumes among others striving for a permanent separation of economic growth and development from the pressure on the natural environment.

As already mentioned, the main direction in the evolution of the idea of sustainable development is so-called sectoral integration of this concept with other research areas. Attempts are also made to identify areas that can complement the basic idea. We can talk about such a supplement in the case of research on the green economy or green growth. These terms function simultaneously in the literature on the subject and strategic documents at various levels. The first of them—the green economy—is used by such organizations as the United Nations Environment Program (UNEP) and European Economic Area (EEA) and the second—green growth—by the Organization for Economic Co-operation and Development (OECD). These terms are most often treated as complementary to each other and at the same time supporting the achievement of sustainable development. According to EEA definition, green economy is "one in which policies and innovations enable society to generate more of value each year, while maintaining the natural systems that sustain us" (EAI (European Environment Agency 2011). The most widely used definition of this term comes from UNEP (United Nations Environment Program) (2011): a green economy is defined by this organization as "one that results in improved human well-being and social; equity, while significantly reducing environmental risks and ecological scarcities." While the green growth which was defined by OECD (OECD (Organization for Economic Co-operation and Development) 2011), means "fostering economic growth and development while ensuring that natural assets continue to provide the resources and environmental services on which our well-being relies." The literature (EAI (European Environment Agency 2011; UNEP (United Nations Environment Program) 2011; OECD (Organization for Economic Co-operation and Development)

2011) also indicates that the concept of green economy or green growth does not replace sustainable development but achieving the purposes of sustainable becomes possible when the economy functioning right.

The key elements of the "green economy" are primarily public and private investments including: renewable energy, clean technologies, energy-efficient (energy-saving) construction, public transport, waste management and recycling, sustainable land use, water, forests, sea fisheries and ecotourism. It is worth noting that monitoring the green growth should be based on indicators that make it possible to distinguish the green economy from the traditional one. These include: green products, services, investments and public procurement as well as green jobs in the green sectors of economy. OECD proposes to divide the indicators monitoring green economy into four basic groups: environmental and resource productivity, natural asset base, the environmental dimension of the quality of life and economic opportunities and policy responses. The measurement of progress in implementing the main assumptions of the green economy, as in the case of sustainable development, should be carried out separately for each distinguished group. Also in this case, we are dealing with the complementarity of these areas, not their substitutability. Hence, the overall measurement concept for examining the relationship between sustainable development and the green economy is based on examining the relationship between all distinguished orders (in the case of sustainable development) and groups (in the case of green economy).

The authors of this work also looked more closely at the problem of a selection of diagnostic features for the study. Usually, often very advanced procedures for selecting diagnostic features for testing are used for this purpose. One of the stages of such selection is the elimination of the so-called collinear features that carry the same information. The literature (Sokołowski and Markowska 2017; Cheba 2017; Sokołowski and Sobolewski 2019) on the subject already contains studies in which it is indicated that in the case of analyses regarding, e.g., the strategic development directions of countries of the world, for which the set list of monitoring indicators is the result of the work of expert teams appointed for this purpose or the use of statistical methods for selection, may distort results research. This problem was also considered in this work. The result is the research conducted using two variants of diagnostic features. In the first of them, all indicators available for each of the analyzed areas of the study were adopted as the final set of features, in the second; the parametrical Hellwig method was used to select diagnostic features.

3 Procedure of the Conducted Research

3.1 Sources of Information on Diagnostic Indicators

In accordance with the adopted assumptions, two datasets were used to study the relationship between sustainable development and green economy. The basis for analyses in the case of sustainable development was the indicators used by the European Commission to monitor progress in the implementation of the latest strategy for sustainable development—The 2030 Agenda for Sustainable Development. The European Commission uses 100 different indicators assigned to the 17 goals of the Agenda for this purpose and not all of them are available for all countries. This applies, for example, to the indicators related to the protection of the sea in countries such as Czechia or Austria, which do not have access to the sea. The result is a set of 64 indicators that enable to carry out comparative analyses of all EU countries. It was also decided that according to the previous research of the Authors (Cheba 2019), in this work these indicators will be assigned to four orders of sustainable development: economic, social, institutional-political and environmental ones. The separation of institutional and political order from the indicators describing social order was decided due to the importance of this area in the case of analyses carried out at the macroeconomic level, in relation to EU countries. The division into orders is the effect of assumptions adopted in the work related to monitoring the implementation of the overall concept of sustainable development and not only the progress in the implementation of the current strategy, which should, however, be treated as the effect of a compromise between the countries of the world. For the same reason, the research did not take into account the target values of indicators indicated in the adopted strategy document, considering that in many cases they are the result of current development opportunities of countries in the world, e.g., technological progress, innovation or social acceptance for restrictions on the use of natural resources (Cheba 2019). Similar assumptions were made by completing indicators available in the OECD database describing progress in implementing the concept of the green growth. In total, 64 indicators assigned to four groups are available in this database for all European Union countries: environmental and resource productivity, natural asset base, environmental dimension of the quality of life and economic opportunities and policy responses. Due to the availability of data on green economy, the results of the research presented in the paper relate to 2015. In the OECD database, indicators for subsequent years are also available, but due to large gaps in the data they cannot be used in further comparative analyses.

3.2 Selection of Diagnostic Features

It should be noted that in further analyses for each indicated area of the study, two sets of diagnostic features were used: variant I ($V1$), in which all available indicators were taken into account, and variant II ($V2$), in which the diagnostic features were selected using the Hellwig parametric method. This method based on a matrix of correlation coefficients between diagnostic features allows to eliminate those features that are too strongly correlated with each other. In the literature on the subject, a description of this method and numerous examples of its use can be found in many works (e.g., Gatnar and Walesiak 2004; Panek 2009).

A detailed list of indicators used in the study, taking into account the division into all indicated orders (in the case of sustainable development) and groups (in the case of green economy), can be presented as follows[1]:

In the area of economic development (E): 18 indicators in variant $V1$ and 15 in variant $V2$: $x_{1.1S}$[2]—agricultural factor income per annual work unit (AWU), $x_{1.2S}$—government support to agricultural research and development, Euro *per capita*, $x_{1.3S}$—area under organic farming, % of utilized agricultural area, $x_{1.4S}$—employment rates of recent graduates, % of population aged 20 to 42 ($V1$), $x_{1.5D}$—inactive population due to caring responsibilities, % of inactive population aged 20 to 73, $x_{1.6S}$—real GDP *per capita*, chain linked volumes (2010 = 100), Euro *per capita*, $x_{1.7D}$—young people neither in employment nor in education, % of population aged 15 to 37, $x_{1.8S}$—employment rate, % of population aged 20 to 72, $x_{1.9D}$—involuntary temporary employment, % of employees aged 20 to 72, $x_{1.10D}$—people killed in accidents at work, number per 100 000 employees, $x_{1.11S}$—gross domestic expenditure on R&D, % of GDP, $x_{1.12S}$—employment in high-and medium-high technology manufacturing sectors and knowledge-intensive service sectors, %of total employment, $x_{1.13S}$—R&D personnel by sector, % of active population, $x_{1.14S}$—purchasing power adjusted GDP *per capita* (in PPS _EU28), ($V1$), $x_{1.15S}$—resource productivity and domestic material consumption (DMC), Euro per kilogram ($V1$), $x_{1.16D}$—volume of freight transport relative to GDP, index (2005 = 100), $x_{1.17D}$—general government gross debt, % of GDP, $x_{1.18S}$—share of labor taxes in total tax revenues, %.

In the area of social development (S): 26 indicators in variant $V1$ and 22 in variant $V2$: $x_{2.1D}$—people at risk of poverty or social exclusion, % ($V1$), $x_{2.2D}$ –people at risk of income poverty after social transfers, %, $x_{2.3D}$—severely materially deprived people, %, $x_{2.4D}$—people living in households with very low work intensity, % of population aged less than 78, $x_{2.5D}$—housing cost overburden rate by poverty status, % of population, $x_{2.6D}$—population living in a dwelling with a leaking roof, damp walls, floors or foundation or rot in window frames of floor, % of population, $x_{2.7S}$—life expectancy at birth, years, $x_{2.8S}$—self-perceived health, very good or good, % of population, $x_{2.9D}$—death rate due to chronic diseases, total, number per 100 000 persons aged less than 65 ($V1$), $x_{2.10D}$—suicide rate, number per 100 000 persons, $x_{2.11D}$—self-reported unmet need for medical care, % of population aged 16 and over, $x_{2.12D}$—early leavers from education and training, % of population aged 18 to 33, $x_{2.13S}$—tertiary educational attainment, % of population aged 30 to 43, $x_{2.14S}$—participation in early childhood education, % of the age group between 4-year-olds and the starting age of compulsory education, $x_{2.15S}$—adult participation in learning, % of population aged 25 to 73, $x_{2.16D}$—final energy consumption in households *per capita*, kg of oil equivalent, $x_{2.17D}$—population unable to keep home adequately

[1]To these indicators which were taken into account only in variant 1, information in brackets ($V1$) was given on their inclusion in second variants of the analyzed features. The rest of indicators were considered in all of analyzed variants.

[2]The indicators were described by 3 symbols as e.g.: $x_{1.1S}$, where the first one means the analysed area, the second one it is the sequential indicator number in this area, and the last one describe the character of indicator: S is dedicated for stimulants, and D—for destimulant.

warm, % of population, $x_{2.18S}$—long-term unemployment rate, % of active population, $x_{2.19D}$—relative median at-risk-of-poverty gap, % distance to poverty threshold, $x_{2.20D}$—Gini coefficient of equalized disposable income, coefficient of 0 (maximal equality) to 100 (maximal inequality), (*V1*), $x_{2.21D}$—income share of the bottom 40% of the population, % of income (*V*2), $x_{2.22D}$—overcrowding rate, % of population, $x_{2.23D}$—population living in households considering that they suffer from noise, % of population, $x_{2.24D}$—people killed in road accidents, rate, $x_{2.25D}$—death rate due to homicide, number per 100,000 persons, and $x_{2.26D}$—population reporting occurrence of crime, violence or vandalism, % of population.

In the area of the institutional and political development (IP): 9 indicators in variant *V*1 and 8 in variant *V*2: $x_{3.1S}$—seats held by women in national parliaments, % of seats, $x_{3.2S}$—seats held by women in national governments, % of seats, $x_{3.3S}$—positions held by women in senior management positions, board members, % of positions, $x_{3.4S}$—general government total expenditure on law courts, Euro per inhabitant, $x_{3.5S}$—population with confidence in EU institutions: European Parliament % of population, $x_{3.6S}$—population with confidence in EU institutions: European Commission, % of population (*V*1), $x_{3.7S}$—population with confidence in EU institutions: European Central Bank, % of population, $x_{3.8S}$—official development assistance as share of gross national income, %, $x_{3.9S}$—EU imports from developing countries by country income groups, million EUR *per capita.*

In the area of environmental development (EN): 11 indicators in variant *V1* and 10 in variant *V*2: $x_{4.1D}$—ammonia emissions from agriculture, kilograms per hectare, $x_{4.2D}$—primary energy consumption, million tonnes of oil equivalent (TOE), $x_{4.3D}$—final energy consumption, million tonnes of oil equivalent (TOE) (*V*1), $x_{4.4S}$—energy productivity, Euro per kilogram of oil equivalent (KGOE), $x_{4.5S}$—share of renewable energy in gross final energy consumption, %, $x_{4.6D}$—energy dependence % of imports in total energy consumption, $x_{4.7S}$—recycling rate of municipal waste, % of total waste generated, $x_{4.8D}$—average CO_2 emissions per km from new passenger cars (source: EEA and EC services), g CO_2 per km, $x_{4.9D}$—greenhouse gas emissions—tonnes *per capita,* $x_{4.10D}$—greenhouse gas emissions intensity of energy consumption (source: EEA and Eurostat), index (2000 = 100), $x_{4.11D}$—shares of environmental taxes in total tax revenues, % of total taxes.

In the similar way, the indicators applying to the description of green economy were divided into four groups. To the first group—environmental and resource productivity (O_1): 12 indicators in variant *V*1 and 7 in variant *V*2 were selected: $y_{1.1D}$—production-based CO_2 productivity, GDP per unit of energy-related CO_2 emissions, $y_{1.2D}$—production-based CO_2 intensity, energy-related CO_2 *per capita* (*V*1), $y_{1.3D}$—demand-based CO_2 productivity, GDP per unit of energy-related CO_2 emissions, $y_{1.4D}$—demand-based CO_2 intensity, energy-related CO_2 *per capita* (*V*1), $y_{1.5S}$—energy intensity, TPES *per capita* (*V*1), $y_{1.6D}$—total primary energy supply, % TPES, $y_{1.7S}$—renewable energy supply, % TPES, $y_{1.8S}$—renewable electricity, % total electricity generation (*V*1), $y_{1.9D}$—energy consumption in agriculture, % total energy consumption, $y_{1.10D}$—energy consumption in industry, % total energy consumption (*V*1), $y_{1.12D}$—energy consumption in transport, % total energy consumption, $y_{1.12D}$—nitrogen balance per hectare.

The second group—natural asset base (O_2)—was represented by 20 indicators in variant $V1$ and 10 in variant $V2$: $y_{2.1S}$—permanent surface water, % total surface, $y_{2.2S}$—seasonal surface water, % total surface ($V1$), $y_{2.3D}$– conversion of permanent water to not-water surface, % permanent water, since 1984, $y_{2.4D}$—conversion of permanent to seasonal water surface, % permanent water, since 1984, $y_{2.5S}$—conversion of not-water to permanent water surface, % permanent water, since 1984 ($V1$), $y_{2.6S}$—conversion of seasonal to permanent water surface, % permanent water, since 1984, $y_{2.7S}$—natural and semi-natural vegetated land, % total ($V1$), $y_{2.8S}$—bare land, % total ($V1$), $y_{2.9D}$—cropland, % total, $y_{2.10D}$—artificial surfaces, % total ($V1$), $y_{2.11S}$—water, % total ($V1$), $y_{2.12D}$—loss of natural and semi-natural vegetated land, % since 1992 ($V1$), $y_{2.13S}$—gain of natural and semi-natural vegetated land, % since 1992 ($V1$), $y_{2.14D}$—conversion from natural and semi-natural land to cropland, % since 1992, $y_{2.15D}$—conversion from natural and semi-natural land to artificial surfaces, % since 1992, $y_{2.16D}$—conversion from cropland land to artificial surfaces, % since 1992 ($V1$), $y_{2.17D}$—built-up area, % total land ($V1$), $y_{2.18D}$—built-up area *per capita*, % total land, $y_{2.19D}$—new built-up area, % since 1990, $y_{2.20S}$—forests under sustainable management certification, % total forest area.

Third group—environmental dimension of quality of life (O_3)—was described by ten indicators in variant $V1$ and by only four in variant $V2$, where $y_{3.1D}$—mean population exposure to PM2.5, micrograms per cubic meter ($V1$), $y_{3.2D}$—percentage of population exposed to more than 10 micrograms/m3, % ($V1$), $y_{3.3D}$—mortality from exposure to ambient PM2.5, per 1,000,000 inhabitants, $y_{3.4D}$—welfare costs of premature mortalities from exposure to ambient PM2.5, GDP equivalent,% ($V1$), $y_{3.5D}$—mortality from exposure to ambient ozone, per 1,000,000 inhabitants, $y_{3.6D}$—welfare costs of premature deaths from exposure to ambient ozone, GDP equivalent, % ($V1$), $y_{3.7D}$—mortality from exposure to lead, per 1,000,000 inhabitants, $y_{3.8D}$—welfare costs of premature deaths from exposure to lead, GDP equivalent, % ($V1$), $y_{3.9S}$—population with access to improved drinking water sources, % total population, $y_{3.10S}$—population with access to improved sanitation, % total population ($V1$).

To the last group—economic opportunities and policy responses (O_4): 14 indicators in variant $V1$ and only 7 in variant $V2$ were selected: $y_{4.1S}$—development of environment-related technologies, % all technologies ($V1$), $y_{4.2S}$—relative advantage in environment-related technology, ratio, $y_{4.3S}$—development of environment-related technologies, % inventions worldwide, $y_{4.4S}$—development of environment-related technologies, inventions *per capita* ($V1$), $y_{4.5S}$—net ODA provided, % GNI ($V1$), $y_{4.6D}$—environmentally related taxes, % GDP, $y_{4.7D}$—environmentally related taxes, % total tax revenue ($V1$), $y_{4.8D}$—energy-related tax revenue, % total environmental tax revenue ($V1$), $y_{4.9D}$—road transport-related tax revenue, % total environmental tax revenue, $y_{4.10D}$—petrol end-user price, USD per liter ($V1$), $y_{4.11D}$—diesel tax, USD per liter ($V1$), $y_{4.12D}$—diesel end-user price, USD per liter, $y_{4.13S}$—mean feed-in tariff for solar PV electricity generation, $y_{4.14S}$—mean feed-in tariff for wind electricity generation.

The lists of indicators used in the work are extensively elaborated. They are prepared according to the requirements formulated by European Commission in the area of sustainable development and by UNEP within green economy. They are developed for measuring the progress in implementation of the main world strategies in these areas of development. Particularly interesting are the results of the selection of green economy indicators. In this case, as a result of applying of Hellwig methods', the list of diagnostic features in the second variant was widely shortened. This may affect the results of ordering of EU countries and significantly change their ranking. In the next part of the paper, the authors' special attention will be paid to this problem.

3.3 Research Method

In the paper, a two-stage research procedure was used to examine the relationships between the analyzed areas of the study. In the first stage, based on diagnostic features indicated in individual variants, EU countries were ordered separately for each area. A taxonomic development measure based on the Weber median was used for this purpose. In the relevant literature, a detailed description of this method and numerous examples of its use can be found in the following papers: Weber (1971), Lira et al. 2002, Andalecio 2009, Młodak et al. (2016), Pulido and Sanchez-Soriano (2009), Pechersky (2015), Adam and Kroupa (2017). The indisputable advantage of this method is its resistance to outliers, which in the case of very diverse results of European Union countries in the analyzed areas is an important factor influencing its choice in the presented study.

The application of this method to ordering EU country according to their results in sustainable development and the green economy can be presented as follows:

1. Normalization of the diagnostic features

The positional option of the linear object assignment takes a different normalization formula,[3] in comparison with the classical approach, based on a quotient of the feature value deviation from the proper coordinate of the Weber median and a weighed absolute median deviation, using the Weber median (Młodak 2006, 2014):

$$z_{ij} = \frac{x_{ij} - \theta_{0j}}{1,4826 \cdot \tilde{\text{mad}}\left(X_j\right)}, \tag{1}$$

where $\theta_0 = (\theta_{01}, \theta_{02}, \ldots, \theta_{0m})$ is the Weber median, $\tilde{\text{mad}}\left(X_j\right)$ is the absolute median deviation, in which the distance from the features to the Weber vector is measured,[4] i.e., $\tilde{\text{mad}}\left(X_j\right) = \underset{i=1,2,\ldots,n}{\text{med}} \left|x_{ij} - \theta_{0j}\right| \; (j = 1, 2, \ldots, m)$.

[3]The median Weber vector was calculated on the basis of features by transforming destimulants into stimulants on the basis of the following formula: $x'_{ij} = \frac{1}{x_{ij}}$, $i = 1, 2, \ldots, n$, $j = 1, 2, \ldots, n$.

[4]The Weber median was calculated in *R program: l1median* of package: *pcaPP*.

2. Calculation of taxonomic measure of development

The synthetic measure μ_i is calculated on the basis of maximum values of normalized features, similarly to the Hellwig (1968) method:

$$\phi_j = \max_{i=1,2,\ldots,n} z_{ij}, \tag{2}$$

according to the following formula:

$$\mu_i = 1 - \frac{d_i}{d_-}, \tag{3}$$

where $d_- = \text{med}(\mathbf{d}) + 2{,}5\text{mad}(\mathbf{d})$, where $\mathbf{d}= (d_1, d_2, \ldots, d_n)$ is a distance vector calculated using the formula: $d_i = \text{med}_{j=1,2,\ldots,m}\left|z_{ij} - \phi_j\right|\ i = 1, 2, \ldots, n$, ϕ_j—the ith coordinate of the development pattern vector, which is constituted of the maximum values of the normalized features.

The division of objects into typological groups is also possible, but in this paper this additional information was not considered.

In the next stage (the second stage of the procedure), based on the obtained results to analyze the relation between values of taxonomic measures of development and the position taken by countries in each rankings, the Kendall τ correlation coefficients were calculated according the following formula (Sanderson 2007):

$$\tau = \frac{P - Q}{\sqrt{P + Q + T) \cdot (P + Q + U)}}, \tag{4}$$

where p—the number of correctly-ordered pairs, Q—the number of incorrectly ordered pairs, T—the number of ties in first ranking, U—the number of ties in second ranking.

The basis of its calculation is the difference between the probability that two variables are arranged in the same order (for observed data) and the probability that their order differs, which was proposed by Kendall (1938) and requires variable values to be ordered (variables must be measured at least on the ordinal scale). This coefficient takes values from the range $\langle -1, 1 \rangle$. The value 1 means full agreement; value 0 does not match the orderings, while the value -1 means the total contradiction. The Kendall factor indicates, therefore, not only the strength, but also the direction of dependence. It is a great tool to describe the similarity of the ordering of a dataset (Lapata 2006; Okazaki et al. 2004).

4 The Study Results

The obtained results of the classification of UE countries within particular orders of sustainable development are presented in the Table 1 (variant I) and in Table 2

Table 1 Results of the ordering of UE countries according their level of development in each considered areas in 2015—*V1*

Country	Order of sustainable development							
	Economic (E_{V1})		Social (S_{V1})		Institutional and political (IP_{V1})		Environmental (EN_{V1})	
	μ_i	Rank	μ_i	Rank	μ_i	Rank	μ_i	Rank
Austria	0.911	10	0.618	5	0.737	7	0.518	7
Belgium	0.865	9	0.191	11	0.547	5	0.359	16
Croatia	0.951	26	0.725	24	0.763	18	0.852	8
Bulgaria	0.897	22	0.547	27	0.391	11	0.696	25
Cyprus	0.912	24	0.552	18	0.600	27	0.299	24
Czechia	0.894	12	0.517	16	0.335	22	0.542	4
Denmark	0.900	1	0.642	3	0.495	4	0.585	1
Estonia	0.855	13	0.265	17	0.285	23	0.481	14
Finland	0.859	6	0.553	4	0.442	2	0.369	6
France	0.912	8	0.661	6	0.482	15	0.621	9
Germany	0.902	7	0.664	15	0.600	6	0.650	28
Greece	0.887	27	0.368	25	0.557	25	0.448	19
Hungary	0.947	15	0.770	23	0.927	28	0.679	2
Ireland	0.868	11	0.323	9	0.463	14	0.423	11
Italy	0.858	16	0.304	21	0.450	10	0.642	21
Latvia	0.879	17	0.657	26	0.228	12	0.554	10
Lithuania	0.934	20	0.658	22	0.585	17	0.320	22
Luxembourg	0.890	4	0.311	7	0.065	9	0.726	26
Malta	0.946	18	0.777	8	0.800	26	0.466	12
Netherlands	0.883	3	0.208	1	0.530	3	0.616	20
Poland	0.868	21	0.491	19	0.511	13	0.316	27
Portugal	0.864	23	0.425	20	0.586	8	0.711	3
Romania	0.824	28	0.044	28	0.425	21	0.508	17
Slovakia	0.892	19	0.618	14	0.479	24	0.535	13
Slovenia	0.878	14	0.552	10	0.294	16	0.551	15
Spain	0.860	25	0.510	13	0.220	19	0.366	23
Sweden	0.923	2	0.716	2	0.845	1	0.677	5
UK	0.928	5	0.600	12	0.433	20	0.497	18

Source Own calculations

(variant II). When analyzing these tables, it should be noted that there are differences in the classification of EU countries due to the variants of diagnostic features adopted for the study, but they do not apply equally to all orders. There were no differences in the rankings for social order. The positions of countries in ranking elaborated on the results in economic order of sustainable development do not differ significantly, and

Table 2 Results of the ordering of UE countries according their level of development in each considered areas in 2015—*V*2

Country	Order of sustainable development							
	Economic (E_{V2})		Social (S_{V2})		Institutional and political (IP_{V2})		Environmental (EN_{V2})	
	μ_i	Rank	μ_i	Rank	μ_i	Rank	μ_i	Rank
Austria	0.913	10	0.550	5	0.707	7	0.329	6
Belgium	0.865	7	0.243	11	0.494	5	0.246	22
Croatia	0.906	26	0.570	24	0.427	18	0.467	9
Bulgaria	0.960	22	0.725	27	0.731	17	0.744	25
Cyprus	0.910	24	0.570	18	0.695	27	0.221	27
Czechia	0.895	11	0.575	16	0.416	21	0.411	12
Denmark	0.900	2	0.683	3	0.540	4	0.491	1
Estonia	0.852	14	0.149	17	0.342	22	0.372	16
Finland	0.856	5	0.575	4	0.450	3	0.299	3
France	0.910	9	0.678	6	0.517	14	0.585	5
Germany	0.909	8	0.625	15	0.648	6	0.531	26
Greece	0.891	27	0.385	25	0.565	25	0.413	18
Hungary	0.961	16	0.781	23	0.961	28	0.688	4
Ireland	0.867	13	0.408	9	0.516	12	0.306	8
Italy	0.855	15	0.320	21	0.474	10	0.478	15
Latvia	0.873	18	0.675	26	0.271	19	0.427	13
Lithuania	0.914	21	0.719	22	0.623	15	0.205	23
Luxembourg	0.888	6	0.342	7	0.163	8	0.607	28
Malta	0.943	19	0.732	8	0.808	26	0.333	14
Netherlands	0.881	3	0.275	1	0.453	2	0.452	21
Poland	0.864	23	0.564	19	0.543	11	0.345	20
Portugal	0.868	20	0.454	20	0.591	9	0.501	7
Romania	0.838	28	0.093	28	0.345	24	0.357	19
Slovakia	0.901	17	0.636	14	0.515	23	0.401	11
Slovenia	0.881	12	0.575	10	0.357	16	0.473	17
Spain	0.857	25	0.475	13	0.261	20	0.208	24
Sweden	0.924	1	0.714	2	0.798	1	0.609	2
UK	0.924	4	0.664	12	0.540	13	0.478	10

Source Own calculations

in 21 countries they do not exceed one position; the biggest difference (3) concerned Portugal.

The situation is similar for the institutional and political order. Also in this case for 21 countries, the difference in positions occupied did not exceed one, except that for two countries (Italy and the UK), it was seven and for Bulgaria—six. The changes in occupied positions can be most frequently seen in the case of environmental order; the maximum difference was eight and concerned two countries: Czechia and the UK. Regardless of the adopted variant of using diagnostic features, some regularities can be noticed: (a) in the case of economic governance, the ranking was headed by Sweden, Denmark and the Netherlands (two first countries are located in Northern part of Europe, and they are usually better classified than others), while the worst situation concerned: Romania, Greece and Croatia (located in Southern Europe and usually perceived as less developed countries), (b) the positions of countries in the case of social order have not changed and the highest positions concerned countries such as the Netherlands, Sweden and Denmark (the same as for economic order), while at the end of the ranking were: Romania, Bulgaria and Latvia, (c) Sweden, the Netherlands and Finland are the leaders in the institutional and political area, while at the end of the ranking were: Hungary, Cyprus and Malta, (d) in the case of environmental governance, Denmark leads the way, with Germany and Poland (variant I) or Luxembourg and Cyprus (variant II), respectively, at the end of the ranking. The results confirmed the high development of the Scandinavian countries for all of the analyzed orders. They are basically the only countries that have managed to decouple economic growth from negative pressure on the environment.

Tables 3 and 4 present rankings of EU countries in terms of green economy. In the case of these rankings, there are significant differences in the positions occupied due to the adopted variants of the selection of diagnostic features. Depending on the group of indicators examined in the field of green economy, the maximum differences in the positions taken are from 4 (O_2—natural asset base) to 22 (O_1—environmental and resource productivity), and it is much more difficult to notice some regularity.

Differences in number of diagnostic features in the analyzed variants are substantial, and it is main reason of the differences of final ordering of the countries. For example, in the case of variant 1 ($V1$) in which all diagnostic features were considered, Hungary, taking into account the results of group I, was only on 23 position and considering option II ($V2$), i.e., the selection of features using the Hellwig parametric method, it came first. The results are also strongly differentiated between the areas. It is much more difficult to find the same leaders in every considered area. We suppose that in the case of O_4 (economic opportunities and policy responses) area, the differences in environmental tax system between the most developed countries located in Northern and Western Europe and less developed countries in Southern and Eastern Europe could be the main reason of significant differences of the results.

One of the main purposes of the research presented in the introduction of the article concerned the mutual relations between sustainable development and green economy. To this end, Pearson and Kendall's τ correlation coefficients between the results of EU countries, taking into account the considered variants of diagnostic features, in all orders of sustainable development and areas of the green economy were determined (Table 5). The high values of coefficients indicate good compliance of the linear ordering of EU countries, although there are discrepancies in the results of some objects. Let us look first at the relationships between individual orders of sustainable development. As it was already mentioned, the highest correlation coefficients (Pearson and Kendall τ) were observed both in the case of economic and

Table 3 Ranking of EU Member States in terms of the green growth level in 2015—$V1$

Country	Areas of the green growth							
	$O1_{V_1}$		$O2_{V_1}$		$O3_{V_1}$		$O4_{V_1}$	
	μ_i	Rank	μ_i	Rank	μ_i	Rank	μ_i	Rank
Austria	0.467	9	0.620	12	0.757	9	0.670	20
Belgium	0.400	18	0.263	27	0.539	20	0.563	22
Bulgaria	0.401	2	0.567	15	0.608	28	0.358	3
Croatia	0.573	6	0.567	22	0.768	27	0.429	7
Cyprus	0.455	19	0.685	5	0.917	14	0.466	23
Czechia	0.569	17	0.880	13	1.000	17	0.485	6
Denmark	0.285	4	0.498	14	0.714	8	0.585	26
Estonia	0.365	13	0.511	7	0.658	3	0.705	4
Finland	0.589	5	0.706	1	0.336	1	0.599	15
France	0.317	25	0.648	21	0.359	10	0.201	16
Germany	0.463	20	0.533	17	0.617	13	0.404	2
Greece	0.285	3	0.765	4	0.905	26	0.535	13
Hungary	0.381	23	0.705	10	0.643	24	0.410	12
Ireland	0.511	24	0.678	3	0.515	4	0.417	14
Italy	0.459	21	0.511	20	0.367	25	0.126	5
Latvia	0.412	11	0.688	9	0.550	18	0.314	24
Lithuania	0.196	8	0.432	8	0.882	21	0.476	1
Luxembourg	0.328	27	0.377	23	0.797	5	0.402	28
Malta	0.442	28	0.367	28	0.514	12	0.518	11
Netherlands	0.460	22	0.670	25	0.585	6	0.240	18
Poland	0.236	16	0.500	6	0.629	19	0.635	10
Portugal	0.515	14	0.427	26	0.714	22	0.331	8
Romania	0.690	12	0.825	18	0.925	23	0.481	27
Slovakia	0.431	10	0.621	16	0.770	16	0.332	9
Slovenia	0.592	7	0.557	24	0.120	11	0.229	17
Spain	0.341	26	0.499	19	0.357	15	0.524	25
Sweden	0.545	1	0.453	2	0.313	2	0.456	21
UK	0.038	15	0.226	11	0.670	7	0.422	19

Source Own calculations

social orders (respectively in $V1$ variant: 0.790 and 0.622, in $V2$: 0.759 and 0.611) as well as economic and institutional and political orders ($V1$: 0.608 and 0.433, $V2$: 0.739 and 0.553). The low values of correlation coefficients between environmental and other orders (regardless of the adopted variant of diagnostic features) confirm the lack of strong relationships between them. It means that the environmental changes are not supported by economic, social, institutional and political development. In the area of green economy, as it results from Tables 3 and 4, the ranking of EU

Table 4 Ranking of EU countries in terms of the green growth level in 2015—$V2$

Country	Areas of the green growth							
	$O1_{V_2}$		$O2_{V_2}$		$O3_{V_2}$		$O4_{V_2}$	
	μ_i	Rank	μ_i	Rank	μ_i	Rank	μ_i	Rank
Austria	0.611	15	0.663	10	0.748	9	0.398	28
Belgium	0.610	16	0.443	23	0.579	18	0.444	24
Croatia	0.761	5	0.534	17	0.321	27	0.741	2
Bulgaria	0.626	13	0.398	26	0.388	25	0.585	13
Cyprus	0.361	26	0.757	5	0.726	11	0.564	15
Czechia	0.538	20	0.547	16	0.680	15	0.688	6
Denmark	0.729	8	0.556	15	0.759	8	0.459	23
Estonia	0.670	9	0.860	2	0.881	3	0.623	12
Finland	0.782	3	0.874	1	0.966	1	0.437	25
France	0.369	25	0.500	20	0.688	13	0.437	26
Germany	0.635	12	0.575	13	0.721	12	0.787	1
Greece	0.755	6	0.733	6	0.396	24	0.707	4
Hungary	0.819	1	0.556	14	0.275	28	0.688	7
Ireland	0.530	21	0.766	4	0.902	2	0.464	22
Italy	0.470	22	0.534	18	0.421	23	0.636	11
Latvia	0.649	10	0.704	7	0.553	20	0.552	16
Lithuania	0.625	14	0.637	11	0.567	19	0.726	3
Luxembourg	0.191	28	0.507	19	0.878	5	0.437	27
Malta	0.218	27	0.335	28	0.686	14	0.578	14
Netherlands	0.413	23	0.447	22	0.830	6	0.548	17
Poland	0.593	17	0.702	8	0.540	21	0.688	8
Portugal	0.569	19	0.335	27	0.495	22	0.651	10
Romania	0.770	4	0.416	24	0.342	26	0.688	9
Slovakia	0.788	2	0.615	12	0.609	17	0.693	5
Slovenia	0.649	11	0.410	25	0.800	7	0.531	18
Spain	0.378	24	0.474	21	0.652	16	0.478	20
Sweden	0.754	7	0.829	3	0.880	4	0.470	21
UK	0.570	18	0.666	9	0.740	10	0.525	19

Source Own calculations

countries when applying individual variants is not the same and in some cases it differs quite significantly. Person and Kendall τ correlation coefficients between synthetic measures are definitely lower than in the case of the correlation between sustainable development orders. The highest values concern: first and second areas in variant 1 ($r = 0.507$; $\tau = 0.275$), and third and fourth areas in variant 2 ($r = -0.611$; $\tau = -0.434$). The attention should be also paid to the negative values of

Table 5 Correlation coefficients between areas of sustainable development and the green economy, including variants of features

Pearson	E_{v1}	E_{v2}	S_{v1}	S_{v2}	IP_{v1}	IP_{v2}	EN_{V1}	EN_{V2}	$O1_{V_1}$	$O1_{V_2}$	$O2_{V_1}$	$O2_{V_2}$	$O3_{V_1}$	$O3_{V_2}$	$O4_{V_1}$	$O4_{V_2}$
E_{V1}	1.000	0.872	0.790	0.734	0.608	0.601	0.277	0.275	−0.370	−0.053	−0.207	−0.009	0.087	−0.095	−0.068	0.111
E_{V2}	0.872	1.000	0.711	0.759	0.536	0.739	0.313	0.497	−0.186	−0.014	−0.128	−0.126	0.158	−0.164	−0.057	0.080
S_{V1}	0.790	0.711	1.000	0.943	0.453	0.513	0.176	0.228	−0.223	0.111	−0.097	0.106	−0.172	−0.018	−0.028	0.061
S_{V2}	0.734	0.759	0.943	1.000	0.377	0.556	0.117	0.295	−0.212	0.094	−0.082	0.051	−0.116	−0.036	−0.087	0.022
IP_{V1}	0.608	0.536	0.453	0.377	1.000	0.851	0.133	0.085	0.025	0.285	−0.024	0.029	0.010	−0.275	0.090	0.220
IP_{V2}	0.601	0.739	0.513	0.556	0.851	1.000	0.039	0.237	0.000	0.181	−0.058	0.032	0.089	−0.211	0.120	0.155
EN_{V1}	0.277	0.313	0.176	0.117	0.133	0.039	1.000	0.825	0.205	0.092	−0.001	−0.363	−0.082	−0.240	−0.511	0.140
EN_{V2}	0.275	0.497	0.228	0.295	0.085	0.237	0.825	1.000	0.085	0.122	−0.005	−0.239	−0.091	−0.198	−0.420	0.063
$O1_{V_1}$	−0.370	−0.186	−0.223	−0.212	0.025	0.000	0.205	0.085	1.000	0.152	0.507	−0.188	−0.132	0.022	−0.141	−0.013
$O1_{V_2}$	−0.053	−0.014	0.111	0.094	0.285	0.181	0.092	0.122	0.152	1.000	0.286	0.341	0.064	−0.307	0.204	0.336
$O2_{V_1}$	−0.207	−0.128	−0.097	−0.082	−0.024	−0.058	−0.001	−0.005	0.507	0.286	1.000	0.186	0.256	−0.167	−0.068	0.233
$O2_{V_2}$	−0.009	−0.126	0.106	0.051	0.029	0.032	−0.363	−0.239	−0.188	0.341	0.186	1.000	0.016	0.398	0.407	−0.098
$O3_{V_1}$	0.087	0.158	−0.172	−0.116	0.010	0.089	−0.082	−0.091	−0.132	0.064	0.256	0.016	1.000	−0.314	0.329	0.427
$O3_{V_2}$	−0.095	−0.164	−0.018	−0.036	−0.275	−0.211	−0.240	−0.198	0.022	−0.307	−0.167	0.398	−0.314	1.000	0.191	−0.611
$O4_{V_1}$	−0.068	−0.057	−0.028	−0.087	0.090	0.120	−0.511	−0.420	−0.141	0.204	−0.068	0.407	0.329	0.191	1.000	−0.112
$O4_{V_2}$	0.111	0.080	0.061	0.022	0.220	0.155	0.140	0.063	−0.013	0.336	0.233	−0.098	0.427	−0.611	−0.112	1.000
Kendall	E_{v1}	E_{v2}	S_{v1}	S_{v2}	IP_{v1}	IP_{v2}	EN_{V1}	EN_{V2}	$O1_{V_1}$	$O1_{V_2}$	$O2_{V_1}$	$O2_{V_2}$	$O3_{V_1}$	$O3_{V_2}$	$O4_{V_1}$	$O4_{V_2}$
E_{V1}	1.000	0.921	0.608	0.608	0.434	0.519	0.172	0.257	−0.090	−0.074	−0.021	0.116	0.545	0.529	−0.196	−0.397
E_{V2}	0.921	1.000	0.614	0.614	0.471	0.545	0.177	0.262	−0.032	−0.058	−0.058	0.079	0.508	0.503	−0.180	−0.381
S_{V1}	0.608	0.614	1.000	1.000	0.370	0.466	0.204	0.246	−0.122	−0.201	−0.063	0.042	0.661	0.635	−0.249	−0.471
S_{V2}	0.608	0.614	1.000	1.000	0.370	0.466	0.204	0.246	−0.122	−0.201	−0.063	0.042	0.661	0.635	−0.249	−0.471

(continued)

Table 5 (continued)

Pearson	E_{v1}	E_{v2}	S_{v1}	S_{v2}	IP_{v1}	IP_{v2}	EN_{V1}	EN_{V2}	$O1_{V_1}$	$O1_{V_2}$	$O2_{V_1}$	$O2_{V_2}$	$O3_{V_1}$	$O3_{V_2}$	$O4_{V_1}$	$O4_{V_2}$
IP_{V1}	0.434	0.471	0.370	0.370	1.000	0.873	0.040	0.103	0.148	−0.016	−0.122	−0.026	0.243	0.259	−0.095	−0.307
IP_{V2}	0.519	0.545	0.466	0.466	0.873	1.000	0.040	0.146	0.085	−0.090	−0.069	0.026	0.317	0.333	−0.085	−0.317
EN_{V1}	0.172	0.177	0.204	0.204	0.040	0.040	1.000	0.759	0.199	0.268	0.087	0.045	0.109	0.061	−0.034	−0.151
EN_{V2}	0.257	0.262	0.246	0.246	0.103	0.146	0.759	1.000	0.230	0.257	0.188	0.156	0.162	0.082	−0.024	−0.151
$O1_{V_1}$	−0.090	−0.032	−0.122	−0.122	0.148	0.085	0.199	0.230	1.000	0.571	0.275	0.201	−0.111	−0.053	0.090	0.132
$O1_{V_2}$	−0.074	−0.058	−0.201	−0.201	−0.016	−0.090	0.268	0.257	0.571	1.000	0.270	0.206	−0.138	−0.164	0.095	0.233
$O2_{V_1}$	−0.021	−0.058	−0.063	−0.063	−0.122	−0.069	0.087	0.188	0.275	0.270	1.000	0.799	0.148	0.122	0.032	0.053
$O2_{V_2}$	0.116	0.079	0.042	0.042	−0.026	0.026	0.045	0.156	0.201	0.206	0.799	1.000	0.286	0.249	0.021	−0.021
$O3_{V_1}$	0.545	0.508	0.661	0.661	0.243	0.317	0.109	0.162	−0.111	−0.138	0.148	0.286	1.000	0.857	−0.238	−0.439
$O3_{V_2}$	0.529	0.503	0.635	0.635	0.259	0.333	0.061	0.082	−0.053	−0.164	0.122	0.249	0.857	1.000	−0.190	−0.434
$O4_{V_1}$	−0.196	−0.180	−0.249	−0.249	−0.095	−0.085	−0.034	−0.024	0.090	0.095	0.032	0.021	−0.238	−0.190	1.000	0.492
$O4_{V_2}$	−0.397	−0.381	−0.471	−0.471	−0.307	−0.317	−0.151	−0.151	0.132	0.233	0.053	−0.021	−0.439	−0.434	0.492	1.000

Source Own calculations

these coefficients in many areas. This means that development in one of them means changes in the opposite direction in the other of these areas.

Taking into account the relationships between sustainable development orders and green economy areas, it should be noticed that these relations are not so strong and only between some of the analyzed orders and areas, the relatively strong or moderate relationships are noticed. It is worth noticing that the higher correlations are observed between the results of ordering of EU countries according the ranks than in the case of relationships measured on the basis of the results of taxonomic measures of development. The visible connections measured by Kendall τ correlation coefficients are observed in the case of economic as well social orders and third area of green economy—environmental dimension of quality of life (in the first case, it is in $V1$: 0.545 and in $V2$: 0.503, and in the second in $V1$: 0.661 and in $V2$: 0.635). The attention should also be paid to the relation between the results considered from the perspective of adopted variants of features. In this case, the very high correlation coefficients both; Person and Kendall τ, are visible. Pearson's correlation coefficients between synthetic measures are at the level from 0.825 (environmental order) to 0.943 (social order) for sustainable development orders, and from -0.112 (O_4) to -0.314 (O_3) for green economy areas.

When analyzing the relationships between the results of ordering due to the variants of features selection, it should be noted that in social order, the positions of EU countries are identical (Kendall τ correlation coefficient equal to 1000).

The highest compliance of ordering results in adopted variants ($V1$ or $V2$) within green economy was obtained for the following areas: the third one (0.857) and the second one (0.799), whereas the lowest was obtained for the fourth area (0.492). Therefore, it is possible to claim that despite the elimination of the diagnostic features, the results obtained for the second and the third areas in the second selection variant do not significantly differ from the results obtained in the first variant where the features were not eliminated.

5 Discussion and Conclusions

Environmental, social and economic problems are closely interrelated and very complex, which is becoming a major challenge for governments, politicians and decision-makers. This creates the need to manage systems such as the economy and society, consisting of many interrelated elements whose behavior cannot be predicted and which have a huge range (Sterman 2012). The European Union wants to create its competitive advantage in the economy, in production, in the development of technology, in research and innovation using the concept of sustainable development. (Frerot 2011). An alternative to the existing economic model based on economic growth is the green economy model as a tool for sustainable development, which using the assets and resources of the natural environment ensures the sustainability of economic processes. For a given country, green economy is a means by which the current economy can move to a sustainable economy (Barbier 2012, Bina 2013).

The main purpose of this article was to examine whether there are links between sustainable development and the green economy in European Union countries. It was assumed that the assessment of the level of sustainable development achieved by a given country should be considered separately for every order. The same approach was used for green economy. This is particularly important in the case of research in which the development of objects is analyzed through the prism of various equally important areas. This is the case when we look at the results of EU countries in terms of sustainable development as well as green economy. In the paper, in accordance with the strong principle of sustainable development, it was assumed that these areas are equally important and the high level of development of the examined objects will also be demonstrated by the high results achieved in each of these areas. Using the taxonomic measure of development based on the Weber median, the EU countries were ordered separately for each area studied, and then the relationships between them were examined using the Pearson and Kendall τ correlation coefficients. It turned out that clear correlations were observed between the third area of green economy (environmental dimension of quality of life) and the two orders of sustainable development: social and economic ones, with the strongest relationship in the case of the social area. The obtained results seem to be profoundly substantiated. The green economy, inextricably linked to green growth is not a substitute for sustainable development—it has a narrower range. It is associated with operational goals that are to lead to specific actions at the interface between the economy and environmental protection by creating the necessary conditions for innovation and investment. The study of green economy includes primarily an assessment of the state of the natural environment and management efficiency. The social aspect is, however, included in a narrower scope—only in the part that is directly related to the environment or the economy. This is directly reflected in the proposed set of indicators. An additional goal of the article was an attempt to verify the impact of the selection of diagnostic features on the classification results. As indicated by the results of research, the rankings of EU countries using different variants of feature selection (without and with a formal and statistical choice) are not the same, and in some cases they differ significantly, especially in areas of green economy. The main reason is different approaches in constructing indicator sets describing sustainable development and the green economy. Indicators of sustainable development are very diverse; each of them is dedicated to a different purpose. In fact, none of them replicates similar information. While the indicators of the green economy describe similar aspects considered in relation to another basis of comparisons, e.g., in relation to the number of inhabitants or GDP. According to the authors, in the case of analyses based on indicators developed for strategic plans, the use of formal and statistical verification may change the results obtained, and therefore, it is postulated to use all the features that may be relevant to achieve the intended goals. As it was indicated in previous part of the paper, the same conclusions were already formulated by other authors especially in the works of: Sokołowski and Markowska (2017) and Sokołowski and Sobolewski (2019).

When answering research questions formulated at the beginning of the paper, it should be emphasized that: (a) there are the compliance between the results of ordering EU countries in the case of individual sustainable development orders and areas of the green economy, considered separately, but they are much more stronger in the case of sustainable development; (b) a consistency of the results of ordering EU countries between individual sustainable development orders and green areas of the economy exists, but it concerns only a few particular orders and areas (economic and social orders with the third area of green economy); (c) the impact of the selection of diagnostic features adopted for the study on the results of ordering EU countries is much more visible in the case of the green economy; in these areas, greater differences between the adopted variants of diagnostic features are also observed.

References

Adam L, Kroupa T (2017) The intermediate set and limiting super differential for coalitional games: between the core and the Weber set. Int J Game Theory 46(4):891–918. https://doi.org/10.1007/s00182-016-0557-3

Aiginger K, Bärenthaler-Sieber S, Vogel J (2013) Competitiveness under new perspectives. Working Paper, 44, www.foreurope.eu. Accessed 10 Aug 2019

Altieri MA (2018) Agroecology: the science of sustainable agriculture. CRC Press, Taylor&Francis Group, Boca Raton

Andalecio MN (2009) Multi-criteria decision models for management of tropical coastal fisheries: a review. Agron Sustain Dev 30(3):557–580. https://doi.org/10.1051/agro/2009051

Barbier EB (2012) The green economy Post Rio + 20. Science 338:887–888

Bina O (2013) The green economy and sustainable development: an uneasy balance? Environ Planning C Govern Policy 31:1023–1047. https://doi.org/10.1068/c1310j

Burget M, Bardone E, Pedaste M (2016) Definitions and conceptual dimensions of responsible research and innovation: a literature review. Science and engineering ethics. Springer, Berlin

Borys T, Czaja S (2009) Badania nad zrównoważonym rozwojem w polskich ośrodkach naukowych. In: Kiełczewski D (ed) Od koncepcji ekorozwoju do ekonomii zrównoważonego rozwoju, Białystok, Wydawnictwo Wyższej Szkoły Ekonomicznej w Białymstoku

Borys T (2011) Zrównoważony rozwój: jak rozpoznać ład zintegrowany. Problemy Ekorozwoju 6(2):75–81

Cheba K (2017) Badanie jednorodności rozwoju w regionach i krajach Unii Europejskiej. Wiadomości Statystyczne 9(676):26–42

Cheba K (2019) Zrównoważona międzynarodowa konkurencyjność krajów Unii Europejskiej. Studium teoretyczno-empiryczne. CeDeWu, Warszawa

Cheba K, Szopik-Depczyńska K (2019) Sustainable competitiveness and responsible innovations—the case of the European Union countries. PN Management and Quality Sciences (in press)

Ciegis R, Ramanauskiene J, Martinkus B (2009) The concept of sustainable development and its use for sustainability scenarios. Eng Econ 62(2):28–37

Cox TF, Cox MAA (2000) A general weighted two-way dissimilarity coefficient. J Classif 17:101–121

Dembińska I (2018) Infrastruktura logistyczna gospodarki w ujęciu środowiskowych uwarunkowań zrównoważonego rozwoju. Wydawnictwo Naukowe Uniwersytetu Szczecińskiego, Szczecin

EAI (European Environment Agency (2011) Europe's environment—an assessment of assessments. Publications Office of the European Union, Luxembourg

Flint RW (2012) basics of sustainable development, practice of sustainable community development. Springer, Berlin, pp 25–54

Frerot A (2011) Unia Europejska a wyzwanie stworzenia zielonej gospodarki. Fundacja Roberta Schumana, Kwestie Europejskie, p 206

Fullwiler ST (2015) Sustainable finance: building a more general theory of finance. Working paper No. 106, Binzagr Institute for Sustainable Prosperity
Gatnar E, Walesiak M (eds) (2004) Metody statystycznej analizy wielowymiarowej w badaniach marketingowych. Wydawnictwo AE we Wrocławiu, Wrocław
Hellwig Z (1968) Zastosowanie metody taksonomicznej do typologicznego podziału krajów ze względu na poziom ich rozwoju oraz zasoby i strukturę wykwalifikowanych kadr. Przegląd Statystyczny. R. XV 4:307–327
Kendall MG (1938) A new measure of rank correlation. Biometrika 30:81–93
Kiba-Janiak M (2015) A Comparative analysis of sustainable city logistics among capital cities in the EU. Appl Mech Mater 708:113–118
Lapata M (2006) Automatic evaluation of information ordering: Kendall's Tau. Comput Linguist 32:471–484
Lira J, Wagner W, Wysocki F (2002) Mediana w zagadnieniach porządkowania obiektów wielocechowych. In: Paradysz J (ed) Statystyka regionalna w służbie samorządu lokalnego i biznesu, Internetowa Oficyna Wydawnicza Centrum Statystyki Regionalnej, Akademia Ekonomiczna w Poznaniu, Poznań, pp 87–99
Młodak A (2006) Analiza taksonomiczna w statystyce regionalnej. Difin, Warszawa
Młodak A (2014) On the construction of an aggregated measure of the development of interval data. Comput Stat 29(5):895–929
Młodak A, Józefowski T, Wawrowski Ł (2016) Zastosowanie metod taksonomicznych w estymacji wskaźników ubóstwa. Wiadomości Statystyczne. R. LXI 2:1–24
OECD (Organization for Economic Co-operation and Development) (2011) Towards green growth, OECD green growth studies. OECD Publishing, Paris. https://doi.org/10.1787/978926411131 8-en
Okazaki N, Yutaka M, Mitsuru I (2004) Improving chronological sentence ordering by precedence relation. In: Proceedings of the 20TH international conference on computational linguistics, Geneva, Switzerland, 23–27 August 2004, pp 750–756
Ostasiewicz W (2004) Ocena i analiza jakości życia. Uniwersytet Ekonomiczny we Wrocławiu, Wrocław
Panek T (2009) Statystyczne metody wielowymiarowej analizy porównawczej. SGH w Warszawie – Oficyna Wydawnicza, Warszawa
Pearce D, Markandya A, Barbier E (1989) Blueprint for a green economy, London. https://doi.org/10.4324/9780203097298
Pechersky S (2015) A note on external angles of the core of convex TU games, marginal worth vectors and the Weber set. Int J Game Theor 44(2):487–498. https://econpapers.repec.org/article/sprjogath/v_3a44_3ay_3a2015_3ai_3a2_3ap_3a487-498.htm. Accessed 10 Aug 2019
Piontek F (2016) Integracja i jej znaczenie dla zarządzania kapitałem ludzkim. Nierówności Społeczne a Wzrost Gospodarczy 46(2):46–59
Pulido M, Sanchez-Soriano J (2009) On the core, the Weber set and convexity in games with a priori unions. Eur J Oper Res 193(2):468–475. https://doi.org/10.1016/j.ejor.2007.11.037
Sanderson M (2007) Problems with Kendall's Tau. In: Proceedings of the SIGIR '07, 30th annual international ACM SIGIR conference on research and development in information retrieval, Amsterdam, The Netherlands, 23–27 July 2007, pp 839–840
Schwab K, Sala-i-Martin X (2012) The global competitiveness report 2012–2013. World Economic Forum, Geneva
Sexton MG, Barrett PS, Lu SL (2008) The evolution of sustainable development. Book Section. University of Sanford, Manchester
Sokołowski A, Markowska M (2017) Iteracyjna metoda liniowego porządkowania obiektów wielocechowych. Przegląd Statystyczny, LXIV(2):153–162
Sokołowski A, Sobolewski M (2019) Jak nie należy analizować danych regionalnych, czyli o błędnym stosowaniu współczynnika zmienności, presentation on Conference: Metodologia Badań Statystycznych, 3–5 July 2019, Warszawa

Stafford-Smith M, Griggs D, Ullah F, Reyers B, Kanie N, Stigson B, Shrivastava P, Leach M, O'Connell D (2017) Integration: the key to implementing the Sustainable Development Goals. Sustain Sci 12(6):911–919

Sterman JD (2012) Sustaining sustainability: creating a systems science in a fragmented academy and polarized world. Springer Science + Business Media, pp 21–58. www.jsterman.scripts.mit.edu [05-10-2019]

The Global Sustainable Competitiveness Index (2012) SolAbility, South Korea

UNEP (United Nations Environment Program) (2011) Towards a green economy: pathways to sustainable development and poverty eradication. http://www.unep.org/greeneconomy. Accessed 10 Aug 2019

Walesiak M (2002) Pomiar podobieństwa obiektów w świetle skal pomiaru i wag zmiennych. Prace Naukowe Akademii Ekonomicznej we Wrocławiu, Ekonometria 10(950):71–85

Wagenmakers EJ, Farrell S (2004) AIC model selection using Akaike weights. Psychon Bull Rev 11:192–196

WCED (World Commission on Environment and Development) (1987) Our common future. Un documents: gathering a body of global agreements has been compiled by the NGO Committee on Education of the Conference of NGOs from United Nations web sites with the invaluable help of information & communications technology, United Nations, New York, USA

Weber A (1909), Reprint (1971) Theory of Location of Industries. Russel & Russel, New York

Zioło M, Filipiak BZ, Bąk I, Cheba K, Tîrca DM, Novo-Corti I (2019) Finance, sustainability and negative externalities. an overview of the european context. Sustainability 11(4249):1–35. https://doi.org/10.3390/su11154249

The Review of Indicators of Data Quality in Intra-Community Trade in Goods. The Choice of an Indicator and Its Effect on the Ranking of Countries

Iwona Markowicz and Paweł Baran

Abstract This article deals with the issue of mirror data concerning intra-Community supplies of goods. Theoretically, in two sources—registers of two countries, trading partners—goods of the same value should be recorded. The observed asymmetries in the declared values were of interest to numerous researchers. The first one to pay special attention to these differences was Morgenstern (1963), who, among other things, dealt with the study of differences in data on world exports and imports. As a result of the literature review, the methods of examining the quality of mirror data in foreign trade (measures and indicators) were systematised. We have also put forward our own proposals. Selected indicators of data asymmetry were used in the study (including aggregated data asymmetry index $_ZW$, aggregated weighted asymmetry index *AER*, symmetric mean absolute discrepancy index *SMADI* and general data asymmetry index $_OW$). Similarities and differences in the obtained results were pointed out (values of indicators and ranking of countries were determined according to the quality of data). The research was conducted using data from the Eurostat COMEXT database—intra-Community supplies (ICS) and acquisitions (ICA) of goods between the EU member states in 2017 were analysed.

Keywords International trade · Mirror data asymmetry · Data accuracy

1 Introduction

Statistical data on international trade in goods are used in a variety of analyses by researchers, policymakers and businesses. Data on intra-Community trade are aggregated and disseminated by Eurostat in the form of COMEXT database, while data at country level are collected through the Intrastat system. Methodologies for

I. Markowicz · P. Baran (✉)
Department of Econometrics and Statistics, Institute of Economics and Finance, University of Szczecin, Szczecin, Poland
e-mail: pawel.baran@usz.edu.pl

I. Markowicz
e-mail: iwona.markowicz@usz.edu.pl

K. Jajuga et al. (eds.), *Classification and Data Analysis*, Studies in Classification, Data Analysis, and Knowledge Organization,
https://doi.org/10.1007/978-3-030-52348-0_12

data collection as well as ways of detecting and correcting errors vary from country to country. Therefore, mirror data (i.e. data on supply from one country and respective data on acquisition from their partner), which should be the same by definition, are divergent to varying degrees. The problem of the quality of international trade data has been addressed in the literature on the subject mostly around the turn of the twentieth and twenty-first centuries (e.g. Parniczky 1980; Federico and Tena 1991; Cate ten 2014).

The aim of the article is to systematise the measures, individual and aggregate indices present in the literature as well as the authors' own proposals used to study the quality of data on international trade in goods. Some of these indices have been calculated in the following empirical example on intra-Community trade in goods. The quality of the data can be compared between countries, and the countries can be ranked according to this quality. Obviously, the choice of a method has a decisive impact on the outcome. The results suggest which of the compared approaches seem to be the most advantageous.

2 Literature Review

Data on foreign trade, including intra-EU trade in goods, are of a bilateral nature and are called mirror data. Eurostat (1998) defines comparing such data as 'bilateral comparison of two basic measures of a trade flow (…) a traditional tool for detecting the causes of asymmetries in statistics'. Numerous researchers carried out analyses of that kind and studied the quality of mirror data on international trade. A synthetic review of the proposed methods of analysing mirror data is presented in Table 1.

According to Guo (2010), the first researchers to observe and describe data asymmetries in foreign trade were Morgenstern (1963) and Tsigas et al. (1992). Morgenstern studied differences in data on world exports and imports. He proposed the determination of absolute and relative differences for the analysis of trade of all countries in general and indices for the analysis of pairs of countries (Table 1, item 1). In Tsigas et al. (1992), the appropriate balance between declared export and import values resulting from differences in the treatment of transport and insurance costs by the declarants is raised. Correct data on exports X_{ij}^* should be equal to correct imports data M_{ij}^*, but they are not. In many cases, two different approaches to the inclusion of transport and insurance costs are being used. These are *free on board* (FOB), i.e. when the buyer is responsible for the freight and insurance cost and *cost, insurance and freight* (CIF), i.e. when the seller is responsible for the costs of transportation of the ordered goods. FOB terms are most common in reporting exports, while CIF terms are most common in reporting imports. The implied formula to account for both CIF and FOB is as follows: $M_{ij}^* = (1 + t)X_{ij}^*$, where t stands for the proportion of these costs. The cited authors considered this proportion in their models, with the aim of searching for other reasons in the data discrepancy. Carrère and Grigoriou (2014) also described the CIF/FOB ratio (Table 1, item 7).

Table 1 An overview of the survey methods for the quality of foreign trade data

No	Authors	Formula	Usage
1	Morgenstern (1963)	Actual difference $M - X$ Relative difference $\frac{M-X}{X}100$ M—imports X—exports	World trade total 1938, 1947–1960 data: Yearbook of International Trade Statistics (New York 1960), Monthly Bulletin of Statistics (1962)
		Ratios $\frac{I_1-E_2}{I_1}$ $\frac{E_1-I_2}{E_1}$ $\frac{b_1-b_2}{b_1}$ I_1—declared imports of country 1 I_2—declared imports of country 2 E_1—declared exports of country 1 E_2—declared exports of country 2 b_1—trade balance of country 1 b_2—trade balance of country 2	For pairs of countries, Different countries, 1909–1913 data: Zuckermann (1921) 1928, 1935, 1938 data: League of Nations, The Network of World Trade (1942) 1948, 1952, 1956, 1960 data: Direction of International Trade
2	Parniczky (1980)	Absolute difference $d_{ij} = \lvert x_{ij} - m_{ij} \rvert$ Ratio $r_{ij} = \frac{x_{ij}}{m_{ij}}$ x_{ij}—exports from country i to country j (declared by the exporter) m_{ij}—imports to country j from country i (declared by the importer)	For pairs of countries Ideal case: $d_{ij} = 0$ $r_{ij} = 1$ Theory only
3	Federico and Tena (1991) - reference to: Morgenstern (1963)	Mirror value indices $M_i = \frac{\sum_{i=1}^{N} M_{ij}}{\sum_{j=1}^{N} X_{ji}}$ $X_i = \frac{\sum_{i=1}^{N} X_{ij}}{\sum_{j=1}^{N} M_{ji}}$ (In numerators summation should also be done over j!) M—imports X—exports i—country under consideration j—partner countries	Comparison of a country with a group of countries (one-to-many), Different countries, 1909–1913, 1928, 1935 data: Zuckermann (1921); League of Nations, The Network of World Trade (1942)

(continued)

Table 1 (continued)

No	Authors	Formula	Usage
4	Ferrantino and Wang (2008)	Relative difference $DIF_{it}^{sr} = 100\frac{M_{it}^{sr}-E_{it}^{sr}}{M_{it}^{sr}}$ Asymmetry index $\mathrm{ER}_{it}^{sr} = 100\frac{M_{it}^{sr}-E_{it}^{sr}}{0{,}5(M_{it}^{sr}+E_{it}^{sr})}$ Aggregated weighted asymmetry index (aggregation with respect to country or commodity group is possible) $\mathrm{AER}_{r}^{sr} = \sum_i w_{it}^{sr}\lvert\mathrm{ER}_{it}^{sr}\rvert$ with weights given by $w_{it}^{sr} = \frac{M_{it}^{sr}+E_{it}^{sr}}{\sum_i(M_{it}^{sr}+E_{it}^{sr})}$ s—country, r—partner, i—commodity, t—period	For pairs of countries, Trade of: China, Hong Kong and the USA with other countries, 1995–2006 data: USITC Oracle database; UN Comtrade
5	Guo (2010)—reference to: Ferrantino and Wang (2008)	Relative difference $\mathrm{DIF}_{st}^{ij} = \frac{\mathrm{Imp}_{st}^{ij}-\mathrm{Exp}_{st}^{ij}}{\mathrm{Imp}_{st}^{ij}}$ Imp_{st}^{ij}—imports to declarant country j from country i of commodity s in year t Exp_{st}^{ij}—exports from declarant country i to country j of commodity s in year t	For pairs of countries, Trade of China with selected countries, 1992–2008 data: IDSB
6	Hamanaka (2012)	Actual difference $A - B$ Mean $\frac{A+B}{2}$ Ratio $\frac{A}{B}$ A—imports to a partner country B—exports from the country under consideration	For pairs of countries, For a country and a group of countries, Cambodia and other countries, Trade totals and divided by HS chapters, 2000–2004, 2008 data: UN Comtrade
7	Carrère and Grigoriou (2014)	CIF/FOB ratio $R_{ijkt}^{cif/fob} = \frac{p_{ijk}^{M}Q_{ijk}^{M}}{p_{ijk}^{X}Q_{ijk}^{X}}$ p—price Q—quantity k—commodity t—period $p_{ijk}^{M}Q_{ijk}^{M}$—value of imports of commodity k to country i from country j $p_{ijk}^{X}Q_{ijk}^{X}$—value of exports of commodity k from country j to country i	For pairs of countries, All countries 2008 data: UN Comtrade, Constant ratio equal to 1 or (1+ τ_{ijk})

(continued)

Table 1 (continued)

No	Authors	Formula	Usage
8	HMRC Trade Statistics (2014)	Absolute asymmetry $\lvert \text{Value}(D) - \text{Value}(P) \rvert$ Relative asymmetry $\frac{\lvert \text{Value}(D) - \text{Value}(P) \rvert}{0{,}5 \lvert \text{Value}(D) + \text{Value}(P) \rvert} 100$ Share in relative asymmetry (%) (e.g. of a group of commodities) $\frac{\lvert \text{Value}(D) - \text{Value}(P) \rvert}{\sum \lvert \text{Value}(D) - \text{Value}(P) \rvert} 100$ *D*—declarant country *P*—partner country	For pairs of countries, UK trade with EU member states 2014 Trade totals and divided into commodity groups (CN chapters) data: COMEXT

UN Comtrade—United Nations Comtrade Database
IDSB—Industrial Demand and Supply Balance Database, UNIDO

The methods of studying differences in mirror data proposed in the literature are used to study the relations between two countries (country–country or one-to-one relations) or between a country and a group of countries (country–countries or one-to-many relations). Proposals for methods and studies for pairs of countries are described by Morgenstern (1963), Parniczky (1980), Ferrantino and Wang (2008), Guo (2010), Hamanaka (2012), Carrère and Grigoriou (2014), HMRC Trade Statistics (2014). In addition to an absolute difference between the values of exports and imports declared in the trading partner countries, growth and intensity rates, and data asymmetry rates were applied in the literature. The latter has been proposed using various formulae, sometimes similar to the data discrepancy index used by the authors (Baran and Markowicz 2018a). The formula proposed by Ferrantino and Wang (2008) (Table 1, item 4, *ER*) has also been adopted by official statistics (Eurostat 2017; GUS 2018). Some surveys focus on the analysis of data differences between a country and a group of countries, i.e. in one-to-many relations (Federico and Tena 1991; Hamanaka 2012). The methodological proposals mainly concerned the intensity index and when aggregation was used, it concerned goods or countries, but the values of exports and imports (and not the differences) were added up.

An alternative approach to the study of the accuracy of mirror data, based on the correlation coefficient between the export and import mirror vector (broken down by commodity groups), is proposed by Fertő and Soós (2009). Although the authors did not present the formula, the text suggests that they calculated the correlation coefficients between the mirrored data vectors for a single country and all its partners separately (in one-to-one relations), and then averaged them, obtaining a measure of the average similarity of the mirror data for all directions combined (i.e. in a one-to-many relation). Next, they created a ranking of 29 selected European countries according to the above-mentioned averaged correlation between the mirror data. The weakness of such an approach is the fact that due to the lack of standardisation of variables or a weighting system, the countries with the highest foreign trade turnover were considered to have the highest data quality.

3 Proposed Methods

The authors analyse mirror (bilateral) data concerning trade between countries. The nature of the data collected by public statistics allows for the identification of differences in the declared values of trade between partner countries. These differences are important for economists using foreign trade information in their research. The problem of data discrepancies can be considered from the point of view of the quality of statistical data (data collection and processing methodology) and from the point of view of tax fraud (Baran and Markowicz 2018b). In both cases, the methods of data discrepancy (asymmetry) analysis can be used. During the literature review and analyses, the authors have developed their own research methodology. These are the following methods:

1. For a pair of countries

- Data asymmetry measure

$$M_E^{AB} = E_{AB} - I_{BA} \tag{1}$$

- Data asymmetry index (taking values from $\langle -2; 2 \rangle$)

$$W_E^{AB} = \frac{E_{AB} - I_{BA}}{K} \tag{2}$$

where:

E_{AB} declared value of exports from country A to country B,
I_{BA} declared value of imports to country B from country A (mirror value),

$K = \frac{E_{AB}+I_{BA}}{2}$ (denominator $K = I_{BA}$ or $K = E_{AB}$ may also be used),

2. For a country and a group of countries (Markowicz and Baran 2019a)

- general data asymmetry index, $\langle -2; 2 \rangle$

$$_{O}W_E^{AU} = \frac{E_{AU} - I_{UA}}{K} \tag{3}$$

where:

E_{AU} declared value of exports from country A to all other EU member states (or a group of countries) total,
I_{UA} declared value of imports to other EU member states UE from country A,

$$K = \frac{E_{AU} + I_{UA}}{2},$$

- aggregated data asymmetry index (aggregation by country), (taking values from $\langle 0; 2\rangle$)

$$_{Z}W_{E}^{AU} = \frac{\sum_{i=1}^{n}\left|E_{AB_i} - I_{B_iA}\right|}{K} \tag{4}$$

where:

E_{AB_i} declared value of exports from country A to country B_i,
I_{B_iA} declared value of imports to country B_i from country A,

$K = \sum_{i=1}^{n} \frac{E_{AB_i}+I_{B_iA}}{2}$ (again, $K = \sum_{i=1}^{n} I_{B_iA}$ or $K = \sum_{i=1}^{n} E_{AB_i}$ may also be used).

The methods presented above concern the formulas used in the one-to-one and one-to-many relations. The methods used for a pair of countries are a reference to the literature proposals presented in Table 1. Data asymmetry measure (1) is related to items 1, 2, 6, 8, and data asymmetry index (2) is related to items 1, 4, 5, 8. Data asymmetry index (3) proposed for use in the analysis of one-to-many relations is analogous to the index for pairs of countries (2). Index of this form has been labelled 'general' as differences between low-level mirror data of different signs are partly compensating, so the measure is rather general in its nature. This is the reason why the authors prefer using the aggregated index (4). Due to the aggregation of absolute values for individual countries, this index is always positive, and its values are higher than the values of the general index (Markowicz and Baran 2019b). In our opinion, the aggregated index better reflects the specificity of trade in goods between a country and its trade partners. Index (4) is an authors' proposal, and we have used it in our previous works, where it proved to be both useful and superior over the general index.

For the empirical part of the work, we also proposed a measure named SMADI (*symmetric mean absolute discrepancy index*); it is calculated similarly to the SMAPE (*symmetric mean absolute percentage error*), an error measure used in forecasting (Armstrong 1986). The following formula was used:

$$\text{SMADI} = \frac{1}{n}\sum_{i=1}^{n}\frac{\left|E_{AB_i} - I_{B_iA}\right|}{\left(E_{AB_i} + I_{B_iA}\right)/2}, \tag{5}$$

where all symbol designations are as in (4). We discuss its characteristics, and we position it among the other indices in the subsequent section.

4 Results

The aim of the research was to compare several ways of determining the data quality for the European Union member states trade in goods data (in one-to-many relations). The analysis was performed using Eurostat's COMEXT data for 2017 (data collected in November 2018). We used data on intra-Community supplies and acquisitions between all pairs of EU member states. The following methods were used in the survey:

1. the aggregated data asymmetry index ${}_{Z}W$ (4), formula proposed by the authors
2. the aggregated weighted asymmetry index *AER* (Table 1, item 4), formula proposed by Ferrantino and Wang (2008)
3. the *SMADI*, formula proposed by the authors
4. the general data asymmetry index ${}_{O}W$ (3), formula proposed by the authors.

The first three indices were calculated with country-by-country aggregation, which is the authors' contribution to the studies on the quality of intra-EU trade data. As an additional measure, we used the general data asymmetry index ${}_{O}W$ (3). When ${}_{O}W$ is applied, the divergences in data between pairs of countries are partly balanced, and therefore, the value of the index is lower than the aggregate index (Markowicz and Baran 2019b).

Table 2 contains only three columns of calculated values of indices because we obtained same values for *AER* and for ${}_{Z}W$. Indeed, a series of uncomplicated transformations lets us derive ${}_{Z}W$ from *AER*, so they are two forms of the same measure, although ${}_{Z}W$ is simpler both in writing and in calculation (designations as in (4)):

$$\begin{aligned}\text{AER} &= \sum_{i=1}^{n} \frac{\left|E_{AB_i} - I_{B_iA}\right|}{\left(E_{AB_i} + I_{B_iA}\right)/2} \cdot \frac{\left(E_{AB_i} + I_{B_iA}\right)}{\sum_{i=1}^{n}\left(E_{AB_i} + I_{B_iA}\right)} \\ &= \sum_{i=1}^{n} \frac{\left|E_{AB_i} - I_{B_iA}\right|}{\sum_{i=1}^{n}\left(E_{AB_i} + I_{B_iA}\right)} \cdot \frac{\left(E_{AB_i} + I_{B_iA}\right)}{\left(E_{AB_i} + I_{B_iA}\right)/2} = \sum_{i=1}^{n} \frac{\left|E_{AB_i} - I_{B_iA}\right|}{\sum_{i=1}^{n}\left(E_{AB_i} + I_{B_iA}\right)/2} \\ &= \frac{\sum_{i=1}^{n}\left|E_{AB_i} - I_{B_iA}\right|}{\sum_{i=1}^{n}\left(E_{AB_i} + I_{B_iA}\right)/2} = {}_{Z}W \qquad (6)\end{aligned}$$

Figure 1 compares the values of data quality indices used in the analysis. The general index ${}_{O}W$ took the lowest values in each case, as the positive and negative discrepancies recorded for different directions often balance (partially) each other. The figure shows the absolute values of ${}_{O}W$ (it is the only measure under consideration to take positive and negative values) so that it could be compared with other methods. On the other hand, the highest values were recorded for the *SMADI* index. Also this result is intuitive, because this time relatively weak trade relations, where the asymmetry values are high, have an excessive impact on the measurement value.

In Fig. 2, these three rankings are presented in a slope graph. This way of displaying information made it possible to distinguish three specific groups of countries. For some countries (Denmark, Latvia, Estonia, dark red lines), negative and

Table 2 Values of calculated data quality indices for EU member states' ICS to all other EU countries in 2017 and rankings of EU member states based on these indices

Country		Index value			Rank*		
		$_ZW, AER$	*SMADI*	$_OW$	$_ZW, AER$	*SMADI*	$_OW$
Germany	DE	0.0550	0.0744	−0.0076	1	1	2
Austria	AT	0.0650	0.1201	−0.0055	2	3	1
Romania	RO	0.0668	0.1481	0.0588	3	7	10
France	FR	0.0704	0.1683	−0.0325	4	11	4
Bulgaria	BG	0.0718	0.1822	0.0550	5	14	9
Belgium	BE	0.0738	0.1193	0.0684	6	2	11
The Netherlands	NL	0.0811	0.1326	0.0706	7	5	14
Spain	ES	0.0838	0.1562	0.0464	8	9	6
UK	GB	0.0864	0.1738	−0.0507	9	13	7
Italy	IT	0.0940	0.1226	0.0872	10	4	16
Hungary	HU	0.0945	0.1491	0.0696	11	8	12
Sweden	SE	0.0948	0.1685	−0.0739	12	12	15
Czechia	CZ	0.1008	0.1586	0.0941	13	10	18
Poland	PL	0.1041	0.1867	0.1025	14	15	19
Slovakia	SK	0.1086	0.3319	0.1080	15	23	20
Lithuania	LT	0.1110	0.1471	0.0937	16	6	17
Denmark	DK	0.1206	0.2106	0.0128	17	17	3
Finland	FI	0.1317	0.2837	−0.0701	18	21	13
Estonia	EE	0.1368	0.2714	0.0510	19	19	8
Greece	GR	0.1430	0.1994	0.1286	20	16	22
Ireland	IE	0.1468	0.3954	−0.1368	21	25	23
Latvia	LV	0.1474	0.2725	0.0405	22	20	5
Slovenia	SI	0.1502	0.3052	0.1410	23	22	24
Portugal	PT	0.1599	0.2600	0.1575	24	18	25
Croatia	HR	0.1976	0.3544	0.1187	25	24	21
Luxemburg	LU	0.2055	0.4517	−0.1837	26	26	26
Malta	MT	0.5196	0.8708	−0.5015	27	28	28
Cyprus	CY	0.5783	0.7417	−0.4635	28	27	27

*The ranking based on $_OW$ was established with the use of the absolute values of the index.
Source Own elaboration

positive discrepancies were balanced out, so that the position according to $_OW$ is much higher than in the case of other two measures. For these countries, in general minor differences exist with different signs between the declared mirror values. Second group (Belgium, the Netherlands and Italy, orange lines) consists of countries with relatively low values of *SMADI* and high values of $_OW$. This means that for these countries, we observed only a few high discrepancies in trade with countries

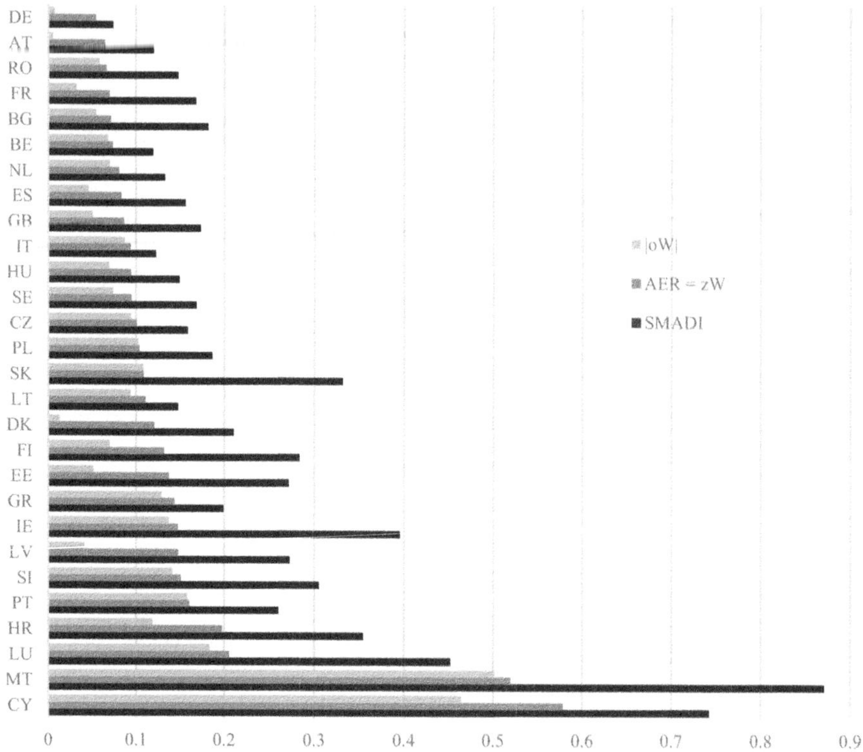

Fig. 1 Values of calculated data quality indices for EU member states' ICS to other EU countries in 2017. *Source* Own calculations

of smaller importance. Another group (Bulgaria, Romania, Slovakia, blue lines) are Eastern European countries with a specific structure of mirror data discrepancies with mostly positive asymmetries (i.e. declared ICS are in most cases greater than respective mirror ICA, which probably indicates a situation where huge exporters send commodities to numerous smaller importers). An example of the discrepancies for Danish ICS (representative of the 'dark red' and 'orange' groups) and for Romanian ICS (representative of the 'blue' group) broken down by country are given in Figs. 3 and 4.

Using the three approaches presented ($_ZW = AER$, *SMADI*, $_OW$) to determine the quality of data on trade between EU countries, different values of indicators and different country rankings were obtained (Fig. 2). As a result, the choice of a method of measuring asymmetry in mirror data is crucial for the survey results. It is important to take account of the discrepancies in data recorded for various product groups in all relations, but also to ensure that the analyses are concise and transparent. The analysis at the level of the Combined Nomenclature chapters and the possible move to four-digit positions where necessary (provided high turnover and strong differentiation of goods within the chapter) seem to strike the right balance between

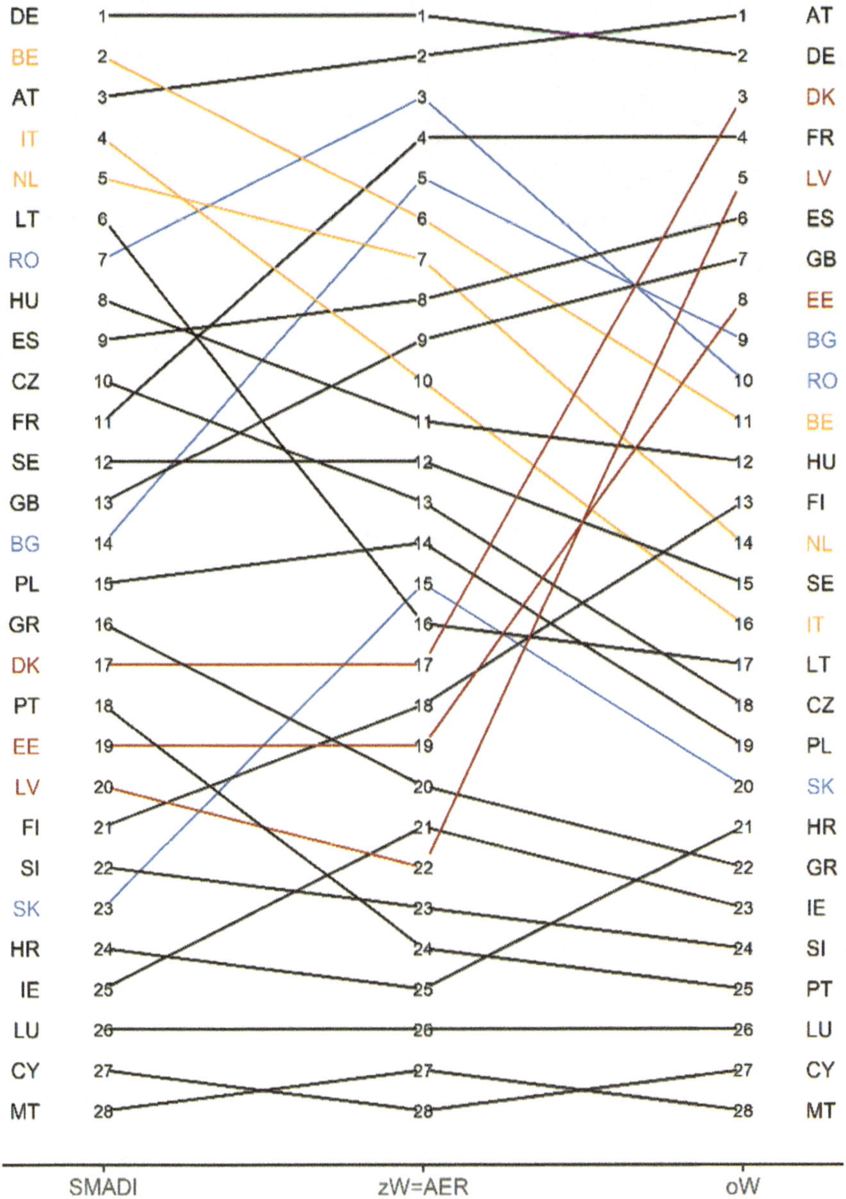

Fig. 2 Rankings of EU member states according to data quality indices for their respective ICS to other EU countries in 2017. *Source* Own calculations

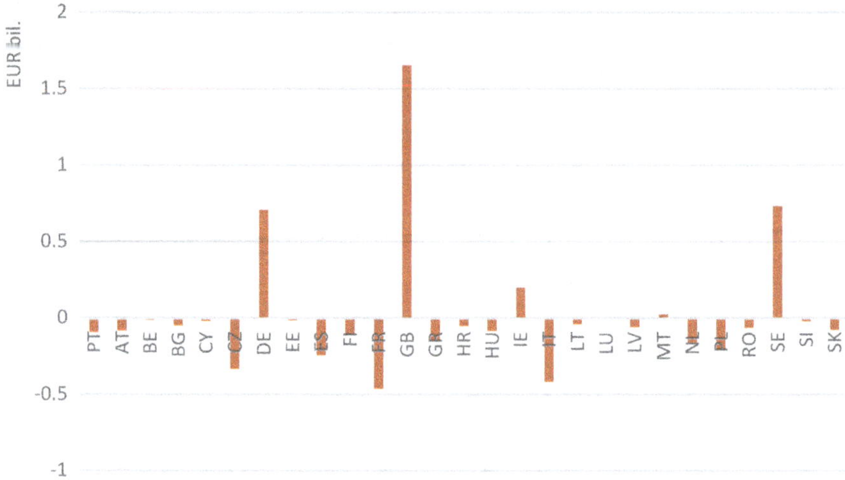

Fig. 3 Aggregate discrepancies in mirror data for Danish ICS broken down by country. *Source* Own calculations

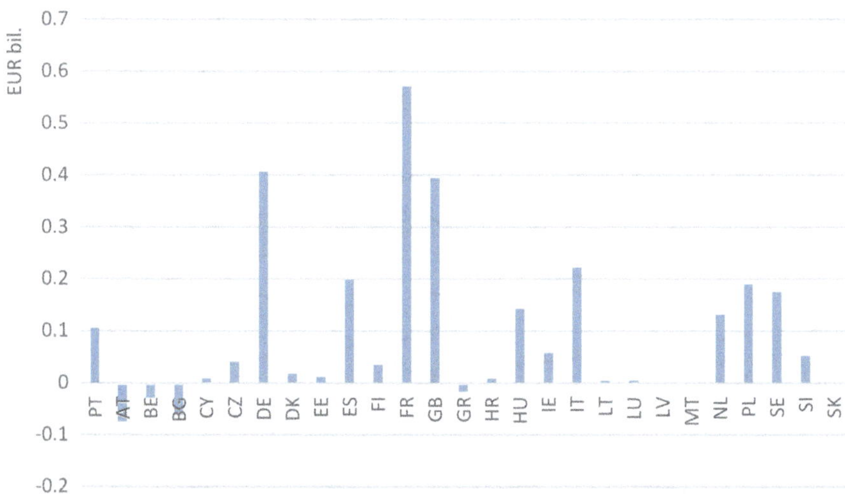

Fig. 4 Aggregate discrepancies in mirror data for Romanian ICS broken down by country. *Source* Own calculations

the complexity of the analysis and the precision of the outcomes received, which leads to the rejection of the $_OW$ measure.

Let us rewrite the *SMADI* in a slightly changed form:

$$\text{SMADI} = \frac{1}{n}\sum_{i=1}^{n}\frac{\left|E_{AB_i} - I_{B_iA}\right|}{\left(E_{AB_i} + I_{B_iA}\right)/2} = \sum_{i=1}^{n}\frac{\left|E_{AB_i} - I_{B_iA}\right|}{\left(E_{AB_i} + I_{B_iA}\right)/2}\cdot\frac{1}{n}$$

and compare it with the *AER*, written as:

$$\mathrm{AER} = \sum_{i=1}^{n} \frac{|E_{AB_i} - I_{B_i A}|}{(E_{AB_i} + I_{B_i A})/2} \cdot \frac{(E_{AB_i} + I_{B_i A})}{\sum_{i=1}^{n}(E_{AB_i} + I_{B_i A})}.$$

If we compare the factors standing on the right side of both formulas, we notice that the *AER* index is the *SMADI* extended by weights, which are the shares of particular directions (partner countries) in the trade of country *A*. This means that in the case of less important trading relations, the value of the *AER* will not increase as rapidly as the *SMADI*.

The *SMADI,* apart from a simpler formula, has no other advantages compared to the *AER*. However, the change in the order of factors in the numerator of the *AER* allows us to notice that in fact, this formula boils down to the formula for the aggregate index $_ZW$ (cf. formula 6 above). Therefore, also the simplicity of calculations is not an argument for using the *SMADI*.

In the opinion of the authors, the proposed aggregate index $_ZW$ is a good measure of data quality. Its formula is consistent with the proposal of the *AER* by Ferrantino and Wang (2008), although these authors did not calculate the index with aggregation by country. At the same time, the decomposition of this index suggests that it is the correct form of a measure of data asymmetry, since it does not only take account of the discrepancy, but also differentiates its significance based on the direction or the product's share in the pattern of trade. The fact that the measure usually takes an intermediate value from among the three tested indices ($|_OW| < {}_ZW < SMADI$) also speaks in favour of the use of this index as it confirms the balanced nature of this measure. In contrast to the *SMADI*, $_ZW$ does not give too high a weight to trade with small countries or countries with minor economic links to the country under examination. And it is precisely these cases that are characterised by high asymmetries in mirror data (as even a single incorrectly declared transaction may significantly affect the measure of relative data asymmetry).

5 Conclusions

Appropriate statistical data are essential for economic research. Foreign trade data are crucial in numerous analyses. Therefore, their quality is extremely important. The article deals with the quality of data on intra-Community trade. Due to the specificity of these data, namely their registration in two sources, there is a possibility to verify their correctness and to measure the quality of the process of collecting such data.

The aim of this article was to systematise the measures, individual and aggregate indices used in the literature to study the quality of data on international trade in goods. Then, the literature and own proposals were used to assess the quality of data on trade in goods between EU member states. Similarities and differences in the obtained values of measures and country rankings were pointed out. A particularly valuable conclusion from the research is the demonstration of equality between the *AER* index (Ferrantino and Wang 2008) and the aggregate index ${}_{Z}W$ (own proposal). Although the authors of the two measures had different ideas and conducted different research, the results are similar. Additionally, the observation leads us to conclusion that we should not use the *SMADI* index in the context of one-to-many trade relations.

Our research shows that ${}_{Z}W$ should be chosen from among the tested measures. This conclusion may be useful not only for other researchers but also for statistical or tax/customs offices looking for an easy way to identify the directions in which the quality of trade data is lower, i.e. the directions that require attention.

References

Armstrong JS (1986) Long-range forecasting. From crystall ball to computer, 2nd edn. Wiley

Baran P, Markowicz I (2018a) Analysis of intra-community supply of goods shipped from Poland. In: Papież M, Śmiech S (eds) The 12th Professor Aleksander Zelias international conference on modelling and forecasting of socio-economic Phenomena. Conference proceedings, Zakopane. Socio-economic modelling and forecasting, 1:12–21

Baran P, Markowicz I (2018b) Behavioral economics and rationality of certain economic activities. The case of intra-community supplies. In: Nermend K, Łatuszyńska M (eds) Problems, methods and tools in experimental and behavioral economics. Proceedings of computational methods in experimental economics (CMEE) 2017 Conference, Springer, Cham, pp 285–299

Carrère C, Grigoriou C (2014) Can mirror data help to capture informal international trade? Policy issues in international trade and commodities research study series No. 65, UNCTAD, New York

Cate AT (2014) The identification of reporting accuracies from mirror data. Jahrbücher für Nationalökonomie und Statistik 234(1). https://doi.org/10.1515/jbnst-2014-0106

Eurostat (1998) Statistics on the trading of goods—user guide. Office for Official Publications of the European Communities, Luxembourg

Eurostat (2017) National requirements for the intrastat system, 2018th edn. Publications Office of the European Union, Luxembourg

Federico G, Tena A (1991) On the accuracy of foreign trade statistics (1909–1935). Morgenstern Revisited. Explor Econ Hist 28(3):259–273

Ferrantino MJ, Wang Z (2008) Accounting for discrepancies in bilateral trade: the case of China, Hong Kong, and the United States. China Econ Rev 19(3):502–520

Fertő I, Soós AK (2008) Treating trade statistics inaccuracies: the case of intra-industry trade. Appl Econ Lett 16(18):1861–1866

Guo D (2010) Mirror statistics of international trade in manufacturing goods: the case of China. UNIDO, Research and statistics branch working paper 19/2009

GUS (2018) Handel zagraniczny. Statystyka lustrzana i statystyka asymetrii, Warszawa

Hamanaka S (2012) Whose trade statistics are correct? Multiple mirror comparison techniques: a test of Cambodia. J Econ Policy Reform 15(1):33–56

HMRC Trade Statistics (2014) A reconciliation of asymmetries in trade-in-goods statistics published by the UK and other European Union Member States. Southend-on-Sea

Markowicz I, Baran P (2019a) ICA and ICS-based rankings of EU countries according to quality of mirror data on intra-Community trade in goods in the years 2014–2017. Oeconomia Copernicana 10(1):55–68. https://doi.org/10.24136/oc.2019.003

Markowicz I, Baran P (2019b) Quality of intrastat data. Comparison between "old" and "new" EU member states. Acta Universitatis Lodziensis. Folia Oeconomica 2(341):69–80. http://dx.doi.org/10.18778/0208-6018.341.05

Morgenstern O (1963) On the accuracy of economic observations, 2nd edn. Princeton University Press, New Jersey
Parniczky G (1980) On the inconsistency of world trade statistics. Int Stat Rev 48(1):43–48
Tsigas ME, Hertel TW, Binkley JK (1992) Estimates of systematic reporting biases in trade statistics. Econ Syst Res 4(4):297–310
Zuckermann S (1921) Statistischer Atlas zum Welthandel, vol 1. Otto Elsner Verlagsgesellschaft MBH, Berlin

Development of ICT in Poland in Comparison with the European Union Countries—Multivariate Statistical Analysis

Małgorzata Misztal and Aleksandra Kupis-Fijałkowska

Abstract The Statistics Poland defines Information and Communication Technologies (ICT) as "a family of technologies that are processing, collecting and sending information in an electronic format." The wide access to the ICT, their ceaseless expansion and extending possibilities of applicability constitute the core for the contemporary society development in Poland, Europe and all over the world. Monitoring and analyzing the changes of and in the ICT area are of great importance in both economic and social dimensions. The ICT expansion is considered to be a stimulant for various processes taking place in the modern economy, significantly affecting the innovation growth in many sectors, as well as increasing competitiveness on both micro- and macroeconomic scales. The study presents and discusses the assessment of the ICT development level in Poland against other European Union countries in the individual users and households perspective. Also, the reasons of the Internet absence at households and types of online activities were investigated. The attention was focused on the identification of subgroups of countries similar in the context of the ICT access and development in the studied societies through the considered years. The analysis was based on the Eurostat and the ITU data for years 2008–2017. The exploratory data analysis methods dealing with three-way data structures (the between and within-class principal component analysis) were applied. Factorial maps (scatterplots and biplots) were presented to summarize the results. The Hellwig method for linear ordering was used to rank the EU-28 countries due to the ICT development level.

Keywords Information and communication technologies · ICT · Multivariate exploratory data analysis · PCA-based methods · EU-28

M. Misztal (✉) · A. Kupis-Fijałkowska
Department of Statistical Methods, Faculty of Economics and Sociology, Institute of Statistics and Demography, University of Lodz, Łódź, Poland
e-mail: malgorzata.misztal@uni.lodz.pl

A. Kupis-Fijałkowska
e-mail: aleksandra.kupis-fijalkowska@uni.lodz.pl

K. Jajuga et al. (eds.), *Classification and Data Analysis*, Studies in Classification, Data Analysis, and Knowledge Organization,
https://doi.org/10.1007/978-3-030-52348-0_13

1 Introduction

Information and Communication Technologies (ICT) are defined by the Statistics Poland as "a family of technologies that are processing, collecting and sending information in an electronic format" (GUS 2018). Eurostat[1] explains that the term ICT "covers all technical means used to handle information and aid communication. This includes both computer and network hardware, as well as their software."

The official statistics bodies as well as non-governmental organizations[2] data about the ICT area for individual users, business and administration, clearly indicate the omnipresence of the ICT. The Internet coverage is spreading, while the digital literacy and the ICT competences are constantly changing and evolving (Vuorikari et al. 2016; Carretero et al. 2017). Many of the official programs at the local and European level were created in aim to reinforce ongoing development of the ICT (e.g., "National Broadband Plan[3]" in Poland) and to exploit the ICT potential for public entities and their services (e.g., "Open data—access, standard, education[4]" or "Digital Sandbox Administration[5]" by The Ministry of Digital Affairs in Poland). The observed ceaseless expansion and widening applicability of these technologies, constitute the core for the modern society and business in Poland, Europe and all over the world.

The ICT development significantly influenced many areas: business (including business digitalization and e-commerce), communication (i.e., instant messengers and social media), education (i.e., e-learning, e-communication platforms with parents), health care services and medicine (including e-medicine and e-health), administration of justice (i.e., e-courts services: e-lawsuits, online monitoring of the court actions), administration (especially e-government—internal and external to specific government entities interactions), finance and banking (mainly e-transactions and e-services), labor market (i.e., job search portals, networks of professionals).

A lot of the research papers can be found about the ICT impact on the economic growth (e.g., Khalili et al. 2014; Próchniak and Witkowski 2016; Jorgenson and Vu 2016; Łaszek et al. 2018) as well as about its supporting role in innovations and competitiveness intensification in many areas, including development of the information society (e.g., Sharafat and Lehr 2017; Mastalerz-Kodzis and Pośpiech 2018). There are articles that focus only on business and enterprises (e.g., Cheba and Saniuk 2016; Becker 2018; Kaliszczak and Pawłowska-Mielech 2019), or mainly on the individuals/households prism (e.g., Miśkiewicz-Nawrocka and Zeug-Żebro 2015; Wojnar 2015; Szczukocka and Pekasiewicz 2017).

---

[1]https://ec.europa.eu/eurostat/statistics-explained/index.php/Glossary:Information_and_communication_technology_(ICT). Accessed 20 Sep 2019.

[2]E.g. International Telecommunication Union (ITU).

[3]https://www.gov.pl/web/cyfryzacja/narodowy-plan-szerokopasmowy. Accessed 20 Sep 2019.

[4]https://www.gov.pl/web/cyfryzacja/otwarte-dane-plus. Accessed 20 Sep 2019.

[5]https://www.gov.pl/web/cyfryzacja/projekt-cyfrowa-piaskownica-administracji-cpa. Accessed 20 Sep 2019.

In conclusion, a lot of scientific attention is paid to the ICT development, its expansion, it is investigated widely how it affects private and public sectors, as well as how far it influences the individual users (life, work and leisure) and their communication models.

Hence, the monitoring and evaluation of the ICT development and accompanying phenomena are of high importance in both economic and social dimensions.

The Polish government research[6] showed that the level of the ICT development is low in Poland in comparison with the vast majority of the UE-28 countries. Moreover, it does not meet all of the objectives formulated in the European strategic documents (e.g., Europe 2020 strategy). Therefore in 2014, the operational program co-financed by the EU structural funds called "Digital Poland for 2014–2020" was formulated and implemented. It is inevitable to undertake activities aimed at expanding the ICT access and improving the digital competences in modern societies. It is also necessary to actively search for new statistical methods that allow to study the ICT development and its components in the multidimensional perspective.

There are three main objectives of this paper:

1. to assess the ICT development level of households in Poland against the European Union 28 member countries (UE-28) in the years from 2008 to 2017;
2. to create the UE-28 ranking of the ICT access based on the Hellwig method for linear ordering and to specify Poland's rank across other considered countries within the years of interest;
3. to identify subgroups of countries within the EU-28 group similar in the perspectives of: the ICT access, reasons for not having Internet access at home and types of Internet activities.

2 Data Sources and Methods

The analyses were based on the Eurostat and the International Telecommunication Union (ITU) data for period from 2008 to 2017. The trend-based single imputation was applied to fill in the few missing data. To assess the differences in the level of development of ICT, taking into account the time of 10 years from 2008 to 2017 and space of 28 European Union countries, an approach using methods based on principal component analysis (PCA) was applied.

Taking into account the years 2008–2017, it is possible to perform 10 separate classical PCAs, one for every year, or one classical PCA after concatenating all datasets; however, the classical PCA carried out for concatenated datasets mixes both the temporal and the spatial typologies (Dufour 2008).

Taking the existence of groups of samples in a data table into consideration, Thioulouse et al. (2018) suggest to use a particular type of analysis, called a between-class analysis, which models the differences between groups by computing the group

[6]i.e. Diagnosis for the Operational Programme Digital Poland for 2014—2020 by The Ministry of Digital Affairs in Poland, https://www.polskacyfrowa.gov.pl/strony/o-programie/dokumenty/program-polska-cyfrowa-2014-2020/. Accessed 20 Sep 2019.

means and analyzes the resulting table. That kind of analysis aims at visually investigating the existence of the groups and describing the main characteristics of the differences between the groups.

In the study, the between-class analysis based on principal component analysis was applied. The basis for the between-class analysis is the table of group means. For groups corresponding to individual countries the values of each variable are averaged across the considered time and then the between-countries PCA can be performed. If the groups correspond to individual years, then the values of each variable are averaged across the considered EU countries and the between-years PCA can be carried out. A more formal description of these methods is presented by Thioulouse et al. (2018).

The between-class analysis may be complemented by the within-class analysis. The within-class analysis operates with the residuals between observed data and the group means, and it aims at looking for structures remaining in the data after removing the differences between groups of samples. The mathematical basis of the within-class PCA can be found in Thioulouse et al. (2018).

Dufour (2008) points out that the within-class and between-class PCA can be regarded as an exploratory generalization of the one-way ANOVA. For each of the considered effects (spatial or temporal), the total inertia (variability) of $\mathbf{X}$ (the matrix containing p variables measured on n objects) can be decomposed in two parts: the inertia of $\mathbf{X}^-$ (the within model after removing the effect of groups) and the inertia of $\mathbf{X}^+$ (the between model, revealing the effect of groups).

To meet the objectives of the study, the between-countries PCA is applied (groups correspond to 28 EU countries, so the spatial effect is taken into account here). Then, the within-countries PCA reveals the temporal effect. All the results are presented graphically with the use of factorial maps: scatterplots and biplots.

The UE-28 ranking due to the level of ICT development was prepared using the Hellwig method for linear ordering multivariate objects (Bąk 2013).

All the calculations were performed with the use of the R-environment, specifically packages: ade4, adegraphics, pllord.

3 Results

3.1 The Assessment of the ICT Access Changes in the Households in Years 2008–2017

The following four variables were taken into consideration for the analysis purposes:

1. Mobile-cellular telephone subscriptions per 100 inhabitants (X_1);
2. Availability of computers on households (X_2);
3. Households with access to the Internet at home (X_3);
4. Households with broadband Internet connection (X_4).

The dynamics of the ICT development in Poland was similar to the average dynamics of the UE-28, with the average rate of change being slightly faster than the average rate in the EU-28 (respectively for each of the considered variables, in Poland: 101.6%, 103.7%, 106.1%, 108.3% and in the EU-28: 100.2%, 102.4%, 104.2%, 106.6%).

The results of the between-class PCA for groups corresponding to the EU-28 countries are presented in Figs. 1, 2 and 3.

The total variability (inertia) in the classical PCA (performed on the combined datasets) equals the number of variables (four in this research). In the presented analysis, the between-class inertia is equal to 2.59, i.e., 64.80% of the total inertia is due to the spatial factor.

The results of the analysis can be summarized with the use of a biplot defined as "a plot of two kinds of information displayed together" by Gower et al. (2015)—Fig. 1. The variables are presented as vectors. The angles between all vectors reflect their

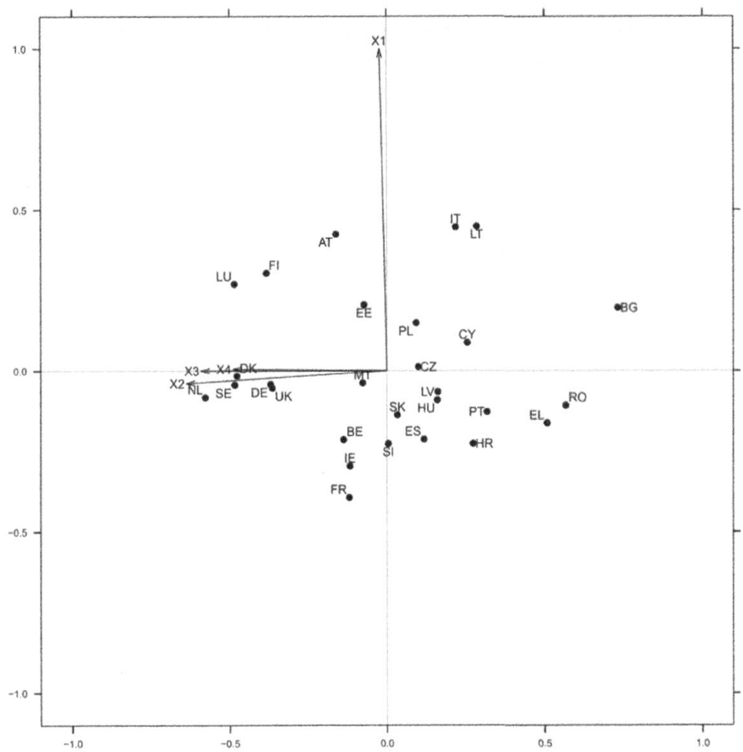

Fig. 1 Between-countries PCA biplot. Legend: BE—Belgium; BG—Bulgaria; CZ—Czechia; DK—Denmark; DE—Germany; EE—Estonia; IE—Ireland; EL—Greece; ES—Spain; FR—France; HR-Croatia; IT—Italy; CY—Cyprus; LV—Latvia; LT—Lithuania; LU—Luxembourg; HU—Hungary; MT—Malta; NL—Netherlands; AT—Austria; PL—Poland; PT—Portugal; RO—Romania; SI—Slovenia; SK—Slovakia; FI—Finland; SE—Sweden; UK—United Kingdom. *Source* Own calculations

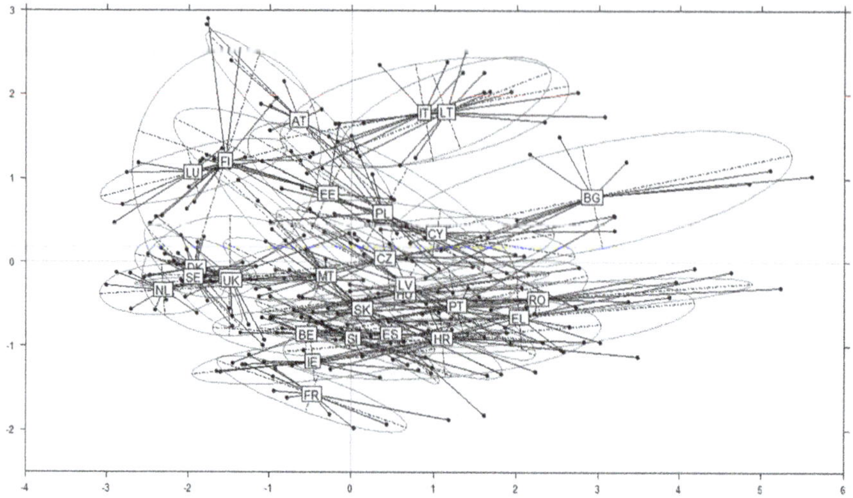

Fig. 2 Diversity of the IT infrastructure development in 2008–2017 for all the EU countries. Legend as in Fig. 1. *Source* Own calculations

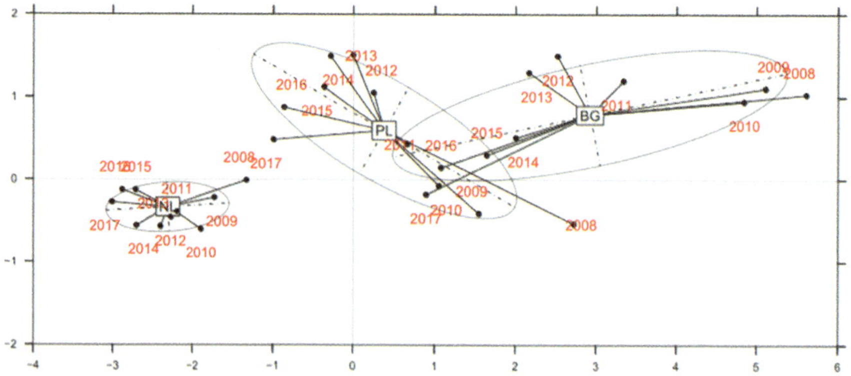

Fig. 3 Star plots with ellipses for Bulgaria (BG), Poland (PL) and the Netherlands (NL). *Source* Own calculations

linear correlations. The direction of the vector corresponds to the direction of the highest variability of a given variable, and its length is proportional to the meaning of this variable.

The angles between the vectors representing the set of variables and the principal components (axes) can be used to assess the linear correlation coefficients.

The first axis in Fig. 1 corresponds to the IT infrastructure (represented mainly by X_2—availability of computers in households and X_3—the percentage of households with access to the Internet at home) with the high level of the IT infrastructure toward the left. The second axis is strongly positively correlated with X_1 (the number of mobile-cellular telephone subscriptions).

The points representing the EU-28 countries can be projected perpendicularly onto the vectors, visualizing the variables to obtain the approximate ordering of the countries in order of the increasing level of the IT infrastructure.

When analyzing the information presented in Fig. 1, it can be observed that Bulgaria (BG), Romania (RO) and Greece (EL) are the countries characterized by the lowest percentage of the households with access to computers and the Internet. The highest percentages of the households equipped with computers, with the Internet access and broadband Internet connection at home are identified in the Netherlands (NL), Sweden (SE), Denmark (DK) and also in Luxembourg (LU) and Finland (FI).

The analysis results show that the last two of the mentioned above are countries with a relatively high number of mobile subscriptions per 100 inhabitants.

The highest number of mobile subscriptions can be observed in Italy (IT), Austria (AT) and Lithuania (LT), while the lowest in France (FR). However, it should be highlighted that at the same time, all these countries are characterized by the average percentage of households equipped with computers and the Internet access.

The countries located close to the origin of the coordinate system on the biplot are characterized by the average level of all considered variables and Poland (PL) belongs this group.

To analyze in more detail the similarities and differences across the EU-28 countries and to assess the diversity of the IT infrastructure individually for each of them, the star plots with ellipses (Fig. 2) are useful. The 10 years for each country are grouped with the ten-pointed star and an ellipse. Each star is labeled with the country identification letters, located at the gravity center of the star.

In addition, Fig. 3 presents star plots for Poland, the Netherlands and Bulgaria, where the respective years are labeled on the rays.

The ellipse shape makes it possible to assess the diversity of the IT infrastructure within each country. The smallest ellipses are observed for the Netherlands (NL) and for Denmark (DK)—this means these countries are characterized by the smallest variability in the IT infrastructure development over the period from 2008 to 2017.

Almost all of the ellipses have elongated horizontal axes; however, they are oriented differently. For example, the ellipses for Bulgaria (BG) and Romania (RO) are oriented more or less in parallel to the first axis of the analysis, representing the IT infrastructure development. This means a high variability with regard to the computer and Internet availability, while the number of the mobile-cellular telephone subscriptions per 100 inhabitants is relatively stable. The ellipse for Poland (PL) is oriented left-up, which means rather high variability in the analyzed years due to the number of the mobile-cellular telephone subscriptions, but there is also a noticeable increase in the percentage of households with access to computer and Internet at home.

To sum up the results of this analysis, it can be also observed that the rays of each star that are corresponding to the individual years can be clustered into three sets. In particular: the most to the right—the years 2008–2011 and the most to the left—the years 2015–2017. It is equivalent to the noticeable improvement in the IT infrastructure access in the analyzed period.

The between-countries PCA was complemented by the within-countries PCA. The within-class PCA is based on the residuals between observed data and the groups' means, and it aims at looking for structures remaining in the data after removing differences between groups of the samples.

The results of the within-countries PCA are presented in Fig. 4. The spatial effect was removed. The within-countries inertia is equal to 1.41, i.e., 35.20% of the total inertia is due to the within PCA.

The annual variations for the three selected years shown in Fig. 4 (the row scores grouped by years) reveal a rather strong temporal structure. This means that after removing the differences related to the spatial effect, there is also fairly clear structure in the data related to the temporal effect. The percentages of households with respectively computers, access to the Internet at home and broadband Internet connection type increase year by year. Also, a decrease in the gap across the EU-28 countries due to the ICT access in the studied years is noticeable.

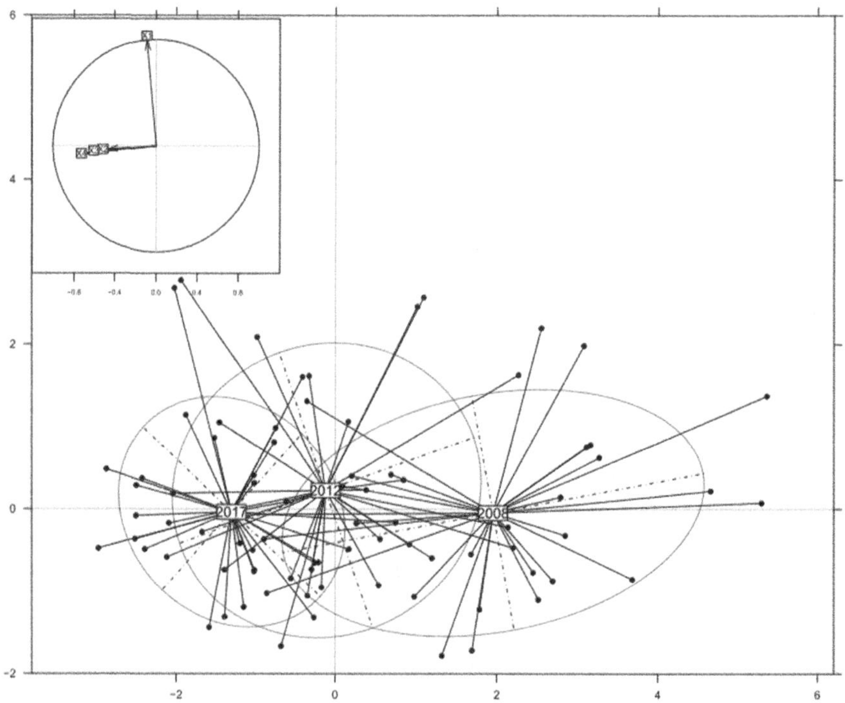

Fig. 4 Within-countries PCA results—the annual variation in 2008, 2012, 2017 after removing the spatial effect; top-left panel: correlation circle. *Source* Own calculations

Table 1 Position of Poland in the ICT access ranking for the UE-28 countries in 2008, 2012 and 2017

Position	2008	2012	2017
1.	Luxembourg	Finland	Austria
2.	Netherlands	Austria	Luxembourg
3.	Finland	Denmark	Germany
…	…	…	…
11.	Lithuania	**Poland**	**Poland**
…	…	…	…
21.	**Poland**	Cyprus	Lithuania
…	…	…	…
26.	Greece	Bulgaria	Croatia
27.	Bulgaria	Greece	Greece
28.	Romania	Romania	Bulgaria

Source Own calculations

In Table 1, selected results of the linear ordering due to the ICT access across the UE-28 countries in the chosen years (2008, 2012, 2017) are shown. The top three, the bottom three and Poland's positions in the ranking were highlighted.

In 2008, Poland was in the 21st position, then moved to the 12th in 2009, and since then was listed in the middle of the rankings between the 11th and 13th positions, respectively, in years: 2010 and 2011—the 12th; 2012 and 2013—the 11th; 2014 and 2015—the 13th and got back in 2016 to the 11th.

Throughout the entire studied period, the leading countries in the ICT access rankings are: Luxembourg, Finland and Austria, while Greece, Bulgaria and Romania the opposite ones.

3.2 *The Analysis of the Lack of Internet Access Reasons in Years 2010–2017*

The following six variables were taken into consideration for the analysis of the lack of Internet access reasons:

1. The access costs are too high (X_1);
2. Have access elsewhere (X_2);
3. The equipment costs are too high (X_3);
4. Access not needed: content is not useful, not interesting, etc. (X_4);
5. Privacy or security concerns (X_5);
6. Lack of skills (X_6).

The complete data were available only for the years 2010–2017; therefore, the analysis was conducted for this period.

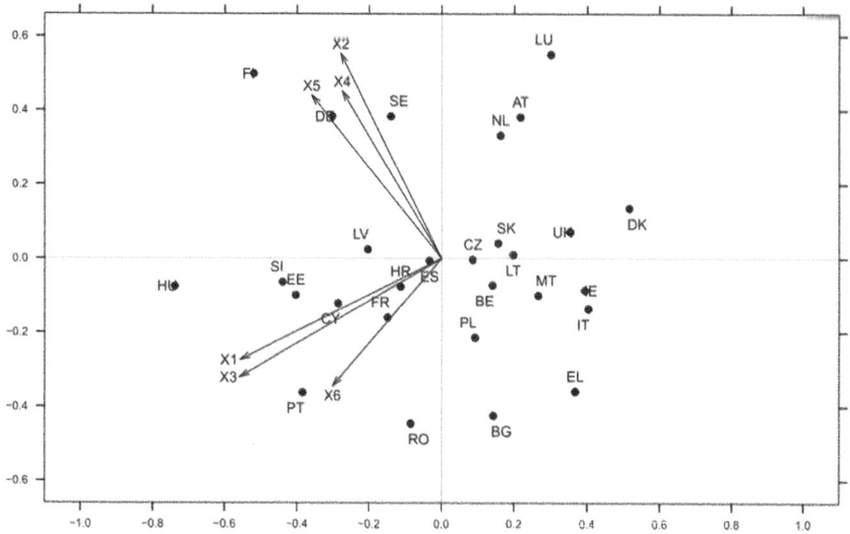

Fig. 5 Between-countries PCA biplot (X_1–X_6: reasons for not having internet at home). Legend as in Fig. 1. *Source* Own calculations

The total inertia in the classical PCA equals six. In the presented analysis, the between-class inertia is equal to 4.01, i.e., 66.89% of the total inertia is due to the spatial factor. The results of the analysis are summarized with the use of a biplot in Fig. 5.

As it is clearly seen, there are two groups of variables on the biplot: X_1 (the access costs are too high), X_3 (the equipment costs are too high) with X_6 (lack of skills) and X_2 (have access elsewhere), X_4 (access not needed), X_5 (privacy or security concerns). Within each of them, strong positive correlation between the variables occurs, and there is no correlation between the groups. The first group of variables (X_1, X_3, X_6) concerns the high costs of Internet access and lack of skills, while the second one (X_2, X_4, X_5) describes the other reasons for the lack of the Internet access at home.

The most strongly correlated with the first PCA axis variables are X_1 and X_3, which means this axis represents the costs of Internet access gradient, with the second axis, respectively X_2 followed by X_4 and X_5.

As it is seen on the biplot, clear clusters of the EU-28 countries related to different reasons of not having the Internet access at home can be observed.

The lack of the Internet access at home due to the high costs and lack of skills relates, inter alia, to Hungary (HU), Portugal (PT) and Romania (RO). Also in the case of Poland (PL), the lack of the Internet access at home is associated with the high costs and lack of skills, while security concerns and access in other places are less important. On the other hand, in countries such as: Finland (FI), Germany (DE), Sweden (SE), Luxembourg (LU), Austria (AT) and the Netherlands (NL), the reasons of not having the Internet access at home are motivated by the Internet

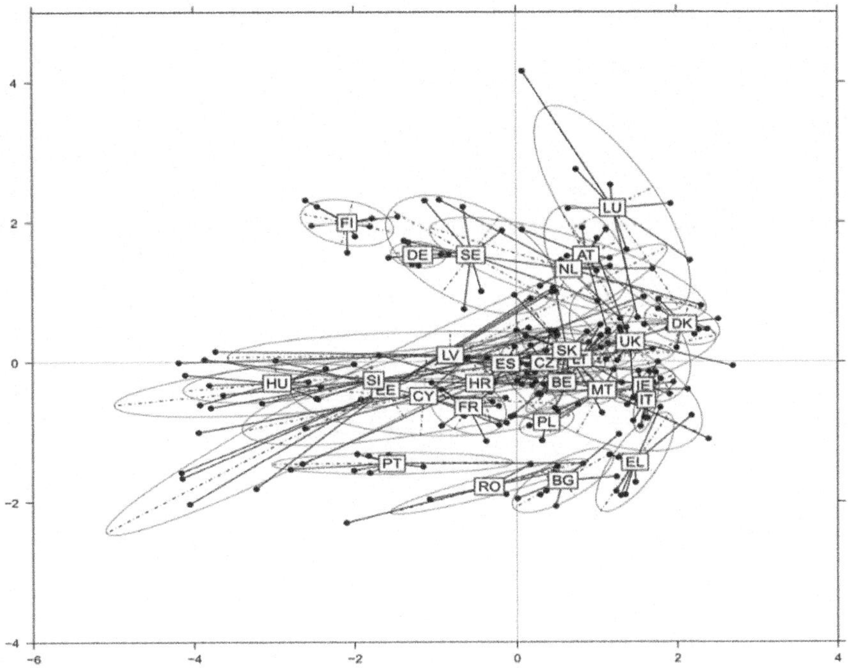

Fig. 6 Diversity of the lack of Internet access reasons in 2010–2017 for all the EU countries. Legend as in Fig. 1. *Source* Own calculations

access availability in other places, the security and privacy concerns and no need of the Internet connection. In this group of countries, costs of the Internet access or lack of skills particularly are not important.

The star plots with ellipses for all EU countries are presented in Fig. 6. The smallest ellipses are observed for Germany (DE), Denmark (DK), Finland (FI), as well as for Poland (PL)—this means the smallest variability in the reasons for not having Internet access at home over the period 2010–2017.

The ellipses for Estonia (EE) and Latvia (LV) with much elongated shape are also very characteristic.

The between-countries PCA was complemented by the within-countries PCA. The within-countries inertia is equal to 1.99, i.e., 33.11% of the total inertia is due to the within PCA.

After removing the differences related to the spatial effect, there is some, not so strong structure in the data related to the temporal effect (see Fig. 7, the row scores are grouped by years for 2010, 2013 and 2017).

When analyzing the results, the conclusion arises that the percentage of the individuals, who do not use the Internet because of the lack of skills (X_6) and to a lesser extent also due to the lack of need to use it (X_4), increases year by year.

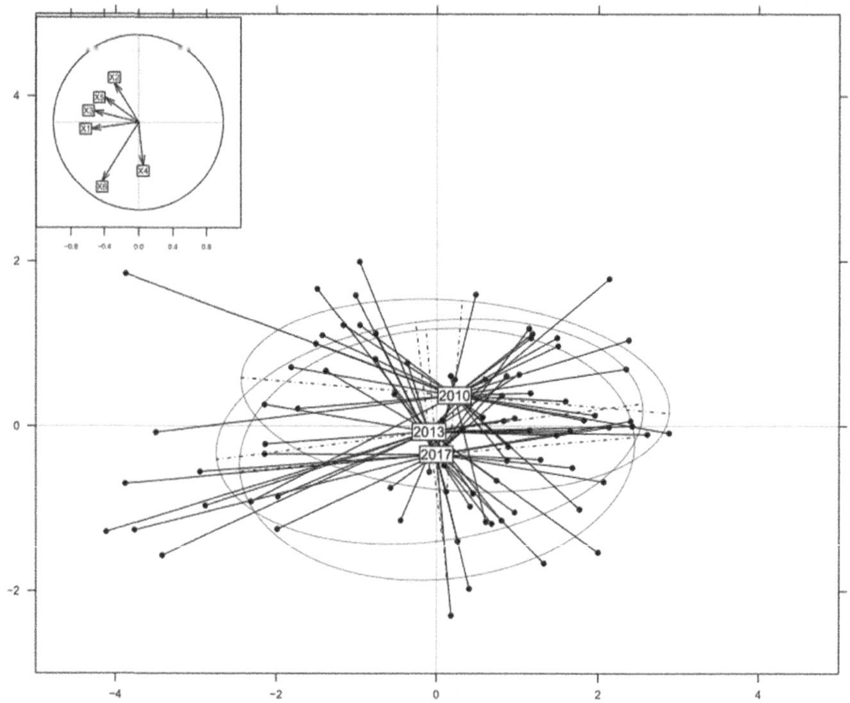

Fig. 7 Within-countries PCA results—the annual variation in 2010, 2013, 2017 after removing the spatial effect; top-left panel: correlation circle. *Source* Own calculations

3.3 The Analysis of the Internet Activities in Years 2010–2017

The following nine variables were taken into consideration for the analysis of the Internet activities:

1. Sending/receiving e-mails (X_1);
2. Telephoning or video calls (X_2);
3. Participating in social networks (X_3);
4. Finding information about goods and services (X_4);
5. Reading online news sites/newspapers/news magazines (X_5);
6. Internet banking (X_6);
7. Travel and accommodation services (X_7);
8. Selling goods or services (X_8);
9. Interaction with public authorities—the last 12 months (X_9).

The complete data were available only for the years 2010–2017, therefore the analysis conducted for this period.

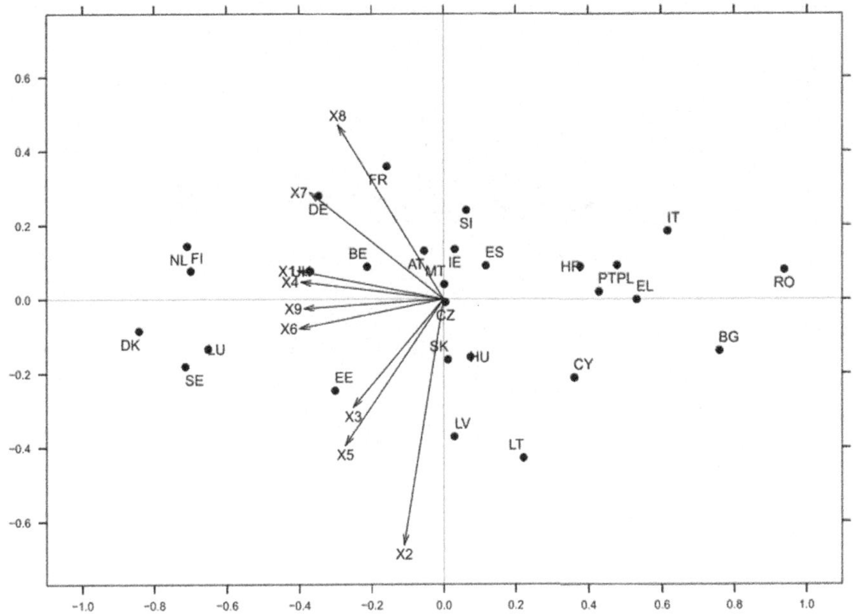

Fig. 8 Between-countries PCA biplot (X_1–X_9: Internet activities). Legend as in Fig. 1. *Source* Own calculations

The total inertia in the classical PCA (performed on the combined datasets) equals nine. In this analysis, the between-class inertia is equal to 7.20, i.e., 80.00% of the total inertia is due to the spatial factor.

The results of the between-class PCA for groups corresponding to the EU-28 countries are summarized using a biplot in Fig. 8.

With the first PCA axis, the most correlated are the following variables: X_1 (sending/receiving e-mails), X_4 (finding information about goods and services), X_6 (Internet banking), X_9 (interaction with public authorities), and respectively with the second one: X_2 (telephoning or video calls) and X_8 (selling goods or services).

Figure 8 clearly presents clusters of the EU-28 countries related to the various activities of the individuals in the Internet. Particularly, the groups of countries ordered along the first PCA axis can be identified. On the left, there are countries in which the most popular performed in the Internet activities are: sending/receiving e-mails (X_1), finding information about goods and services (X_4), Internet banking (X_6) and interaction with public authorities (X_9). In this group, the following countries are included: Denmark (DK), Sweden (SE), the Netherlands (NL), Finland (FI), Luxembourg (LU) (Scandinavian countries and Benelux). Moreover, users in the mentioned cluster are very engaged in all considered Internet activities.

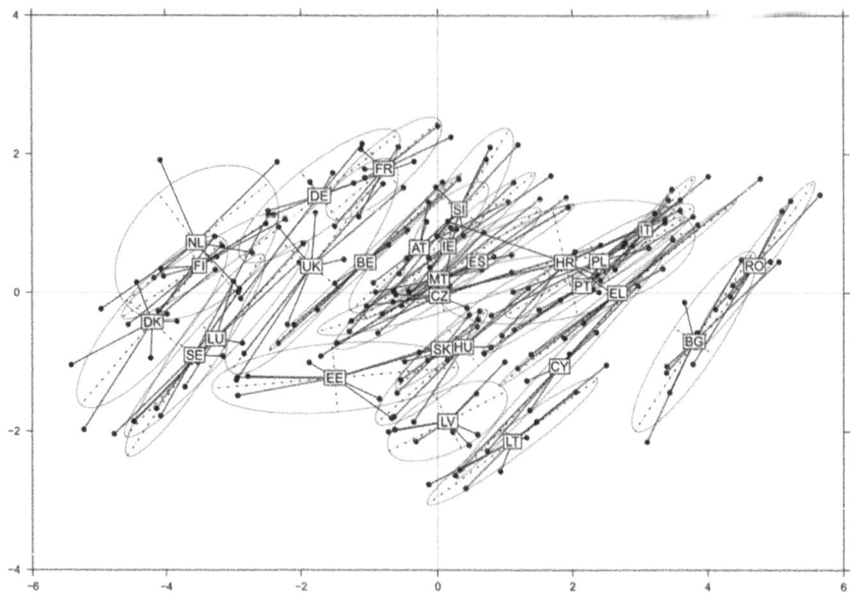

Fig. 9 Diversity of the internet activities in 2010–2017 for all the EU countries. Legend as in Fig. 1. *Source* Own calculations

On the right of the first PCA axis, the group of countries with users characterized by the lowest involvement in the Internet-analyzed activities is observed. This cluster includes Poland (PL) with Romania (RO), Bulgaria (BG), Italy (IT), Greece (EL), Portugal (PT), Hungary (HU) and Cyprus (CY). It should be noticed that the telephoning or video calls (X_2) are the most popular Internet activities in the Baltic countries: Lithuania (LT), Latvia (LV) and Estonia (EE).

The star plots with ellipses for all EU countries are presented in Fig. 9.

Almost all of the ellipses are right-up oriented. This indicates that in the analyzed years, a high variability with regard to telephoning or video calls (X_2) and selling goods or services (X_8) is observed.

The between-countries PCA was complemented by the within-countries PCA. The within-countries inertia is equal to 1.80, i.e., 20.00% of the total inertia is due to the within PCA. The annual variation reveals some temporal structure so after removing the differences related to the spatial effect, there is some, not so strong structure in the data related to the temporal effect (see Fig. 10 for the three selected years).

When analyzing the results, it can be stated that year by year, the percentage of the individuals using Internet for more and more various forms of activities increases.

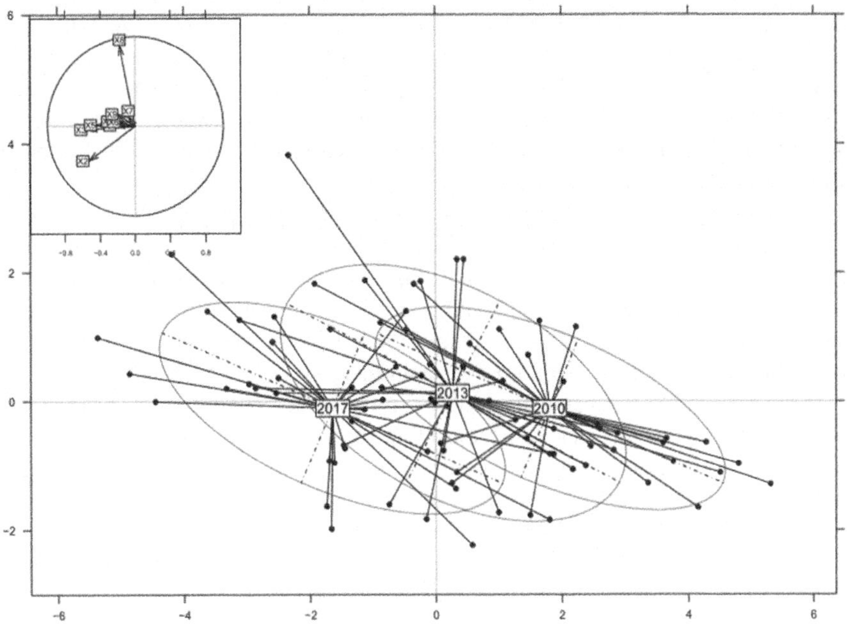

Fig. 10 Within-countries PCA results—the annual variation in 2010, 2013, 2017 after removing the spatial effect; top-left panel: correlation circle. *Source* Own calculations

4 Conclusions

The obtained results and their graphical representations allowed to successfully meet the research goals formulated in the Introduction of the paper. The chosen multivariate statistical analysis methods provided the assessment of the ICT development across space of the UE-28 countries and over the selected time from 2008 to 2017. Also, the ICT development course in Poland in comparison to other European Union countries and the UE-28 average level could be investigated.

To the best knowledge of the authors, there are no studies about the ICT development and no analyses concerning the reasons of not having the Internet connection or types of online activities, which are based on the between and within PCA. The PCA-based methods seem to have wide potential for use in the analysis of the ICT development and related issues. The factorial maps analysis enabled to identify groups of the EU countries similar due to the ICT development, reasons for not having access to the Internet at home and types of Internet activities, as well as to determine the relationships between the analyzed variables.

The dynamics of the ICT development in Poland was similar to the average dynamics of the UE-28, with the average rate of change being slightly faster than in the EU-28. Poland in the ranking of the ICT development level was on the 21st position in 2008, but in the years 2009–2017 was always placed not higher than the 11th and not lower than the 13th rank, so it can be concluded that regarding the

ICT access and the IT infrastructure, Poland was also at the average level within the EU-28 in the studied years. The most developed countries due to the ICT access are: Luxembourg, Finland and Austria, and respectively the least: Bulgaria, Romania, Portugal, Croatia and Spain. Regarding the IT infrastructure, Sweden, Denmark, the Netherlands, Luxembourg, Finland and UK are opening the ranking, while Romania, Bulgaria and Greece are at the end of it.

In all European Union countries, over the considered period, the ICT access expanded and the individual users activities performed in the Internet evolved a lot—all users were more active in the Internet. Also, significant changes were observed in the structure of the reasons of not having access to the Internet at home. The security and privacy concerns are more important for users in countries placed in the top of the rankings than in those with more distant ranks.

In general, the less developed countries were characterized by higher variability than the developed ones regarding the IT infrastructure and the ICT access over the studied years. Concerning the mobile-cellular telephone subscriptions, the individuals in Austria, Italy and Lithuania are leading, while French and Irish users in comparison to the UE-28 individuals use them the least.

A visible breakdown occurs in reasons of not having Internet access, clearly two groups can be identified. The first one includes the beliefs that the cost of access and equipment is too high and the lack of skills cause, the second—privacy and digital security concerns and no need at all motive. It is worth to mention that the percentage of individuals not having the Internet access because of the lack of skills, increases year by year.

The structure of the Internet activities varies a lot in different countries. Polish users are among the least active ones, this group also includes users from: Bulgaria, Romania, Spain, Portugal, Italy, Cyprus and Hungary. The most active in the Internet are individuals from Denmark, Sweden, the Netherlands, Finland and Luxembourg, and they most frequently use e-mail accounts, interact with the e-administration and e-banking, look for the information about goods and services. In Germany and France, the tourist services bookings and information in accompanying with e-shopping are the most popular Internet activities. Lithuania, Latvia and Estonia are characterized by big interest in the phone and video connections via Internet. The individuals from Denmark, Sweden and Luxembourg are the most active in the social media and information services. In all of the EU-28 countries, ways of using Internet and its resources are expanding year by year.

The UE-28 countries are undoubtedly at different levels of the ICT development, but the growth of the ICT access and usage by individuals is noticeable in all of member countries. On the basis of the conducted analyses, Poland is a country which progressed a lot in 2009—by making up nearly the half of the distance to the ranking leaders (from rank 21 to 12), however since then stands still in the rankings. In other words, in the considered period in comparison the UE-28, the ICT development in Poland was not fast and intense. In published by the European Commission report on the Digital Economy and Society Index (DESI)[7] for 2018, Poland's general rank (based on the following five categories/areas: connectivity, human capital, use of Internet services, integration of digital technology, digital public services) is very low, namely the 24th. Regarding the connectivity, Poland is on the 21st position and in the terms of Use of Internet Services—on the 25th. It is worth to mention that in the DESI Report, out of the UE-28, Poland is outstanding in Mobile Broadband Take-up: the 2nd rank and performs the worst in the terms of Fixed Broadband Coverage (% of households): the last rank.

The ICT development is perceived as a key to the development of the Polish society and modern business. The government, especially The Ministry of Digital Affairs as well as many different non-governmental organizations and independent experts, highlight the need of developing the broadband infrastructure, enhancing the e-administration access and improving digital competences in Poland. In this context, it is also important to delve into advanced statistical methods and search for most effective ones to study thoroughly the ICT development in a multidimensional perspective.

It should be noted that several approaches to analyze the spatial-temporal data can be considered. The authors opted for the methods of exploratory data analysis, with the awareness of their numerous advantages and disadvantages. The possibility of graphical presentation of the results with the use of scatterplots and biplots and the potentiality of drawing conclusions directly from the investigation of the factorial maps, where the use of formal mathematics is minimal, can be listed as the strongest advantages. On the other hand, as discussed above, the applied between-countries PCA is based on the table of group means. Averaging the data over time may lead to the loss of relevant information. It is a serious disadvantage of this approach. Therefore, more formal methods of spatial-temporal data analysis should also be considered to be applied, including the PCA for functional data (see e.g. Górecki and Krzyśko 2012) or the nonlinear PCA (see, e.g., Krzyśko et al. 2018).

The presented paper is an introduction to the further multivariate statistical research on the ICT development and related issues—in both, individual users and enterprises perspectives, within Poland as well as Poland across Europe.

References

Bąk A (2013) Metody porządkowania liniowego w polskiej taksonomii—pakiet pllord. Pr Nauk Uniw Ekon Wroc 287:54–62

Becker A (2018) Wykorzystanie technologii informacyjno-telekomunikacyjnych w przedsiębiorstwach w ujęciu wojewódzkim. Wiad Stat 3:69–82

[7] Digital Economy and Society Index 2018, Country Report Poland, http://ec.europa.eu/information_society/newsroom/image/document/2018-20/pl-desi_2018_-_country_profile_eng_B440E0DD-F8E8-B007-4A97A5E2BE427B1F_52233.pdf. Accessed 20 Sep 2019.

Carretero S, Vuorikari R, Punie Y (2017) DigComp 2.1: the digital competence framework for citizens with eight proficiency levels and examples of use. JRC Report, EUR 28558 EN. https://doi.org/10.2760/38842

Cheba K, Saniuk S (2016) Wielowymiarowa analiza wykorzystania technologii ICT w przedsiębiorstwach w Polsce. Przedsiębiorczość i Zarządzanie 12(1):41–53

Dufour AB (2008) Within PCA and between PCA. http://pbil.univ-lyon1.fr/R/pdf/course4.pdf. Accessed 25 Mar 2019

Gower JC, Le Roux NC, Gardner-Lubbe S (2015) Biplots: quantitative data. WIREs Comput Stat 7:42–62

Górecki T, Krzyśko M (2012) Functional principal components analysis. In: Pociecha J, Decker R (eds) Data Analysis Methods and its Applications. C.H. Beck, Warszawa, pp 71–87

GUS (2018) Społeczeństwo informacyjne w Polsce. Wyniki badań statystycznych z lat 2014–2018. GUS, Warszawa, Szczecin

Jorgenson DW, Vu KM (2016) The ICT revolution, world economic growth and policy issues. Telecommun Policy 40:383–397

Kaliszczak L, Pawłowska-Mielech J (2019) Nowoczesne technologie informacyjno-komunikacyjne jako determinanta rozwoju MSP. Nierówności Społeczne a Wzrost Gospodarczy 58:129–140

Khalili F, Lau W, Cheong K (2014) ICT as a source of economic growth in the information age: empirical evidence from ICT leading countries. Res J Econ Bus ICT 9(1):1–26

Krzyśko M, Łukaszonek W, Ratajczak W, Wołyński W (2018) Nonlinear principal component analysis for geographically weighted temporal-spatial data. Folia Oeconomica. Acta Universitatis Lodziensis 4(337):169–181

Łaszek A, Borchmann Ł, Husiatyński M et al (2018) E-rozwój. Cyfrowe technologie a gospodarka, Forum Obywatelskiego Rozwoju, Warszawa

Mastalerz-Kodzis A, Pośpiech E (2018) Ekonomiczno-społeczne uwarunkowania rozwoju społeczeństwa informacyjnego krajów Unii Europejskiej. Nierówności Społeczne a Wzrost Gospodarczy 53:109–118

Miśkiewicz- Nawrocka M, Zeug-Żebro K (2015) Zastosowanie statystyk przestrzennych w analizie dostępności do infrastruktury ICT w Polsce. Nierówności Społeczne a Wzrost Gospodarczy 44:43–55

Próchniak M, Witkowski B (2016) Digitalizacja i internetyzacja a wzrost gospodarczy. DELab UW Working Paper 4. Uniwersytet Warszawski

Sharafat AR, Lehr WH (eds) (2017) ICT-centric economic growth, innovation and job creation. ITU Publications, Geneva, Switzerland

Szczukocka A, Pekasiewicz D (2017) Analiza rozwoju nowych technologii w gospodarstwach domowych w Polsce. Nierówności Społeczne a Wzrost Gospodarczy 52:247–257

Thioulouse J, Dray S, Dufour A-B et al (2018) Multivariate analysis of ecological data with ade4. Springer, New York

Wojnar J (2015) Tempo rozwoju ICT w Polsce oraz syntetyczna ocena dystansu Polski od krajów Unii Europejskiej w zakresie wykorzystania technologii informacyjnych. Nierówności Społeczne a Wzrost Gospodarczy 44:372–381

Vuorikari R, Punie Y, Carretero Gomez S, Van den Brande G (2016) DigComp 2.0: the digital competence framework for citizens. Update phase 1: the conceptual reference model. JRC Report, Luxembourg Publication Office of the European Union, EUR 27948 EN. https://doi.org/10.2791/11517

Sensitivity Analysis in Causal Mediation Effects for TAM Model

Adam Sagan and Mariusz Grabowski

Abstract One of the goals of scientific research is to identify cause-effect relationships, which in many cases are made in non-experimental research design, based on correlation measures or using regression methods. A special case is a structural equation model (SEM) that is often and incorrectly labeled "causal" models. The aim of the paper is to identify causal relationships in relation to technology acceptance models (TAM) (Davis in MIS Q 13:319–340, 1989; Davis et al. in Manage Sci 35:982–1003, 1989) using the analysis of mediation effects and causal dependencies that stem from Markov's causal rule. Identification of causal relationships is made using d-separation (Pearl in Stat Surv 3:96–146, 2009) and sensitivity analysis (Imai et al. in Stat Sci 1:51–71, 2010; Tingley et al. in J Stat Soft 59:1–38, 2013). The aim of this article is to assess the impact of unknown disturbing variables (confounders) affecting both the mediation and focal-dependent variables. The analysis allowed for simulations of correlated disturbances effect of dependent variables in the TAM model on the degree of average causal mediation effect bias. The TAM model was built on the basis of research conducted on a quota sample of 150 students of the Cracow University of Economics.

Keywords Technology acceptance model · Causal mediation · Sensitivity analysis

1 The Concept of Causality

The concept of causality as an essential element of metaphysics was present in science from the beginning of ancient times (Leucippus, Democritus, Aristotle), addressed in the works of Descartes and questioned in more recent times by Hume. Over time, many concepts of causality mutually incomparable and heterogeneous were created

A. Sagan (✉) · M. Grabowski
Cracow University of Economics, Kraków, Poland
e-mail: sagana@uek.krakow.pl

M. Grabowski
e-mail: grabowsm@uek.krakow.pl

K. Jajuga et al. (eds.), *Classification and Data Analysis*, Studies in Classification, Data Analysis, and Knowledge Organization,
https://doi.org/10.1007/978-3-030-52348-0_14

and developed (Osińska 2008). The causality may be considered on various levels of abstraction starting from ontology, through epistemology and finally by being expressed in quantitative terms. The last one is the level of abstraction which is the object of interest presented in this article.

The term "causality" is defined in Meriam-Webster dictionary (https://www.merriam-webster.com/dictionary/causality) as: "the relation between a cause and its effect or between regularly correlated events or phenomena." More formally, when reflecting the causal relationship as a relation, we should consider the following attributes (Osińska 2008): transversality, asymmetry and transitivity. The first attribute indicates that no event can be a cause for itself. Asymmetry means that in the relation, one event is a cause and the second is an effect. The cause logically precedes the effect. Transitivity is related to treating the relations of causes and effects in terms of logically connected chains of events.

At the end of second decade of twenty-first century, the relationships of cause and effects are the objects of research in many fields and disciplines. In psychology and social psychology, causality is related mostly to the concept of potential/counterfactual states (Hume 2000; Holland 1986; Rubin 2005). Sociology and disciplines dealing with highly institutional social problems represent contextual, multilevel approach to identification of causal relationships (Raudenbusch and Bryk 2001; Goldstein 1999; Snijders and Bosker 2012). The economics and econometrics are not the exceptions. In these disciplines, a special consideration is devoted to the definition of causality proposed by Granger (1969). C. W. J. Granger concept of causality is related to time series. It tests the hypothesis whether the previous value of time series is able to predict the following one. C. W. J. Granger causality is being criticized for equating causality with prediction, and for this reason, it is restricted to "predictive causality." Throughout this paper, the authors adapt the terms "parent" and "child" within the meaning of J. Pearls' acyclic graph (DAG) terminology (a node can have two parent nodes).

2 Technology Acceptance Model

Technology acceptance model (TAM) (Davis 1989; Davis et al. 1989) is the most frequently used framework explaining the causal nature of Information and Communication Technology (ICT) users' behavior. The theoretical background for this model constitutes the Theory of Reasoned Action (TRA) (Fishbein and Ajzen 1975; Ajzen and Fishbein 1980) and the Theory of Planned Behavior (TPB) (Ajzen 1985, 1991), (Fig. 1).

According to this theory, the Intention is an essential predictor of Behavior and is shaped by three interlinked factors: Subjective Norm, Perceived Behavioral Control and Attitude. When Perceived Behavioral Control reflects factual influence of a person on the situation, it may be treated as a direct predictor of behavior.

In TAM (Fig. 2), similarly like in TPB, the Actual System Use is determined by the Intention to Use. The Intention, however, is shaped directly by Attitude toward Using

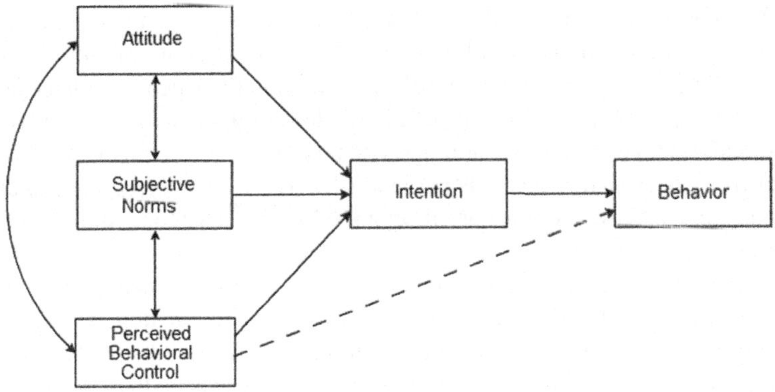

Fig. 1 Theory of planned behavior. *Source* Ajzen (1991, p. 182)

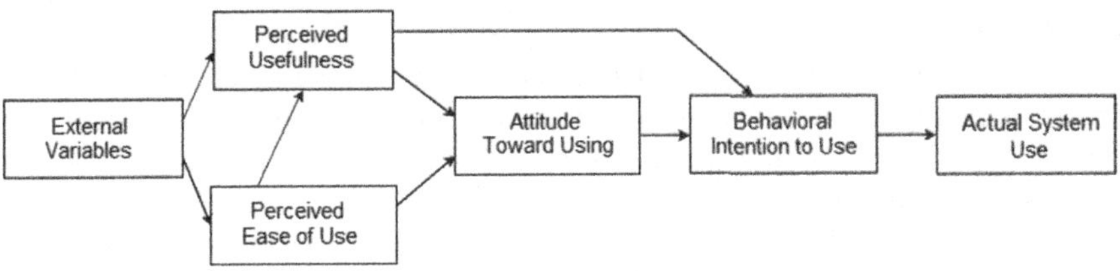

Fig. 2 Technology acceptance model. *Source* Davis et al. (1989, p. 985)

and Perceived Usefulness, and indirectly by Perceived Ease of Use. Additionally, Perceived Ease of Use is a predictor of Perceived Usefulness and both, Perceived Ease of Use and Perceived Usefulness can be influenced by external variables.

Although TAM is rooted in causal in their nature, TRA and TPB, it is rarely analyzed in cause-and-effect terms. In particular there is a lack of published research which focuses on the analysis of the strength of causal relationships between the constructs.

3 Identification of Causal Effects

The identification of causal relationships in TAM is based on the idea of Markov condition, directed separation approach (d-separation), (Pearl 2000, 2009) and sensitivity analysis (Imai et al. 2010; Tingley et al. 2013). The Markov causal rule is frequently used for identification of causal relations in path models with mediators. Using the directed acyclic graph (DAG) terminology, variables presented in Fig. 2, Perceived Usefulness and Perceived Ease of Use, are parents of Attitude Toward Using. Therefore, Attitude Toward Using is a child of Perceived Usefulness and Perceived Ease of Use. Similarly, Behavioral Intention to Use is a child of Attitude

Toward Using. Perceived Usefulness, Perceived Ease of Use and Attitude Toward Using are ancestors of Behavioral Intention to Use. The Markov condition is defined as follows: "every variable with parents is unconditionally dependent on its parents" (Mulaik 2009, p. 114; Pearl 2000, p. 19). In models with mediators, each variable is conditionally independent of its non-descendants, given its parents ("child is independent from its ancestors given its parents"). Therefore, Perceived Usefulness and Perceived Ease of Use variables are d-separated from Behavioral Intention to Use by Attitude Toward Using.

Sensitivity analysis facilitates an assessment of the influence of unknown disturbing variables (confounders) affecting both the mediation and the focal-dependent variables in SEM model. The confounding effect in the mediation results from the assumption that the unobservable common cause can affect both the mediation and focal-dependent variables in the model. Correlation of disturbances (residuals in the model) allows to control the influence of unobserved common cause. However, in regular (i.e., most commonly used) applications, such a model is non-identifiable (negative number of degrees of freedom). The correlation coefficient between disturbances (ρ) for mediating and dependent variables determines the degree of influence of the unknown common cause and model parameter bias. If $\rho = 0$, then there is no disturbance of the causal effect (no correlation between residuals/disturbances in the model).

The sensitivity analysis is an attempt to understand the mechanism of causal effect (through mediation mechanism). It is related to concept of counterfactual/potential state (Rubin 2005). The causality concept based on counterfactual state on individual level relates existence of individual in factual state [i.e., in experimental group (1)] with its counterfactual state [as if were in control group (0)]. Individual treatment effect Y_i is therefore the difference between causal (treatment) effect in factual and counterfactual state:

$$ITE = Y_i(1) - Y_i(0) \tag{1}$$

If assumptions of Stable Unit Treatment Value Assumption (SUTVA) and Conditional Independence Assumption (CIA) in experimental settings with random assignment of the participants are valid, one can calculate average treatment effect for group level.

$$ATE = E[Y(1) - Y(0)] \tag{2}$$

In order to control the causal effect, the Markov causal condition allows to introduce the mediator (m) and decompose total causal effect into direct causal effect and indirect causal effect. Pure natural direct effect can be expressed as:

$$DE = E[Y(1, M(0)) - Y(0, M(0))|C = c] \tag{3}$$

In case of direct effect, M is changing with respect to control group only and total indirect effect is given as:

$$TIE = E[Y(1, M(1)) - Y(1, M(0))|C = c] \quad (4)$$

Pure indirect effect is given as:

$$PIE = E[Y(0, M(1)) - Y(0, M(0))|C = c] \quad (5)$$

Total effect is a sum of both direct and indirect effects.

$$TE = E[Y((1) - Y(0))|C = c] \quad (6)$$

The evaluation of the causal effect using the evaluation of total indirect effect is performed in the simulation analysis for various fixed levels of ρ coefficient. It allows comparing the effect of mediation for $\rho = 0$ with the coefficient ρ when total indirect effect (TIE) = 0. If the values of the ρ coefficients significantly exceed TIE value, for which coefficient $\rho = 0$, then it can be assumed that the total indirect effect is significant and there is no impact of disturbance correlation on the causal mediation effect. On the other hand, if the values of the ρ coefficients insignificantly exceed TIE value, for which coefficient $\rho = 0$, then it can be inferred that the total indirect effect is irrelevant, and the disturbance correlation influences the causal mediation effect (Muthen and Asparouhov 2015). Sensitivity analysis can be performed by many statistical packages like SENSITIVITY command in Mplus software, *mediation* library in R package, *medsens* library in STATA and PROC CAUSALMED procedure in SAS.

4 Mediation Effects in TAM

Empirical TAM is presented in Fig. 3. This model is a modification of the original TAM and was utilized for an assessment of e-learning Moodle platform used in

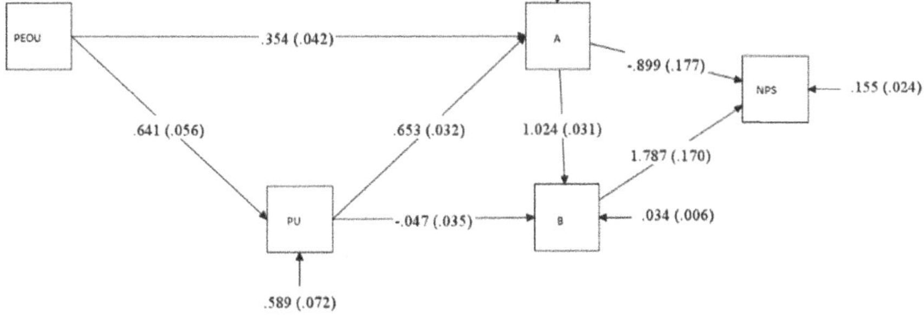

Fig. 3 TAM path coefficients. *Source* Own

Cracow University of Economics, Poland. The quota sample of 149 undergraduate and graduate students (users of Moodle platform) was selected for the study. The measurement of construct indicators under the study was based on 5-point Likert scales. The final set of variables in the path model was operationalized as an optimally weighted sum of scores for underlying constructs in TAM model.

Three sets of mediation dependencies should be distinguished in the model depicted in Fig. 3. The first set of variables determines the relationship between Perceived Ease of Use of the Moodle platform (PEOU), Perceived Usefulness (PU) and Attitudes Toward the Platform Use (A). The second set of variables shapes the relationship between Perceived Usefulness (PU), Attitudes Toward the Platform Use (A) and the Intention to Use the Platform in the Future (B). The third set of variables includes Attitudes Toward the Platform Use (A), Intention to Use the Platform (B) and Overall Platform Recommendations measured by Net Promoter Score (NPS).

Model fit is acceptable. The value of the χ^2 statistic is 10.661 ($df = 3, p = 0.014$). The relative fit (χ^2/df) is 3.55. Statistics describing the incremental fit are NFI = 0.991, RFI = 0.969, TLI = 0.977 and CFI = 0.993, respectively. The value of the root mean square approximation error (RMSEA) is 0.131 (90% p.u = 0.052–0.221). The high RMSEA value is associated with a large random error of the model resulting from the relatively small sample size ($n = 150$ and the small number of degrees of freedom ($df = 3$). This is also demonstrated by wide RMSEA confidence interval and the relatively high probability of a close fit (PCLOSE = 0.46). As indicated by simulation analysis, for models with a small number of degrees of freedom, the RMSEA indicator incorrectly reflects the fit of the model to population data and the other indicators of model fit should be taken into account (Kenny et al. 2014).

Standardized path coefficients are presented in Table 1, whereas correlation coefficients between latent variables are illustrated in Table 2. All correlation coefficients between latent variables are positive and statistically significant. Also, all model parameters, except for the PU-B path, are statistically significant. Comparing regression parameters with appropriate bivariate correlations, one should notice the compatibility of coefficients reflecting the relationship between Perceived Ease of Use (PEOU), Perceived Usefulness (PU) and Attitude Toward the Moodle Platform (A). In the second set of variables, there is a mediation effect—there are insignificant negative direct relationships between PU and B in the path model, but a positive pairwise correlation between factors (0.85).

Table 1 Standardized paths coefficients

	PEOU	PU	A	B
PEOU				
PU	0.641			
A	0.354	0.653		
B		−0.047	1.024	
NPS			−0.899	1.787

Source Own based on AMOS

Table 2 Correlation coefficients between latent variables

	PEOU	PU	A	B	NPS
PEOU	1.00				
PU	.641	1.00			
A	.772	.879	1.00		
B	.749	.854	.983	1.00	
NPS	.677	.693	.857	.904	1.00

Source Own based on AMOS

It means that the direct relationship between PU and B was fully explained by attitudes toward Moodle Platform Use (A). The third set of variables is characterized by a suppression effect resulting from significant negative direct path dependencies between attitudes (A) and recommendations (NPS) but a positive correlation between them (0.857). It can be concluded that the Intentions of Using Moodle Platform in the Future (B) modifies the relationship between Attitudes Toward the Platform Use (A) and Overall Platform Recommendation (NPS).

The direct, indirect and total mediation effects in the relationship between latent variables are presented in Table 3.

All indirect effects are statistically significant as indicated in Table 3. Further analysis of Table 3 points out to the significant role of PU, A and B latent variables as mediators in causal explanation of the relationships in TAM model.

The calculations for the Sobel, Aroian and Goodman significance tests of mediation effects are given in Figs. 4, 5 and 6 obtained by means of test calculator (Preacher and Leonardelli 2019). The Aroian version of the Sobel test is recommended that

Table 3 Partial mediation effects of the TAM model

Paths	Direct effect	Indirect effect	Total effect
PEOU—PU—A	0.354*	0.418*	0.772*
PU—A—B	−0.047	0.668*	0.621*
A—B—NPS	−0.899*	1.830*	0.931*

Source Own based on AMOS.
* p< 0.05

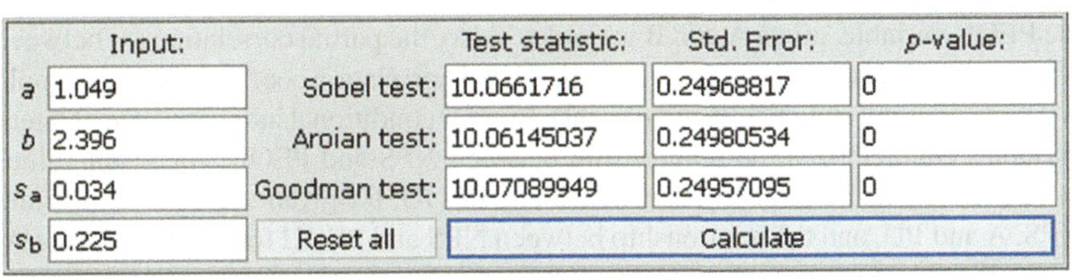

Fig. 4 Mediation tests of A-B-NPS relationship. *Source* Own

Input:			Test statistic:	Std. Error:	p-value:
a	0.605	Sobel test:	4.49254714	0.14126619	0.00000704
b	1.049	Aroian test:	4.4851802	0.14149822	0.00000729
s_a	0.036	Goodman test:	4.4999505	0.14103377	0.0000068
s_b	0.225	Reset all	Calculate		

Fig. 5 Mediation tests of PU-A-B relationship. *Source* Own

Input:			Test statistic:	Std. Error:	p-value:
a	0.993	Sobel test:	8.42392678	0.0713165	0
b	0.605	Aroian test:	8.41278262	0.07141097	0
s_a	0.102	Goodman test:	8.43511535	0.07122191	0
s_b	0.036	Reset all	Calculate		

Fig. 6 Mediation tests of PEOU-PU-A relationship. *Source* Own

is suggested in Baron and Kenny (1986) because it does not make the unnecessary assumption that the product of s_a and s_b is vanishingly small.

The size of the quota sample (150 students) determined the use of the calculation method, given the Monte Carlo study suggests that the Sobel and Aroian test (MacKinnon et al. 1995) converges with sample sizes greater than 50.

5 D-Separation of Causal Relationships in TAM

The principle of directed separation (d-separation) presented above is the basis for identifying causal relationships arising from non-cyclic directed graph theory (DAG).

Taking into account the Markov causality principle, the DAG analysis allowed the identification of individual d-separation conditions. These conditions are shown in Table 4.

The results of the analysis of d-separation of causal relationships in the TAM are presented in Table 4. We observe from Table 4 that NPS variable is independent of the PEOU variable, given A and B variables. Also, the partial correlation (r) between NPS and PEOU is 0 in the population (the partial correlation coefficient in the sample is 0.113 and is not statistically significant). Similar conditional independence of latent variables characterizes the relationship between NPS and PEOU, for given values of A and PU, the relationship between B and PEOU, for given values of A, PU and NPS, A and PU, and the relationship between NPS and PEOU for given values of A, B and PU.

Table 4 Conditions of d-separation

D-separation conditions	*r*	*t*	*p*-level
NPS ∥ PEOU ∣ A, B	0.113	1.550	0.123
PU ∥ NPS ∣ A, B	−0.226	−2.729	0.006
PU ∥ NPS ∣ PEOU, A, B	−0.212	−2.605	0.010
NPS ∥ PEOU ∣ A, PU	0.015	0.175	0.861
B ∥ PEOU ∣ A, PU	−0.970	−1.180	0.240
B ∥ PEOU ∣ NPS, A, PU	−0.139	−1.681	0.095
NPS ∥ PEOU ∣ A, B, PU	0.100	1.209	0.226

Source Own based on DAGitty in AMOS

6 Sensitivity Analysis of TAM

The isolated causal dependencies based on Markov's causal rule and DAG theory rely on the effects of mediation to control the relationship between the independent and dependent variables in a given analytical system. These relationships do not take into account the potential impact of disturbing variables that can affect both the mediator and the dependent variable. In order to control the confounding effect, a direct natural sensitivity analysis (pure natural direct effect) and a total indirect effect (total natural indirect effect) were estimated because of potential bias resulting from the covariance between unobserved disturbances (ρ). The results of a simulation analysis of the impact of bias on the values of direct and indirect effects for the first set of mediation relationships (PEOU-PU-A) can be seen in Figs. 7 and 8.

Figure 7 shows that the pure natural direct effect of PEOU on A, for a set factor $\rho = 0$, is −2.11. The lower bound of the confidence interval of the ρ coefficient when the direct effect is 0 is −0.87. Comparing the direct effect for $\rho = 0$ with the

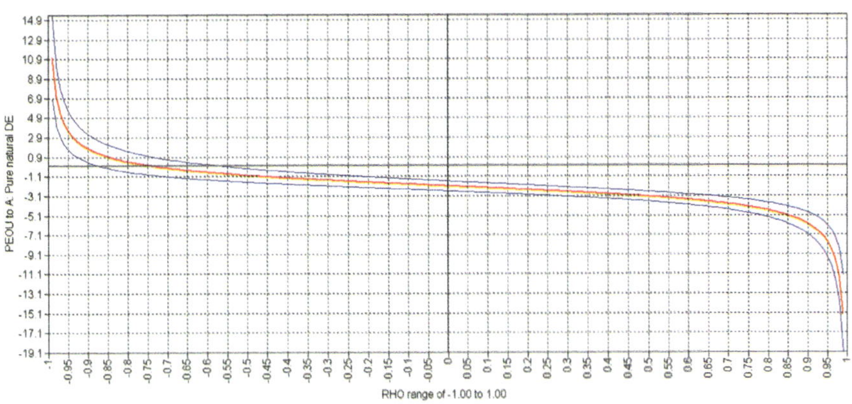

Fig. 7 Sensitivity analysis of the PEOU-PU-A set for direct effect. *Source* Own based on Mplus 8.2

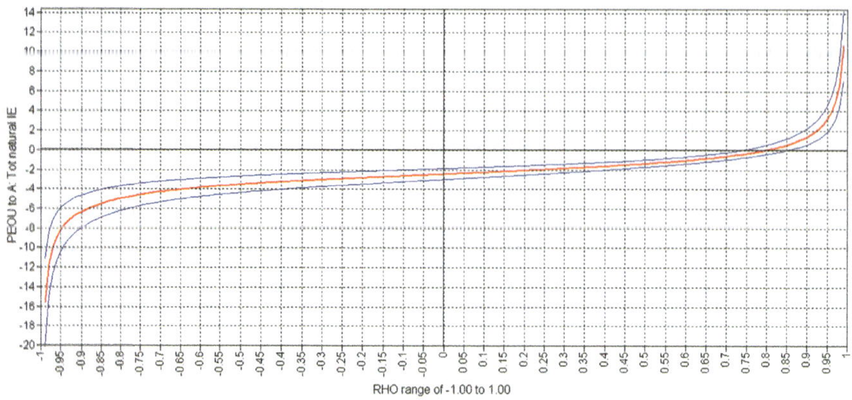

Fig. 8 Sensitivity analysis of the PEOU-PU-A set for indirect effect. *Source* Own based on Mplus 8.2

coefficient ρ when direct effect $= 0$, it can be concluded that the pure direct effect is devoid of influence of disturbances correlation.

Figure 8 shows that the total indirect effect of PEOU on A through the PU mediator variable, for fix $\rho = 0$, is -2.26. The lower bound of the confidence interval of the ρ factor when the indirect effect is 0 is 0.85. Comparing the indirect effect for $\rho = 0$ with the coefficient ρ when indirect effect $= 0$, it can be assumed that the total indirect effect is devoid of disturbances correlation.

In summary, the analysis of d-separation and sensitivity indicates a causal relationship between the perceived easiness of use (PEOU) the Moodle system on positive attitudes towards the platform use (A).

Figures 9 and 10 present the results of a simulation analysis of the impact of bias on the value of direct and indirect effects for the second set of mediation relationships (PU-A-B).

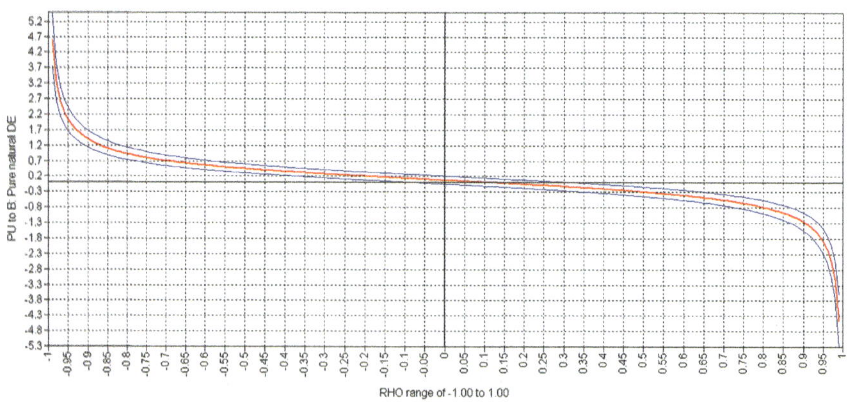

Fig. 9 Sensitivity analysis of the PU-A-B variable set for direct effect. *Source* Own based on Mplus 8.2

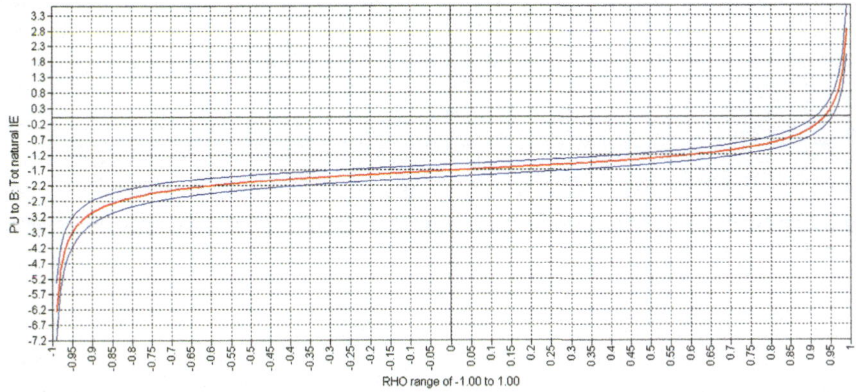

Fig. 10 Sensitivity analysis of the PU-A-B variable system for indirect effect. *Source* Own based on Mplus 8.2

The pure natural direct effect of PU on B, for $\rho = 0$, is 0.12 (Fig. 9). The lower bound of the confidence interval of the ρ coefficient when the direct effect is 0 is only -0.13. Comparing the direct effect for $\rho = 0$ with the coefficient ρ when direct effect $= 0$, it can be concluded that the pure direct effect is strongly "contaminated" by the influence of correlation of disturbances.

The total indirect effect of PU on B through the mediation variable A, for given $\rho = 0$, is -2.1 (Fig. 10). The lower bound of the confidence interval of ρ coefficient when the direct effect is 0 is 0.97. Comparing the indirect effect for $\rho = 0$ with the coefficient ρ when indirect effect $= 0$. Therefore, it can be assumed that the total indirect effect is devoid of disturbance correlation.

In summary, the analysis of d-separation and sensitivity indicates no direct relationship between PU and B. This relationship is explained by unbiased mediation effects of PU-A and A-B

Figures 11 and 12 present the results of a simulation analysis of the impact of bias effect on the value of direct and indirect effects for the third set of mediation relationships (A-B-NPS).

The pure natural direct effect of A on NPS, for given $\rho = 0$, is 4.24 (Fig. 11). The lower bound of the confidence interval of the ρ coefficient when the direct effect is 0 is only -0.25. Comparing the direct effect for $\rho = 0$ with the factor ρ when direct effect $= 0$, it can be concluded that the pure direct effect is "contaminated" by the influence of correlation of disturbances.

Figure 12 shows that the total indirect effect of PU on B through the mediation variable A, for a set factor $\rho = 0$, is -4.7. The lower bound of the confidence interval of the ρ coefficient when the indirect effect is 0 is 0.75. Comparing the indirect effect for $\rho = 0$ with the coefficient ρ when indirect effect $= 0$, it can be assumed that the total indirect effect is devoid of disturbance correlation.

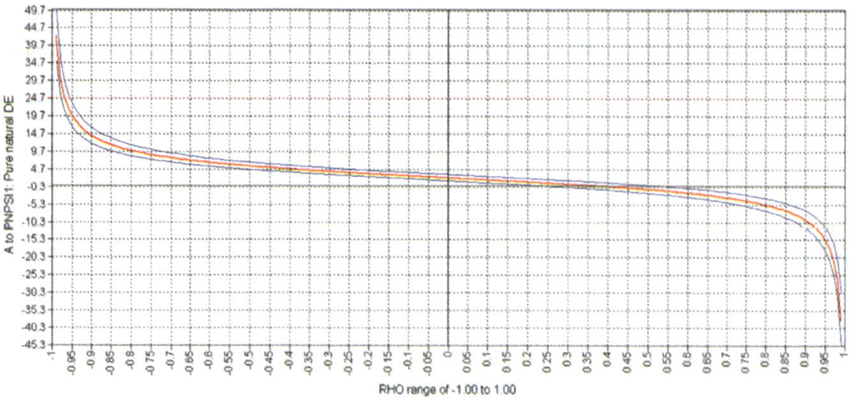

Fig. 11 Sensitivity analysis of the A-B-NPS variables set for direct effect. *Source* Own based on Mplus 8.2

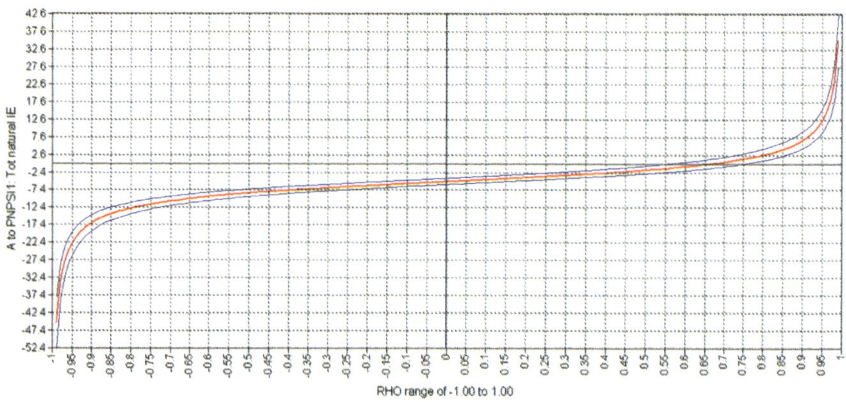

Fig. 12 Sensitivity analysis of the A-B-NPS variables set for indirect effect. *Source* Own based on Mplus 8.2

In summary, the analysis of d-separation and sensitivity indicates no direct relationship between A and NPS. This relationship is explained by the unbiased mediation effects of A-B and B-NPS, which are not affected by correlated disturbances.

7 Conclusions

In summary, the analysis of d-separation and sensitivity check proves a causal relationship between perceived easiness of use of the Moodle system (PEOU) and positive attitudes (A) toward the platform.

On the other hand, it indicates no direct relationship between Perceived Usefulness (PU) and the Intention to Use the Platform in the Future (B). This relationship is explained by the unbiased mediation effects of PU-A and A-B. So, A is a true cause of B.

Also, there is no direct relationship between A and Net Promoter Score (NPS). This relationship is explained by the mediation effects of A-B and B-NPS, which are not affected by correlated disturbances.

The overall causality analysis shows the appropriate specification of TAM; however, the path coefficients cannot be regarded within the causal framework without taking into account mediation effects that help to explain the spurious regressions between constructs of TAM model.

References

Ajzen I (1985) From intentions to actions: a theory of planned behavior. In: Kuhl J, Beckmann J (eds) Action control: from cognition to behavior. Springer, Berlin, Heidelberg, New York

Ajzen I (1991) The theory of planned behavior. Org Beh Hum Dec Proc 50:179–211

Ajzen I, Fishbein M (1980) Understanding attitudes and predicting social behavior. Prentice-Hall, Englewood Cliffs, NJ

Baron RM, Kenny DA (1986) The moderator-mediator variable distinction in social psychological research: conceptual, strategic, and statistical considerations. J Pers Soc Psychol 51(6):1173–1182. https://doi.org/10.1037/0022-3514.51.6.1173

Davis FD (1989) Perceived usefulness, perceived ease of use, and user acceptance of information technology. MIS Q 13(3):319–340. https://doi.org/10.2307/249008

Davis FD, Bagozzi RP, Warshaw PR (1989) User acceptance of computer technology: a comparison of two theoretical models. Manage Sci 35(8):982–1003. https://doi.org/10.1287/mnsc.35.8.982

Fishbein M, Ajzen I (1975) Belief, attitude, intention, and behavior: an introduction to theory and research. Addison-Wesley, Reading, MA

Goldstein H (1999) Multilevel statistical models. Institute of Education, London

Granger CWJ (1969) Investigating causal relations by econometric models and cross-spectral methods. Econometrica 37(3):424–438. https://doi.org/10.2307/1912791

Holland PW (1986) Statistics and causal inference. J Am Stat Assoc 81(396):945–960

Hume D (2000) An enquiry concerning human understanding. Clarendon Press, Oxford, UK

Imai K, Keele L, Yamamoto T (2010) Identification, inference and sensitivity analysis for causal mediation effects. Stat Sci 1:51–71

Kenny DA, Kaniskan B, McCoach B (2014) The performance of RMSEA in models with small degrees of freedom. Soc Meth Res 44(3):486–507. https://doi.org/10.1177/0049124114543236

MacKinnon DP, Warsi G, Dwyer JH (1995) A simulation study of mediated effect measures. Multivar Beh Res 30(1):1–23. https://doi.org/10.1207/s15327906mbr3001_3

Mulaik SA (2009) Linear causal modeling with structural equations. Chapman and Hall/CRC, Boca Raton

Muthen B, Asparouhov T (2015) Causal effects in mediation modeling: an introduction with applications to latent variables. Struct Equ Mod Multi J 22(1):12–23. https://doi.org/10.1080/10705511.2014.935843

Osińska M (2008) Ekonometryczna analiza zależności przyczynowych. Wydawnictwo Naukowe Uniwersytetu Mikołaja Kopernika, Toruń

Pearl J (2000) Causality. Cambridge University Press, Cambridge

Pearl J (2009) Causal inference in statistics: An overview. Stat Surv 3:96–146. https://doi.org/10.1214/09-SS057
Preacher KJ, Leonardelli GJ (2019) An interactive calculation tool for mediation tests. http://quantpsy.org/sobel/sobel.htm. Accessed 18 Sept 2019
Raudenbusch SW, Bryk AS (2001) Hierarchical linear models: applications and data analysis methods. Thousand Oaks, Sage Pub
Rubin DB (2005) Causal inference using potential outcomes. Design, modeling, decisions. J Am Stat Assoc 100:322–331. https://doi.org/10.1198/016214504000001880
Snijders TAB, Bosker RJ (2012) Multilevel analysis: an introduction to basic and advanced multilevel modeling. Sage Pub, London
Tingley D, Yamamoto T, Hirose K, Keele L, Imai K (2013) Mediation: R package for causal mediation analysis. J Stat Soft 59(5):1–38. https://doi.org/10.18637/jss.v059.05

Applications in Social Problems

Prentice–Williams–Peterson Models in the Assessment of the Influence of the Characteristics of the Unemployed on the Intensity of Subsequent Registrations in the Labour Office

Beata Bieszk-Stolorz

Abstract In the analysis of the duration of socio-economic phenomena, events subject to the study may occur more than once. They are called recurring or multiple events. Most analyses focus only on the first event and ignore the next one. In many cases, the risk of the next event occurring depends on the previous events. The aim of the paper is to analyse risk of subsequent registrations in the labour office depending on the characteristics of the unemployed (gender, age, education, seniority) using Prentice–Williams–Peterson's conditional models. Two types of models for multi-dimensional survival data were used in this paper. The first one (PWP-CP model) considers the time until the event occurs from the beginning of observation, and the second one (PWP-GP model) considers the time from the previous event. The basis of these models is the stratified Cox proportional hazards model, in which the strata are created by subsequent events. These models are an extension of the classical approach to survival analysis. In the study, individual data of persons registered in the Poviat Labour Office in Szczecin were used. The research revealed that age and education influenced the risk of multiple registrations in the office, while gender and seniority did not have a significant impact. In a similar way, the characteristics of the unemployed affected the risk of first return to office. However, they did not affect subsequent registrations.

Keywords Survival analysis · Recurrent events · Prentice–Williams–Peterson models · Registered unemployment

1 Introduction

In the survival analysis, the duration of the unit in a given state, up to the moment of the occurrence of a specific event, is examined. It is also possible to analyse processes during which the unit can be in this state several times. In this case, events ending the process may occur several times. The majority of analyses focus only on time to

B. Bieszk-Stolorz (✉)
University of Szczecin, Szczecin, Poland
e-mail: beata.bieszk-stolorz@usz.edu.pl

K. Jajuga et al. (eds.), *Classification and Data Analysis*, Studies in Classification, Data Analysis, and Knowledge Organization,
https://doi.org/10.1007/978-3-030-52348-0_15

the first event, ignoring the subsequent events. Most of the methods in the literature describe the statistical analysis of single-spell duration data, because most of the regression models for durations were developed in biostatistics by analysing survival times of patients in a clinical trial. In other areas such as economics, sociology, ecology, psychology or industrial engineering, study subjects can experience more than one event or failure as time elapses and, moreover, these events or failures may be of various kinds (Hamerle 1989). Several statistical models have been proposed for analysing multiple events. Processes of recurrent events are defined as processes that repeatedly generate specific events (Cook and Lawless 2007). They can be analysed using models of recurring events, also called multiple events in the literature. In this way, the technical sciences analyse assembly line downtimes and the processes for detecting and correcting software errors. Most often, such models are used in medical science to study the time to relapse (Sagara et al. 2014). Many diseases and other clinical outcomes may recur in the same patient. Examples include asthma attacks, skin cancers, myocardial infarctions, injuries, migraines, seizures in epileptics and admissions to hospital. In economic and social sciences, it is possible to analyse the time of entering and leaving the sphere of poverty (Sączewska-Piotrowska 2015), for assessment of the credit risk (Chen et al. 2012; Watkins et al. 2014), as well as the time of subsequent guarantee or insurance claims (Keiding et al. 1998).

There are many articles in the literature devoted to the application of the survival analysis in the study of socio-economic phenomena. However, only a small part of them describe the use of multiple event models in the labour market research. They address issues related to professional mobility (Blossfeld and Hamerle 1989), the duration of subsequent episodes of unemployment depending on the time and level of benefits received (Hamerle 1989; Kovačević and Roberts 2007) or the likelihood of reemployment of young people (Trivedi and Alexander 1989). Similar topics using multiple event models are discussed only in a few articles on the Polish labour market (Bieszk-Stolorz 2018; Gałecka-Burdziak 2016; Gałecka-Burdziak and Góra 2017; Grzenda 2019).

The literature deals with the problem of the duration of unemployment with the use of survival analysis. However, the subject of the analysis of the intensity of subsequent registrations in the office is not mentioned. The aim of the paper is to analyse the intensity of subsequent registrations in the labour office depending on selected characteristics of the unemployed (gender, age, education, seniority) using conditional Prentice–Williams–Peterson models. Two types of models for multidimensional survival data will be used in this paper. In the first one (PWP-CP model), the time from the beginning of observation to the moment of event occurrence is analysed, and in the second one (PWP-GP model) the time from the previous event is analysed. The basis of these models is the stratified Cox proportional hazards model, in which the strata create subsequent events. These models are an extension of the classical approach to survival analysis. They are examples of conditional models in which the risk of an event occurring is determined by the risk of previous events occurring. In the study, individual data of persons registered in the Poviat Labour Office in Szczecin were used.

2 Multiple Registrations with the Employment Office

In order to benefit from the services and support provided by the labour office, a person must register as unemployed or as a job seeker. If the person is unemployed, he/she can only register with the relevant employment office in his/her home country. Such person can take full advantage of the help offered by the office, but at the same time he/she has more responsibilities. A job seeker may register with several labour offices in order to have access to the widest possible range of services. He/she does not have the right to unemployment status (receiving unemployment benefit) because for example he/she is employed or earns income from rent or disability benefits. A job seeker is not registered for health insurance and is not paid for by health insurance, but can benefit from certain forms of assistance. The term "registered unemployment" therefore refers to persons who have the status of an unemployed person.

In Poland, the registered unemployment rate since 2013 has fallen from 13.4 to 5.8% (2018). Analysing the unemployment rate and the inflow and outflow of the unemployed from the labour offices, a certain regularity can be observed (Fig. 1). In the good economic situation, the outflow of the unemployed is smaller than the inflow, and the unemployment rate decreases. This was the case in 2004–2008 and 2013–2018. During the crisis, the situation is reversed, resulting in a rising unemployment rate (2009–2013). Among the persons registered in the labour offices, a large part of them are persons registered once again. They constituted from 68% (in 2004) to 83% (in 2016). Despite favourable labour market conditions and improving labour market outcomes, the share of such persons remains relatively high. It follows from this that a large number of people using the intermediation of labour offices do not take up permanent employment.

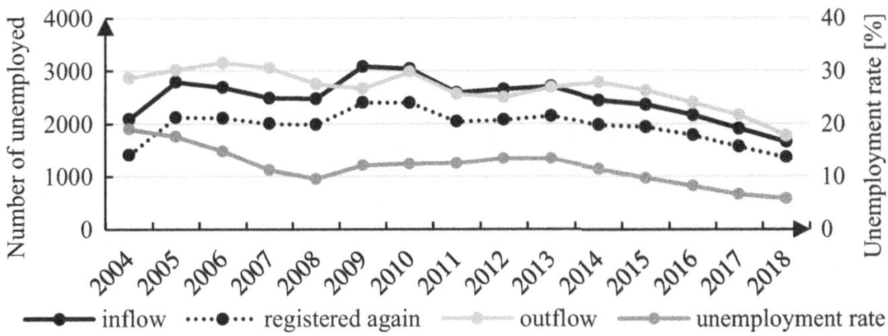

Fig. 1 Inflow and outflow of the registered unemployed (in thousands) and the unemployment rate registered in Poland in the years 2004–2018

3 Research Methodology

Selected methods of survival analysis were used in the study. The duration of the unit in a given condition, which is a random variable T, is observed. The basis for this type of analysis is the survival function, defined as follows:

$$S(t) = P(t < T) = 1 - F(t) \tag{1}$$

where
T—the duration of the phenomenon,
$F(t)$—cumulative distribution of random variable T.

The survival function informs about the probability that an event will not occur at least until time t. If the distribution of the survival time of the analysed occurrence is unknown, the survival function is mostly estimated by means of the Kaplan–Meier estimator (Kaplan and Meier 1958):

$$\hat{S}(t) = \prod_{j:t_j \le t} \left(1 - \frac{d_j}{n_j}\right) \tag{2}$$

where
d_j—the number of events at the moment t_j,
n_j—the number of individuals at risk by the moment t_j.

The second function used in the survival analysis is the hazard function. It describes the intensity of occurrence of an event at the moment t under the condition of survival until time t and is defined as follows (Kleinbaum and Klein 2005):

$$h(t) = \lim_{\Delta t \to 0} \frac{P(t \le T < t + \Delta t | T \ge t)}{\Delta t} \tag{3}$$

The Cox proportional hazards model expressed in the following formula (Cox 1972) is often used to study the intensity of events:

$$h(t, X) = h_0(t)\exp\left(\sum_{i=1}^{n} \beta_i X_i\right), \tag{4}$$

where
t—time,
$X = [X_1, X_2, \ldots, X_n]$—vector of variables,
$\beta_1, \beta_2, \ldots, \beta_n$—model parameters,
$h_0(t)$—basic hazard.

In 1981, Prentice, Williams and Peterson proposed two models for the analysis of recurrent events which are considered the first extension of the Cox proportional hazards model (Yadav et al. 2018). These models are included in the group of conditional models (Hosmer and Lemeshow 1999, pp. 308–311; Machin et al. 2006,

p. 247; Aalen et al. 2008, p. 473). It is assumed that recurrent events in a given unit are interrelated and that the basic risks are different for them. Examples of objects with multiple events are shown in Fig. 2. The first object was affected by one event, the second by two events, the third by three events and the fourth by two events before censorship. The fifth entity did not survive any event before censorship.

The occurrence of recurrent events is particularly evident in medical research. The risk of a subsequent heart attack is always higher than risk of a previous one. The two models proposed have a common feature: the basic risk is different depending on the event and depends on the occurrence of the previous event. The difference between them lies in the different determination of the intervals of risk of an event occurring. The risk intervals indicate when the unit is exposed to an event along a given time scale. In this study, time will be counted in two ways: the counting process method and the time gap method (Fig. 3). The first method uses calendar time and counts the

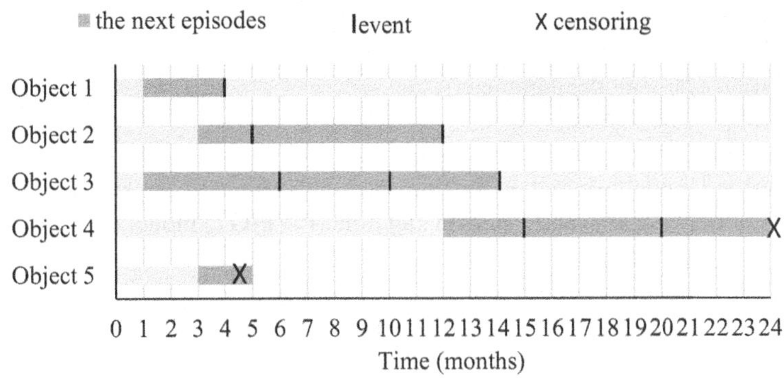

Fig. 2 Example of objects with multiple events

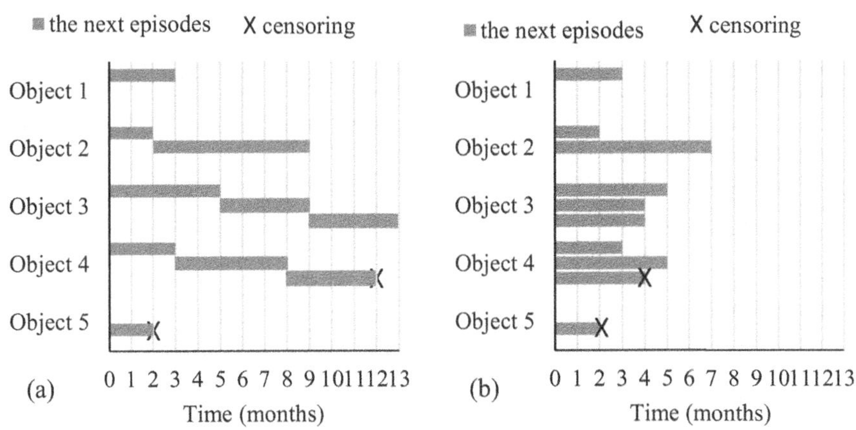

Fig. 3 Risk intervals counting process (**a**) and gap time (**b**)

risk intervals preceding the event. The second method consists in calculation of the time between subsequent events.

The first model, known as the Prentice, Williams and Peterson counting process (PWP-CP) model, counts the intervals. Counting process requires consideration of calendar time and time gaps, which means that it is a double-indexed process. The counting process uses the same scale of time as the total time, but takes into account the fact that an entity may have a delayed entry or a censorised period before it becomes exposed to the risk of an event occurring. (Sączewska-Piotrowska 2015). It is very useful for modelling the full process of cyclic events. It is a model.

$$h_g(t, X) = h_{0g}(t)\exp\left(\sum_{i=1}^{n} \beta_i X_i\right), \tag{5}$$

where

t—time elapsing from the end of previous interval,
$\mathbf{X} = [X_1, X_2, \ldots, X_n]$—vector of variables,
$\beta_1, \beta_2, \ldots, \beta_n$—model parameters,
$g = 1, 2, \ldots, k$—strata,
$h_{0g}(t)$—baseline hazard in stratum g.

The second model is based on the time interval between two events and is known as the Prentice, Williams and Peterson gap time (PWP-GT) model. It is a model

$$h_g(t, X) = h_{0g}\left(t - t_{g-1}\right)\exp\left(\sum_{i=1}^{n} \beta_i X_i\right), \tag{6}$$

where

$t - t_{g-1}$—time elapsing from the moment of the previous event,
$\mathbf{X} = [X_1, X_2, \ldots, X_n]$—vector of variables,
$\beta_1, \beta_2, \ldots, \beta_n$—model parameters,
$g = 1, 2, \ldots, k$—strata,
$h_{0g}(t)$—baseline hazard in stratum g.

Hazard ratio (HR), which expresses the relative intensity of an event occurring, is determined for both models using the formula:

$$HR = \exp\beta_i \tag{7}$$

where $\beta_1, \beta_2, \ldots, \beta_n$ are model parameters.

The PWP-CP model can be used if someone wants to interpret the effect of an intervention on a variable from the start of the study, while the PWP-GT model should be used if someone is interested in learning about the effect of previous events. Although both models are very suitable for the analysis of cyclical events, they have certain limitations. The main one is that they can give unreliable results for events of a higher order, because as the sequence of events increases, and the number of objects in the risk interval decreases (Cai and Schaubel 2004).

4 Data Used in the Study

In the study, anonymous unit data on unemployed persons from the Poviat Labour Office in Szczecin were used. The observation period covered the years 2016–2017. Among the persons registered in the office in 2016, those for whom it was the first registration were selected. Each person registered in the labour office can be assigned a registration history. The history of successive registrations until the end of 2017 was observed. As a starting point for the observation of each person ($t = 0$), the first registration was assumed. Each history consists of events, i.e. successive registrations in the office. These are recurrent events. After a preliminary analysis of the number of events in the history of registrations, it was decided to divide them into four groups: with one, two, three and four or more events. The separation of the latter group resulted from the small number of people registered with at least four events. A total of 3644 persons were analysed. Most of the observed persons were the unemployed, who did not register again in the analysed period (2808 persons), and there were 836 returning persons. Once again, 648 persons registered and twice—144 persons. The number of groups decreased with the number of subsequent registrations. The study included five groups of the unemployed due to the number of subsequent registrations, of which the fifth group included the unemployed who returned to the labour office four times or more by the end of 2017. The sizes of groups of persons according to subsequent registrations are included in Table 1.

If a given unit has suffered a k-th event (the k-th subsequent registration in the office for the $k = 1, 2, 3, 4$) and not suffered, by the end of 2017, an event with the number $k + 1$, such observation was considered to be censored. The observation period ended at the end of 2017, however, in the future the observed persons may register again at the labour office. In the case of the research, all observations shall be of the same character as object 4 in Fig. 2.

The following characteristics of the unemployed were taken into account in the study: gender, age, education and seniority. Gender is a dichotomous variable in which men are the reference group. Six age groups were distinguished: 18–24, 25–34, 35–44, 45–54, 55–59 and 60 or more, while the youngest group was the reference group. Education is divided into five levels: at the most lower secondary, basic vocational, general secondary, vocational secondary and higher. The lowest level

Table Sizes of groups of persons by number of events

Number of events (subsequent registrations)	Number of persons
0	2808
1	648
2	144
3	33
4 and more	11
Total	3644

Table Sizes of groups of persons by number of events

Characteristics	Size	Percentage (%)
Gender		
Males	1853	50.85
Females	1791	49.15
Education		
At most lower secondary	481	13.20
Basic vocational	416	11.42
General secondary	601	16.49
Vocational secondary	664	18.22
Higher	1482	40.67
Age		
18–24	1295	35.54
25–34	1302	35.73
35–44	456	12.51
45–54	296	8.12
55–59	189	5.19
60+	106	2.91
Seniority		
Without seniority	1967	53.98
With seniority	1677	46.02

of education was taken as the reference group. The unemployed were divided into two groups according to their seniority. The reference group consisted of people without seniority. The second group was made up of people with seniority. Structure of analysed persons is presented in Table 2. Among analysed unemployed persons registered for the first time, young persons (18–34 years old) with higher education dominated. The unemployed without seniority constituted a slightly larger group than persons with seniority.

Gender and seniority are the dichotomous variables. The multi-state variables: age groups and education were transformed into dummy variables. The 0-1 coding was used. All variables in the models therefore became dichotomous.

5 Analysis of Multiple Registrations in the Labour Office

The main part of the analysis was preceded by the determination of the median time until the re-registration and the maximum value of this time among people who experienced the k-th event ($k = 1, 2, 3, 4$). Both values decreased with the next registration. This shows that the time between the subsequent returns to the employment office was reduced (Table 3). Half of the observed persons who returned

Table Sizes of groups of persons by number of events

Number of subsequent registrations	Median (months)	Maximum (months)
1	6.7	21.3
2	5.1	18.3
3	3.6	12.0
4 and more	1.6	9.2

to the register for the first time did so within 6.7 months. The shortening time between registrations indicates that people who have been repeatedly registered have problems with maintaining employment.

Next, Kaplan–Meier estimators for subsequent events were determined. In the presented study, the event was defined as another return to the register of the labour office. In such a situation, the Kaplan–Meier estimator informs about the probability of the lack of subsequent registration. In this case, it is more convenient to interpret the opposite event, i.e. the probability of subsequent registration in the office after the time t. The probability of survival decreases as the recurrent events are observed. Reciprocally, it can be said that these events present an increasing risk of occurrence. Hence, the differences between the curves clarify the importance of considering methods for multiple events analysis. The course of Kaplan–Meier estimators confirmed the existence of differences in the probability of successive returns to the labour office. It also confirms the value of the test for many samples being a generalisation of Gehan's test (Gehan 1965), Peto's and Peto's test (Peto and Peto 1972) and log-rank test ($\chi^2 = 119.396, p = 0.0000$). The course of the survival curves in Fig. 4 indicates that only less than 25% of the persons registered for the first time returned to the labour office in the analysed period. Of these, about 31% returned for the second time. This probability increased with the subsequent registration, and for the subsequent events it was equal to 40% and 61%, respectively.

Medians and Kaplan–Meier estimators indicate that if people return to the labour office again, they do it more and more quickly and with greater probability. It follows that the intensity of these returns increases. Using the Cox regression model, the relative intensity of subsequent events was assessed. The parameters estimates and hazard ratios are presented in Table 4. It shows that these intensities were significantly different from each other. The risk of second, third, fourth and further return to the office in comparison with the first one was higher by 64%, 172% and 699%, respectively. The last high figure may be due to the fact that for the fourth return there was a low group size.

The main study was divided into two stages. It was examined whether the characteristics of the unemployed: gender, age, education and professional experience influenced the intensity of subsequent registrations in the office. In the first stage, the parameters of the PWP-CP and PWP-GT models were estimated for all episodes together and the appropriate hazard ratios were determined. All the variables in the models were dichotomous. Gender and seniority had two categories. Multi-category

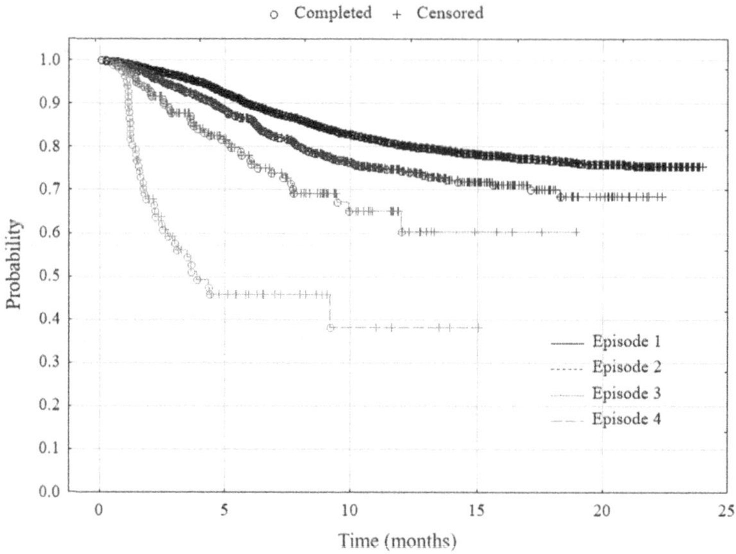

Fig. 4 Kaplan–Meier estimators for subsequent registrations

Table 4 Parameters estimates and hazard ratios for subsequent registrations in the labour office ($\chi^2 = 144.588$, $p = 0.0000$)

Variables	Parameter	Standard error	p-value	Hazard ratio (HR)
Second registration	0.4972	0.0818	0.0000	1.64
Third registration	1.0005	0.1564	0.0000	2.72
Fourth registration	2.0779	0.1672	0.0000	7.99

variables: age group and education were converted into dummy variables 0-1. It was examined whether the characteristics of unemployed persons influence the risk of subsequent registrations in the office. Both models gave similar results. Gender and seniority did not determine the risk of re-registration in the office. People with at most lower secondary education or up to 24 years of age were mostly at risk of being re-registered in the office. This risk diminished as education levels increased. However, for people with higher education, its growth was noted. The lowest risk was in the case of people with vocational secondary education—by 40% (PWP-CP) and 39% (PWP-GP) lower than the reference group. In the case of age groups, the lowest risk was observed in the 55–59 age group. In both models, the risk was 38% lower than for people aged 18–24. For people aged 60+, the parameters of the model were statistically insignificant (Table 5).

The second stage of the study consisted in determining the estimators of parameters of PWP-CP and PWP-GT models for subsequent events (Table 6). It was examined whether the characteristics of unemployed people influence the risk of k-th

Table 5 Hazard ratio (HR) estimated on the basis of PWP-CP and PWP-GT models

Variables	PWP-CP		PWP-GT	
	HR	*p*-value	HR	*p*-value
Gender				
Males	1.00		1.00	
Females	0.97	0.6919	0.97	0.6938
Education				
At most lower secondary	1.00		1.00	
Basic vocational	0.71	0.0030	0.70	0.0023
General secondary	0.67	0.0000	0.68	0.0001
Vocational secondary	0.60	0.0000	0.61	0.0000
Higher	0.81	0.0188	0.82	0.0251
Age				
18–24	1.00		1.00	
25–34	0.90	0.1763	0.90	0.2119
35–44	0.70	0.0018	0.70	0.0019
45–54	0.76	0.0385	0.77	0.0504
55–59	0.62	0.0069	0.62	0.0071
60+	0.85	0.4713	0.82	0.3874
Seniority				
Without seniority	1.00			
With seniority	0.96	0.5468	0.96	0.5886
χ^2	54.94		53.02	
p-value	0.0000		0.0000	

registration in the office. According to the assumptions, the estimators for the first event are the same. Women did not differ significantly from men in terms of the risk of re-registration for the first, second and third events. Only in the fourth event, the risk was significantly lower for women than for men (in the PWP-CP model at the significance level 0.1, and in the PWP-GT model—0.05). Education differentiated the unemployed only in the case of the first event. The age for both models differed between the unemployed in the case of the first event and in the PWP-GP model also for the fourth event.

6 Conclusions

The paper presents a review of two methods useful in the analysis of data on multiple events: PWP-CP and PWP-GT. Both models assume that risk of occurrence of each event depends on the observation of the previous event. It is also assumed that

Table 6 Hazard ratios for PWP-CP and PWP-GT models for subsequent events

Variables	PWP-CP				PWP-GT			
	1	2	3	4	1	2	3	4
Gender								
Males	1.00	1.00	1.00	1.00	1.00	1.00	1.00	1.00
Females	1.01	1.00	1.36	0.39***	1.01	1.01	1.18	0.27*
Education								
At most lower secondary	1.00	1.00	1.00	1.00	1.00	1.00	1.00	1.00
Basic vocational	0.62*	1.05	0.63	2.21	0.62*	1.02	0.45	2.46
General secondary	0.58*	1.09	0.99	0.58	0.58*	1.11	0.97	0.68
Vocational secondary	0.54*	0.95	0.85	0.00	0.54*	0.95	0.79	0.00
Higher	0.71*	1.01	1.15	1.72	0.71*	1.06	1.09	1.27
Age								
18–24	1.00	1.00	1.00	1.00	1.00	1.00	1.00	1.00
25–34	0.83**	1.09	1.38	3.26	0.83**	1.09	1.55	5.03**
35–44	0.67*	0.86	0.89	0.00	0.67*	0.86	0.84	0.00
45–54	0.71**	0.87	1.84	4.43	0.71**	0.83	1.95	8.90**
55–59	0.56**	0.88	1.24	0.51	0.56*	0.91	1.44	1.22
60+	0.64***	1.54	3.08	3.10	0.64***	1.52	2.56	2.34
Seniority								
Without seniority	1.00	1.00	1.00	1.00	1.00	1.00	1.00	1.00
With seniority	0.97	0.93	1.18	0.80	0.97	0.94	1.38	0.74
χ^2	64.73	2.83	5.98	29.17	64.73	3.09	7.36	30.25
p-value	0.0000	0.9928	0.8750	0.0021	0.0000	0.9895	0.7696	0.0014

*significance level 0.01, **significance level 0.05, ***significance level 0.1

there are differences in base risk for subsequent events. They should therefore be considered according to the order in which they occur, i.e. in strata. Ultimately, the choice of the right model depends on the aims of the researcher and the specificity of the data. The declining median and the maximum value of time until the next registration indicate that people who repeatedly register at the office have problems with finding a permanent job or are not interested in it. The study showed that in the analysed period, only age and education influenced the risk of multiple registrations at the Poviat Labour Office in Szczecin. Gender and seniority did not have a significant impact on this. The analysis performed in each stratum, i.e. for subsequent registrations, confirmed the impact of the same features in the first stratum, i.e. on the first subsequent registration. In general, it can be stated that the characteristics of the unemployed did not have a significant impact on the second and subsequent returns to the labour office. The risk of further registrations was the highest in the case of people with low education and up to 24 years of age. The result may be affected by the number of subsequent strata. With the next event, the number of observations decreases. As Cai and Schaubel (2004) point out, this may affect the reliability of the estimates in the last strata. Therefore, the aim of further research will be to estimate models for a smaller number of strata and to use other models of multiple events [e.g. Wei, Lin and Weissfeld models (1989), Andersen and Gill models (1982)].

References

Aalen OO, Borgan O, Gjessing HK (2008) Survival and event history analysis. A process point of view. Springer, New York

Andersen PK, Gill RD (1982) Cox's regression model for counting processes: a large sample study. Ann Stat 10(4):1100–1120

Bieszk-Stolorz B (2018) Stratified Cox model with interactions in analysis of recurrent events. Acta Universitatis Lodziensis. Folia Oeconomica 3(335):207–218. https://doi.org/10.18778/0208-6018.335.14

Blossfeld HP, Hamerle A (1989) Using Cox models to study multiepisode processes. Sociol Methods Res 17(4):432–448

Cai J, Schaubel DE (2004) Analysis of recurrent event data. In: Balakrishnan N, Rao CR (eds) Handbook of statistics, vol 23. North Holland, Elsevier BV, pp 603–623. https://doi.org/10.1016/s0169-7161(03)23034-0

Chen YS, Ho PH, Lin CY, Tsai WC (2012) Applying recurrent event analysis to understand the causes of changes in firm credit ratings. Appl Finan Econ 22(12):977–988. https://doi.org/10.1080/09603107.2011.633888

Cook RJ, Lawless JF (2007) The statistical analysis of recurrent events. Springer, New York

Cox DR (1972) Regression models and life-tables. J Roy Stat Soc B 34(2):187–220

Gałecka-Burdziak E (2016) Multiple unemployment spells duration in Poland, Collegium of Economic Analysis SGH—Working Papers 19(10)

Gałecka-Burdziak E, Góra M (2017) How do unemployed workers behave prior to retirement? A multi-state multiple-spell approach. Discussion Paper Series, IZA DP 10680, ftp.iza.org/dp10680.pdf. Accessed 15 Jan 2018

Gehan EA (1965) A generalized Wilcoxon test for comparing arbitrarily single-censored samples. Biometrika 52:203–223

Grzenda W (2019) Survival Modelling of Repeated Events Using the Example of Changes in the Place of Employment. Acta Universitatis Lodziensis. Folia Oeconomica 3(342):183–197. https://doi.org/10.18778/0208-6018.342.10

Hamerle A (1989) Multiple-spell regression models for duration data. Appl Stat 38(1):127–138

Hosmer DW, Lemeshow S (1999) Applied survival analysis. Regression modeling of time to event data. Wiley, New York

Kaplan EL, Meier P (1958) Non-parametric estimation from incomplete observations. J Am Stat Assoc 53:457–481

Keiding N, Andersen Ch, Fledelius P (1998) The Cox regression model for claims data in nonlife insurance. Astin Bull 28(1):95–118. https://doi.org/10.2143/AST.28.1.519081

Kleinbaum D, Klein M (2005) Survival analysis. A self-learning text. Springer, New York

Kovačević MS, Roberts G (2007) Modelling durations of multiple spells from longitudinal survey data. Surv Methodol 33(1):13–22

Machin D, Cheung YB, Parmar MKB (2006) Survival analysis. A practical approach, 2nd edn. Wiley, Chichester

Peto R, Peto J (1972) Asymptotically efficient rank invariant test procedures. J Roy Stat Soc 135(2):185–207

Prentice RL, Williams BJ, Peterson AV (1981) On the regression analysis of multivariate failure time data. Biometrika 68(2):373–379

Sagara I, Giorgi R, Doumbo OK, Piarroux R, Gaudart J (2014) Modelling recurrent events: comparison of statistical models with continuous and discontinuous risk intervals on recurrent malaria episodes data. Malaria J 13:293. https://doi.org/10.1186/1475-2875-13-293

Sączewska-Piotrowska A (2015) Badanie ubóstwa z zastosowaniem nieparametrycznej estymacji funkcji przeżycia dla zdarzeń powtarzających się. Przegląd Statystyczny LXII(1):29–51

Trivedi PK, Alexander JN (1989) Reemployment probability and multiple unemployment spells: a partial-likelihood approach. J Bus Econ Stat 7(3):395–401

Watkins JGT, Vasnev A, Gerlach R (2014) Multiple event incidence and duration analysis for credit data incorporating non-stochastic loan maturity. J Appl Econometrics 29(4):627–648. https://doi.org/10.1002/jae.2329

Wei L, Lin D, Weissfeld L (1989) Regression analysis of multivariate incomplete failure time data by modeling marginal distributions. J Am Stat Assoc 84(408):1065–1073. https://doi.org/10.2307/2290084

Yadav CP, Sreenivas V, Khan MA, Pandey RM (2018) An overview of statistical models for recurrent events analysis: a review. Epidemiology (Sunnyvale) 8:354. https://doi.org/10.4172/2327-4972.1000354

Right-Skewed Distribution of Features and the Identification Problem of the Financial Autonomy of Local Administrative Units

Romana Głowicka-Wołoszyn and Feliks Wysocki

Abstract Linear ordering methods with ideal solutions may sometimes suffer from identification problems when determining the development levels of examined objects. These problems manifest themselves in the form of inconsistencies between the range of the constructed synthetic measure and the development level of the complex phenomenon it should depict—especially, when the measure's low values are assigned to objects with obviously high level of development. Such inconsistencies often arise when simple features are strongly skewed—which is the case of the financial autonomy of the second-level local administrative units (communes) that is described by a number of asymmetric financial indicators. The aim of the research was to pose the problem of identifying levels of financial autonomy of Polish communes—assessed by synthetic measures constructed with ideal solution methods such as Hellwig's and TOPSIS—and to present proposals to resolve it. Two variants of the classical Hellwig's and TOPSIS methods were analyzed: standard and with correction of ideal values by the quartile criterion. Additionally, the positional TOPSIS method was also considered. It was found that with the standard classical methods or the positional TOPSIS, the prevalence of asymmetric simple features would reduce the range of the synthetic measure and shift it toward lower values. The variants with correction of ideal values had the range much broader and more centered, the broadest in the case of the corrected TOPSIS method. This contributed to the improvement of consistency between the identified levels of the communes financial autonomy and the synthetic measure values assigned to them.

Keywords Identification of development levels · Asymmetry of simple features · TOPSIS method · Hellwig's method · Financial autonomy of communes

R. Głowicka-Wołoszyn (✉) · F. Wysocki
Poznań University of Life Sciences, Poznań, Poland
e-mail: romana.glowicka@up.poznan.pl

F. Wysocki
e-mail: wysocki@up.poznan.pl

K. Jajuga et al. (eds.), *Classification and Data Analysis*, Studies in Classification, Data Analysis, and Knowledge Organization,
https://doi.org/10.1007/978-3-030-52348-0_16

1 Introduction

Linear ordering methods with ideal solutions may suffer from identification problems when determining the development level of examined objects. This is often the case when the simple features display strong asymmetry or have many outliers, as happens with various financial indicators of the households, companies or local administrative units. As a result, the values of a constructed synthetic index may cover only a small part of the potential area of its variability, which means that low values of the index may be assigned to objects with a high level of development of the studied phenomenon. This problem was presented in empirical studies (Glowicka-Woloszyn and Wysocki 2018a) that illustrated lack of correspondence between the classes of Polish communes with identified level of development and the values of the synthetic index. Specifically, the class of high financial autonomy in 2014 was formed by communes with synthetic index ranging between 0.136 and 0.583 (Hellwig's method).

The synthetic index is constructed by ideal solution methods that use ideal objects as a reference. Coordinates of the ideal objects may be exogenous[1] or endogenous—in the latter case usually set to be the maxima (minima) of the ranges of simple features.[2] However, with strong asymmetry of the features, it may be misleading to adopt for ideal objects the extreme values, as required by the classical ideal solution methods (either Hellwig's or TOPSIS). Such classical approaches would lead to

- excessive distance between the typical and ideal values,
- significant reduction in the range of the synthetic index,
- strong concentration of the index values in the lower part of the distribution (which corresponds to the right asymmetry of simple features).[3]

As a result, full and correct identification of development classes based on the calculated index values may be difficult. The problem particularly affects the attempts to classify large collections of objects, because if the applied approach is arbitrary—empty classes may appear, and if statistical—objects with high level of development may be assigned relatively low values of the synthetic index (Wysocki 2010). Often,

---

[1] Externally imposed ideal values are usually some widely accepted norms. They cannot or should not be computed from the observed sample, so the discussed problem is in such cases moot. Examples of these norms include 100% of the population's access to the water supply, or the environmental standards adopted through emission regulations.

[2] Other choices for the ideal object coordinates include *corrected* ideal values, that are smaller than maximum (or larger than minimum) (see Kozera and Wysocki 2016; Roszkowska et al. 2017; Glowicka-Woloszyn and Wysocki 2018a). They are used when the extremes are unusual and/or represent exceptional events. In some cases, the maximum may be replaced even by a mean, as in Yue (2011), where it was described as "maximum compromise among all individual decisions of a group".

[3] Simulation studies (Glowicka-Woloszyn and Wysocki 2018a) showed that predominance of moderately or strongly asymmetric features leads to a decreasing range of the synthetic index with a shift toward lower values.

the standard procedure to adopt for ideal objects the extreme values of simple features is essentially unjustified.[4]

The literature on the identification problem in the presence of strong asymmetry (or outliers) mentions a number of possible solutions. Some empirical studies suggest positional methods based on the robust L1 median (Lira et al. 2002; Luczak and Wysocki 2013); others propose alternative settings of the ideal values (Kozera and Wysocki 2016; Roszkowska et al. 2017; Glowicka-Woloszyn and Wysocki 2018a). One simulation study (Glowicka-Woloszyn and Wysocki 2018b) found that with classical ideal solution methods (Hellwig's or TOPSIS) more asymmetric features and stronger asymmetry exacerbate the identification problem, reducing the range of the synthetic index and concentrating its values at the bottom of the distribution. The study advocates the use of the quartile criterion as a remedy to the problem, or even positional TOPSIS, but in the latter case only when moderately or strongly skewed features do not form majority.

The aim of the research was to present the problem of identifying the levels of financial autonomy of Polish communes—assessed by a synthetic index constructed with ideal solution methods—and to propose solutions to the problem. Synthetic assessment of financial autonomy was based on simple features (financial indicators), whose distributions were mostly moderately or strongly right-skewed. Two variants of the classical Hellwig's and TOPSIS methods were analyzed: one standard and one with a correction of the ideals that applied the quartile criterion for individual simple features. Besides those classical approaches, positional TOPSIS method was also considered.

2 Research Methods

2.1 Source Material

The presentation of the research problem was illustrated by the case of synthetic assessment of financial autonomy of Polish communes (gminas)[5] in 2017. The study

---

[4]Communes' own income per capita may be one example of highly skewed features where extreme values should not serve as the ideals' coordinates. It is doubtful, for instance, whether Kleszczów, the most affluent commune in Poland, should be accepted as an ideal object. Its own income is 29 times higher than the median, 1.7 times higher than the second commune in the ranking (Rewal) and five times higher than the third (Krynica Morska). Normalization of the feature would yield the value of 1 for Kleszczów, while the median would only be 0.022. The high level of the commune's own income is due to the presence of a large lignite mine with a power plant—not a result that can or should be emulated by others. Also, own income of Rewal and other affluent coastal communes is largely derived from tourism-driven property income, which is essentially transient in its nature. It is not materially justified to take such levels of own income as models of development for other communes.

[5]Gminas—lower level of local administrative units (LAU 2 according to Classification of Territorial Units for Statistics).

drew on data from the Local Data Bank/Public Finance, maintained by the Central Statistical Office (www.stat.gov.pl/bdl), and on data published by the Ministry of Finance (www.mf.gov.pl). The research covered 2478 Polish communes, counting 1555 rural, 621 mixed urban–rural and 302 urban communes (including 66 cities with county (powiat)[6] status).

2.2 The Methods

The research was carried out in two stages. First, different methods of synthetic index construction were used to assess the financial autonomy of Polish communes in 2017.

Then, in the second stage, these results were compared and evaluated on the range and concentration of the values of the computed indices, as well as class delimitations determined by those values.

The methods of stage I included

1. Classical Hellwig's (Hellwig 1968):
   - without correction of ideal values,
   - with correction of ideal values,
2. TOPSIS:
   - classical[7] (Hwang and Yoon 1981):
     - without correction of ideal values,
     - with correction of ideal values,
   - positional (Lira et al. 2002; Wysocki 2010).

Ideal values were corrected by the quartile criterion (Tukey 1977; Trzęsiok 2014; Oliveira et al. 2016; Roszkowska et al. 2017; Kozera and Wysocki 2016; Głowicka-Wołoszyn and Wysocki 2018a, b). The criterion calculates the extreme values (minimum and maximum) not over the whole range of every simple feature, but rather over a restricted range of the form:

$$[Q_1 - a \cdot IQR; Q_3 + a \cdot IQR], \tag{1}$$

where Q_1 and Q_3 are the first and the third sample quartile, IQR is the interquartile range, and $a = 1.5$ is the correction parameter. Smaller value of the parameter $a = 1.0$. was also considered, which corresponded to tightening of the quartile

[6]Powiats—upper level of local administrative units (LAU 1 level according to Classification of Territorial Units for Statistics).

[7]The groundwork for classical TOPSIS can be found in the papers by Hellwig (1969, 1972a, b), and especially in Hellwig (1981), where synthetic measure construction uses both ideal and anti-ideal (see Walesiak 2016, 2017).

criterion, as well as larger values of $a = 2, 3, 5, 10$ that amounted to relaxing of the criterion.

Stage I—construction of synthetic indices—was conducted in six following steps:

Step 1—the selection of simple features (financial indicators) followed literature studies of the subject (Jastrzębska 2004; Heller 2006; Surówka 2013, 2018; Głowicka-Wołoszyn and Wysocki 2014) and statistical procedures of eliminating features that showed little variation (with coefficient of variation under 10%) or excessive correlation (with the diagonal coefficient of the inverse correlation matrix above 20) (Malina and Zeliaś 1997).

Step 2—normalization of the remaining features—included transformation of destimulants into stimulants and min–max scaling procedure (Kukuła 2000) for classical methods or L1 standardization (Lira et al. 2002; Młodak 2006; Łuczak and Wysocki 2013) for the positional TOPSIS.

The min–max scaling for stimulant features is given by

$$z_{ik} = \frac{x_{ik} - \min_i\{x_{ik}\}}{\max_i\{x_{ik}\} - \min_i\{x_{ik}\}}, \tag{2}$$

and for destimulant features by

$$z_{ik} = \frac{\max_i\{x_{ik}\} - x_{ik}}{\max_i\{x_{ik}\} - \min_i\{x_{ik}\}}, \tag{3}$$

where x_{ik} is the value of the kth feature for ith object, $k = 1, \ldots, K; i = 1, \ldots, N$.

L1 standardization (Lira et al. 2002; Młodak 2006) follows the formula:

$$z_{ik} = \frac{x_{ik} - m\tilde{e}d_k}{14,826 \cdot m\tilde{a}d_k}, \tag{4}$$

where $m\tilde{e}d_k$ is the L1 median component corresponding to the kth feature, $m\tilde{a}d_k = med_i|x_{ik} - m\tilde{e}d_k|$—median absolute deviation of the kth feature values from the median component, 14,826 constant scaling factor corresponding to normally distributed data ($\sigma \approx E(14,826 \cdot m\tilde{a}d_k(X_1, X_2, \ldots, X_K))$, σ—standard deviation).

Step 3—calculation of the ideal (A^+) and anti-ideal (A^-) values—as maxima and minima over the whole ranges of feature values (for classical variants) or over restricted ranges (for corrected variants):

$$A^+ = \left(\max_i(z_{i1}), \max_i(z_{i2}), \ldots, \max_i(z_{iK})\right) = \left(z_1^+, z_2^+, \ldots, z_K^+\right) \tag{5}$$

$$A^- = \left(\min_i(z_{i1}), \min_i(z_{i2}), \ldots, \min_i(z_{iK})\right) = \left(z_1^-, z_2^-, \ldots, z_K^-\right) \tag{6}$$

In the corrected variants, the ranges of every simple feature were restricted independently from one another. Observations that fell outside of the restricted range were winsorized, i.e., given the respective value of the ideal or the anti-ideal.

Step 4—calculation of the distance of each object (commune) from the ideal (for the Hellwig's and classical TOPSIS methods):

$$d_i^+ = \sqrt{\sum_{k=1}^{K} \left(z_{ik} - z_k^+\right)^2} \tag{7}$$

and the distance from the anti-ideal (classical TOPSIS):

$$d_i^- = \sqrt{\sum_{k=1}^{K} \left(z_{ik} - z_k^-\right)^2}. \tag{8}$$

The distances were calculated differently for positional TOPSIS:

$$d_i^+ = med_k\left(\left|z_{ik} - z_k^+\right|\right) \tag{9}$$

$$d_i^- = med_k\left(\left|z_{ik} - z_k^-\right|\right) \tag{10}$$

where $i = 1, \ldots, N$, and med_k is the marginal median of the kth simple feature.

Step 5—calculation of the synthetic index q_i:

- by the Hellwig's method

$$q_i = 1 - \frac{d_i^+}{\overline{d}^+ + 2s_{d^+}} \tag{11}$$

where $\overline{d}^+$ and s_{d^+} are the mean and standard deviation of the distance of objects from the ideal,

- by TOPSIS

$$q_i = \frac{d_i^-}{d_i^+ + d_i^-} \tag{12}$$

Step 6—identification of development types based on the mean ($\overline{q}$) and standard deviation $\left(s_q\right)$ of the empirical values q_i of the synthetic index:

- class I(high): $q_i \geq \overline{q} + s_q$,
- class II(medium high): $\overline{q} + s_q > q_i \geq \overline{q}$,
- class III(medium low): $\overline{q} > q_i \geq \overline{q} - s_q$,
- class IV(low financial autonomy): $q_i < \overline{q} - s_q$.

Genesis of the identification problem of development types, its consequences and proposals for solutions is presented in Fig. 1.

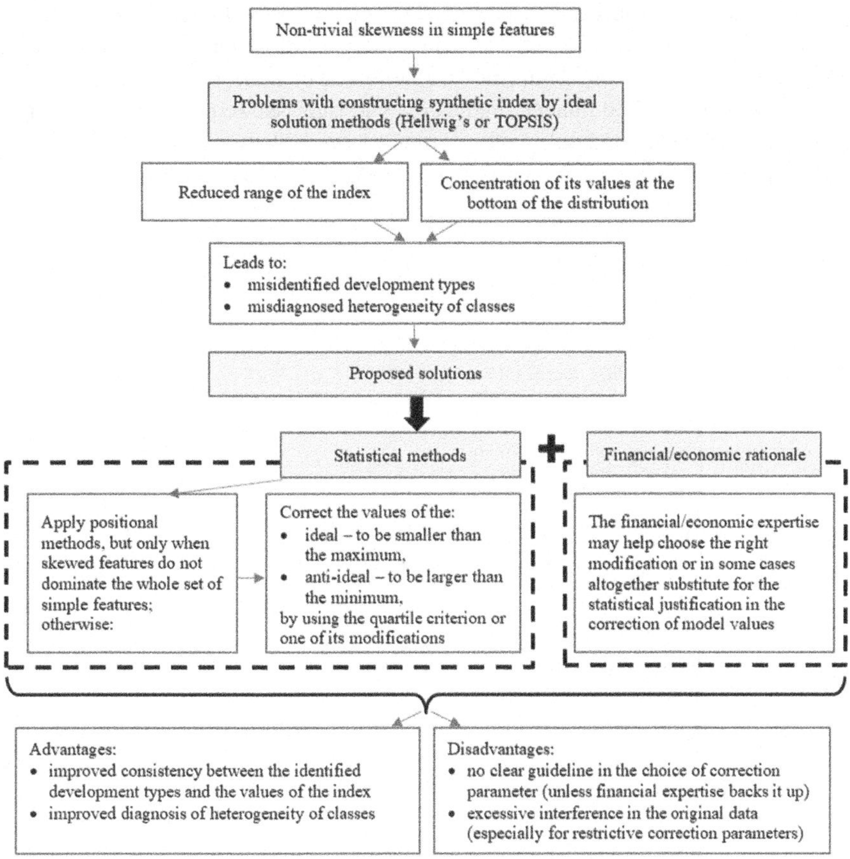

Fig. 1 Genesis of the identification problem of development types, its consequences and proposals for solutions

3 Results

3.1 Characteristics of the Simple Features

To characterize financial autonomy of Polish communes, a set of ten financial indicators was initially proposed, of which three did not meet statistical criteria and were discarded. These were tax income *per capita*, transfer income *per capita* and the share of operating surplus and property income in total expenditures. Thus, the financial autonomy of the communes was jointly described by remaining seven simple features, whose names and definitions are presented in Table 1.

The distribution of all financial indicators was right-skewed (Table 2), and while the W2 and W7 exhibited only mild asymmetry (with skewness below 1), and W3 moderately mild (with 1.6), the other four indicators were strongly asymmetric,

Table 1 Names and definitions of financial indicators

Financial indicator	Definition
W_1—own income *per capita* (PLN/cap.)	$W_1 = \frac{DW}{L}$
W_2—financial autonomy first degree (%)	$W_2 = \frac{DW}{DO}$
W_3—fiscal autonomy (%)	$W_3 = \frac{DP}{DB}$
W—operating surplus *per capita* (PLN/cap.)	$W_4 = \frac{NO}{L}$
W_5—share of operating surplus in total income (%)	$W_5 = \frac{NO}{DO}$
W_6—property expenditures *per capita* (PLN/cap.)	$W_6 = \frac{WM}{L}$
W_7—share of investment expenditures in total expenditures (%)	$W_7 = \frac{WMI}{WO}$

where
DW—communes' own income, *NO*—operating surplus, *DO*—total income, *WO*—total expenditures, *DP*—tax revenues, *WM*—property expenditures, *DB*—current income, *WMI*—investment expenditures, *L*—population

Table 2 Statistics of the financial indicators that describe the financial autonomy of Polish communes in 2017

Statistic	W1	W2	W3	W4	W5	W6	W7
Minimum	541.5	12.4	4.4	−1809.3	−33.8	10.0	0.3
Maximum	43,048.9	92.7	140.7	16,054.6	210.9	12,808.7	41.2
Mean	1716.9	37.7	29.3	395.3	9.7	547.0	13.2
Median	1491.8	35.6	27.0	334.0	8.8	483.0	12.5
CV	74.7	35.4	47.6	119.9	84.0	73.4	42.8
Skewness	16.7	0.6	1.6	19.0	13.1	12.9	0.8
Kurtosis	480.8	−0.2	735.7	4.8	541.3	281.9	471.9

with skewness ranging from 13 to 19. Moreover, all indicators showed rather high variability: four of them (W1, W4, W5, W6) had coefficients of variation exceeding 50% and W4 reaching 120%.

3.2 *Identification of Financial Autonomy Levels of Polish Communes*

Synthetic indices constructed with uncorrected classical Hellwig's and TOPSIS methods, or positional TOPSIS differed from the indices constructed with corrected classical methods in three points (Table 3; Fig. 2):

- lower range of values—for uncorrected classical variants, the synthetic indices reached at best 0.721 and their mean varied between 0.096 and 0.199, while for

Table 3 Range of the synthetic indices of financial autonomy of Polish communes constructed by the Hellwig's and TOPSIS methods without or with ideal values correction

Construction of the index			Min	Mean	Max	Standard deviation
Hellwig's method	Classical approach	Uncorrected	−0.009	0.096	0.650	0.048
		Corrected ($a = 1.5$)	−0.036	0.344	1.000	0.172
TOPSIS method	Classical approach	Uncorrected	0.082	0.199	0.721	0.059
		Corrected ($a = 1.5$)	0.083	0.435	1.000	0.151
	Positional approach		0.008	0.166	1.000	0.074

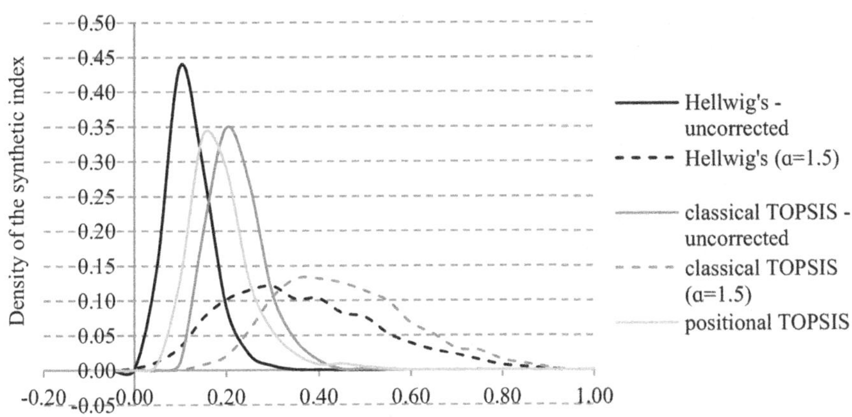

Fig. 2 Distribution of the synthetic indices of communes' financial autonomy constructed with different methods and variants

corrected variants covered the whole range of possible values, and their mean fell in the 0.344–0.433 interval;

- value shift toward the bottom of the distribution—uncorrected variants were much more affected, as seen in Fig. 2;
- smaller variation—the standard deviation of the synthetic indices for uncorrected variants ranged between 0.048 and 0.074, while for the corrected ones between 0.151 and 0.172.

Correction parameter of $a = 1.5$ was not the only one considered in the study. Modifications included tightening of the quartile criterion by setting stricter $a = 1.0$ or relaxing it with milder $a = 2, 3, 5, 10$. Distributions of the synthetic indices that corresponded to those correction parameters are illustrated in Figs. 3 and 4 for Hellwig's and TOPSIS methods, respectively.

Tightening of the criterion can be seen to yield more symmetric distributions with means closer to 0.5—the center of the range of possible values. However, that improvement was paid for by winsorizing of up to 9% of all values for highly skewed features, while the usual criterion of $a = 1.5$ called for winsorization of only 5%.

Synthetic indices constructed with uncorrected classical Hellwig's and TOPSIS, and positional TOPSIS methods can be seen to identify the class of high financial autonomy very poorly (Table 4). The class was formed by communes with low index values—some as low as 0.145 in the Hellwig's method and 0.239–0.257 in TOPSIS. Moreover, this class seems to be extremely heterogeneous, as the range of the index values is much larger here than in the other classes.

Application of the usual quartile criterion ($a = 1.5$) and its stricter version ($a = 1.0$) to the Hellwig's and TOPSIS methods contributed to the resolution of

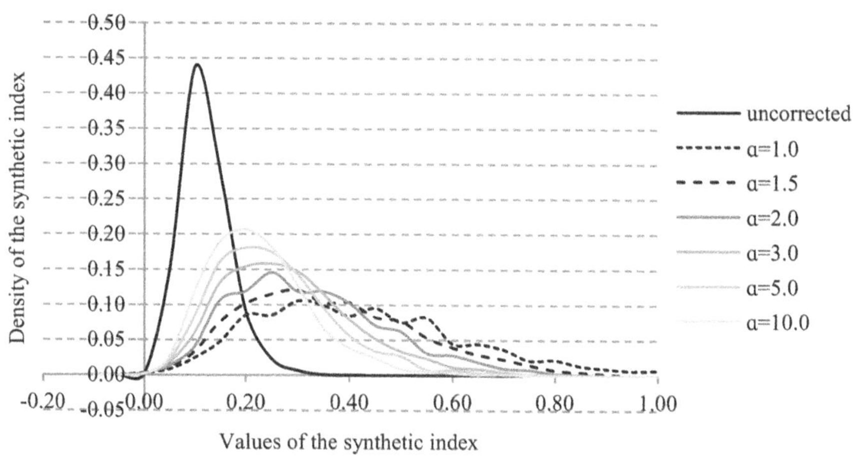

Fig. 3 Distribution of the synthetic indices of communes' financial autonomy constructed with Hellwig's method and different correction parameters

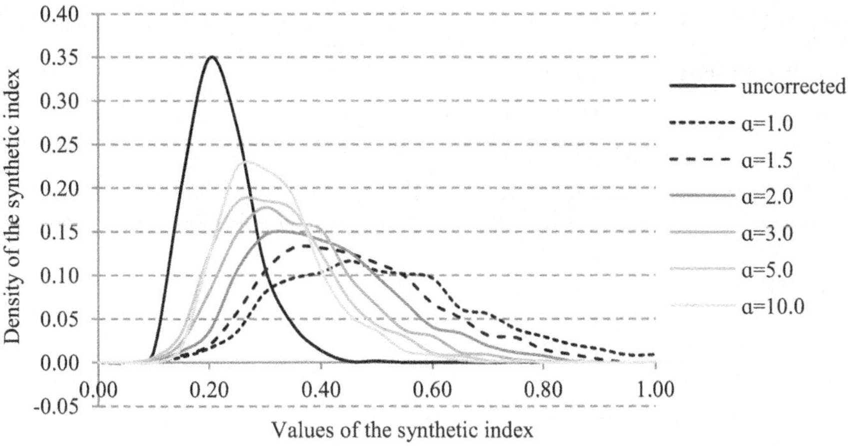

Fig. 4 Distribution of the synthetic indices of communes' financial autonomy constructed with TOPSIS method and different correction parameters

Table 4 Identification of communes' level of financial autonomy through classification based on the mean and standard deviation of the synthetic index constructed with classical Hellwig's and TOPSIS and positional TOPSIS methods (without correction)

Class	Classical methods				Positional TOPSIS	
	TOPSIS		Hellwig's			
	Range	%	Range	%	Range	%
IV (low)	[0.082, 0.140)	14.4	[−0.01, 0.048)	15.4	[0.008, 0.092)	12.0
III (medium low)	[0.140, 0.199)	30.4	[0.048, 0.096)	37.0	[0.092, 0.166)	28.1
II (medium high)	[0.199, 0.257)	40.6	[0.096, 0.145)	33.6	[0.166, 0.239)	49.5
I (high)	[0.257, 0.721]	14.6	[0.145, 0.565]	14.0	[0.239, 1.000]	10.4

inconsistency between identified high level of financial autonomy and the corresponding range of the synthetic index (Tables 5 and 6). That level matched now considerably higher values of at least 0.516 (Hellwig's) and 0.586 (TOPSIS), instead of 0.145 in the uncorrected Hellwig's method.

4 Conclusions

In the ideal solution methods (Hellwig's and TOPSIS), the correction of ideal values allowed to extend the empirical range of the index, so as to cover a much larger part of its potential range and to shift its values toward the middle of the unit interval.

Table 5 Identification of communes' level of financial autonomy through classification based on the mean and standard deviation of the synthetic index constructed with classical Hellwig's and TOPSIS methods (with correction)

Class	Correction parameter ($a = 1.5$)			
	Hellwig's method		TOPSIS	
	Range	%	Range	%
IV (low)	[−0.04, 0.172)	16.2	[0.083, 0.284)	15.9
III (medium low)	[0.172, 0.344)	29.4	[0.284, 0.435)	29.9
II (medium high)	[0.344, 0.516)	38.3	[0.435, 0.586)	38.4
I (high)	[0.516, 1.000]	16.1	[0.586, 1.000]	15.8

Table 6 Identification of communes' level of financial autonomy through classification based on the mean and standard deviation of the synthetic index constructed with classical Hellwig's and TOPSIS methods (with stricter correction)

Class	Correction parameter ($a = 1.0$)			
	Hellwig's method		TOPSIS	
	Range	%	Range	%
IV (low)	[−0.05, 0.198)	16.5	[0.067, 0.313)	16.5
III (medium low)	[0.198, 0.396)	29.5	[0.313, 0.487)	29.5
II (medium high)	[0.396, 0.594)	37.3	[0.487, 0.660)	37.8
I (high)	[0.594, 1.000]	16.6	[0.660, 1.000]	16.2

In the financial autonomy application, the quartile criterion used to correct the ideal values helped identify typological classes determined by the ranges of synthetic index far more consistent with the names of their corresponding classes.

Tightening of the criterion led to a broader range of the index, but for highly skewed features up to 9% of observations had to be winsorized, while with the usual criterion only 5%. It is still debatable at what level should the ideal values be set so that the interference in the original dataset is not too great. It seems that an appropriate approach to establishing ideal values should be based both on expert methods (following a financial or economic rationale) and statistical application of the quartile criterion.

References

Głowicka-Wołoszyn R, Wysocki F (2014) Uwarunkowania społeczno-ekonomiczne samodzielności finansowej gmin województwa wielkopolskiego. Prace Naukowe Uniwersytetu Ekonomicznego we Wrocławiu 346:34–44

Głowicka-Wołoszyn R, Wysocki F (2018a) Problem identyfikacji poziomów rozwoju w zagadnieniu konstrukcji cechy syntetycznej. Prace Naukowe Uniwersytetu Ekonomicznego we Wrocławiu 508:56–65

Głowicka-Wołoszyn R, Wysocki F (2018b) Symulacyjne badania wpływu asymetrii rozkładu cech na zakres zmienności wartości konstruowanego miernika syntetycznego, Materiały konferencyjne XXVII Konferencja Sekcji Klasyfikacji i Analizy Danych Polskiego Towarzystwa Statystycznego, XXXII Konferencja Taksonomiczna „Klasyfikacja i analiza danych - teoria i zastosowania" Ciechocinek, 9–12 września 2018. https://skad2018.wsb.torun.pl/public/conferences/3/schedConfs/2/program-pl_PL.pdf. Accessed 1 Sept 2019

Heller J (2006) Samodzielność finansowa samorządów terytorialnych w Polsce. Studia Regionalne i Lokalne 2(24):137–151

Hellwig Z (1968) Zastosowania metody taksonomicznej do typologicznego podziału krajów ze względu na poziom ich rozwoju i strukturę wykwalifikowanych kadr. Przegląd Statystyczny 4:307–327

Hellwig Z (1969) On the problem of weighting in international comparisons. In: Study VII of the UNESCO statistical office, towards a system of quantitative indicators of components of human resources indicators development, UNESCO, Paris

Hellwig Z (1972a) Approximative methods of selection of an optimal set of predictors. In: Study XVI of the UNESCO Statistical Office, towards a system of quantitative indicators of components of human resources indicators development, UNESCO, Paris

Hellwig Z (1972b) On optimal choice of predictors. In: Gostkowski Z (ed) Towards a system of human resources indicators for less developed countries, UNESCO, Ossolineum, Paris, Wrocław, pp 69–90

Hellwig Z (1981) Wielowymiarowa analiza porównawcza i jej zastosowanie w badaniach wielocechowych obiektów gospodarczych. In: Welfe W (ed) Metody i modele ekonomiczno-matematyczne w doskonaleniu zarządzania gospodarką socjalistyczną. PWE, Warszawa, pp 46–68

Hwang CL, Yoon K (1981) Multiple attribute decision-making: methods and applications. Springer, Berlin

Jastrzębska M (2004) Samodzielność ekonomiczna i finansowa jednostek samorządu terytorialnego. Ekonomia/Uniwersytet Warszawski 13:100–112

Kozera A, Wysocki F (2016) Problem ustalania współrzędnych obiektów modelowych w metodach porządkowania liniowego obiektów. Prace Naukowe Uniwersytetu Ekonomicznego we Wrocławiu 427:131–142

Kukuła K (2000) Metoda unitaryzacji zerowanej. Wydawnictwo Naukowe PWN, Warszawa

Lira J, Wagner W, Wysocki F (2002) Mediana w zagadnieniach porządkowania obiektów wielocechowych. In: Paradysz J (ed) Statystyka regionalna w służbie samorządu lokalnego i biznesu. Internetowa Oficyna Wydawnicza Centrum Statystyki Regionalnej, Poznań, pp 87–99

Łuczak A, Wysocki F (2013) Zastosowanie mediany przestrzennej Webera i metody TOPSIS w ujęciu pozycyjnym do konstrukcji syntetycznego miernika poziomu życia. Prace Naukowe Uniwersytetu Ekonomicznego we Wrocławiu 278:63–73

Malina A, Zeliaś A (1997) Taksonomiczna analiza przestrzennego zróżnicowania jakości zycia ludności w Polsce w 1994 r. Przegląd Statystyczny 1(44):11–27

Młodak A (2006) Analiza taksonomiczna w statystyce regionalnej. Difin, Warszawa

Oliveira EC, Faro AO, Anderson LF (2016) Comparison of different approaches for detection and treatment of outliers in meter proving factors determination. Flow Meas Instrum 48:29–35

Roszkowska E, Filipowicz-Chomko M, Wachowicz T (2017) Wykorzystanie metody TOPSIS do oceny zróżnicowania rozwoju województw Polski w latach 2010-2014 w kontekście kształtowania się ładu instytucjonalnego. Prace Naukowe Uniwersytetu Ekonomicznego we Wrocławiu 469:149–158

Surówka K (2013) Samodzielność finansowa samorządu terytorialnego w Polsce. Polskie Wydawnictwo Ekonomiczne, Warszawa

Surówka K (2018) Sources of income and financial autonomy of local self-government. Econ World 6(1):22–33. https://doi.org/10.17265/2328-7144/2018.01.003

Trzęsiok M (2014) Wybrane metody identyfikacji obserwacji oddalonych. Prace Naukowe Uniwersytetu Ekonomicznego we Wrocławiu 327:157–166

Tukey JW (1977) Exploratory data analysis. Addison-Wesley, Boston

Walesiak M (2016) Visualization of linear ordering results for metric data with the application of multidimensional scaling. Ekonometria 2(52):9–21

Walesiak M (2017) The application of multidimensional scaling to measure and assess changes in the level of social cohesion of the Lower Silesia region in the period 2005–2015. Ekonometria 3(57):9–25

Wysocki F (2010) Metody taksonomiczne w rozpoznawaniu typów ekonomicznych rolnictwa i obszarów wiejskich. Wydawnictwo Uniwersytetu Przyrodniczego w Poznaniu, Poznań

Yue Z (2011) An extended TOPSIS for determining weights of decision makers with interval numbers. Knowl Based Syst 24(1):146–153

Multi-criteria Rankings with Interdependent Criteria: Case of EU Countries on Their Way to Healthy Lives and Well-Being

Iwona Konarzewska

Abstract One of the main assumptions when making multi-criteria rankings or using multivariate statistical analysis methods in comparative surveys is independence of considered diagnostic criteria. However, in practical research, we meet problems with statistical independence of chosen essential properties of objects under comparison. The other problem is the choice of adequate weights for criteria. Weights are usually chosen subjectively (as in the AHP method), as equal (if there are no reasons justifying diversification) or as statistically justified, taking into account discriminant capability and information capacity of the criteria. Our proposition is to choose weights according to the values of variance inflation factors (VIFs)—diagonals of the inverse to the matrix of correlation coefficients across criteria. Greater VIF means smaller information capacity of the criterion and smaller weight is imposed in consequence. The other proposition is choosing weights basing on principal component analysis (PCA) of the covariance matrix—reduction of criteria space dimension. We compare these proposals for classical simple average weighting method (SAW). Another proposition is the method multi-criteria principal components (MCPC) in which weights are assigned to principal components. Our rankings base on EUROSTAT indices for 28 EU countries measuring their achievements of targets of the UN 2030 Agenda for Sustainable Development Goal 3: "Ensure healthy lives and promote well-being for all at all ages" for the year 2017 (or closest).

Keywords Multi-criteria rankings · Interdependent criteria · Criteria weighting · VIF · PCA · Sustainable development Goal 3

1 Introduction

The main tool which is widely used when making multi-criteria rankings of decision alternatives, objects, countries, etc., is composite indicator (CI). Composite indicators are the aggregated indices developed using individual performance indicators. They

I. Konarzewska (✉)
Department of Operational Research, University of Łódź, Łódź, Poland
e-mail: iwona.konarzewska@uni.lodz.pl

K. Jajuga et al. (eds.), *Classification and Data Analysis*, Studies in Classification, Data Analysis, and Knowledge Organization,
https://doi.org/10.1007/978-3-030-52348-0_17

are used to measure the complex phenomena which cannot be captured by one single criterion (indicator). There are many arguments for constructing composite indicators, among them (see Saisana and Tarantola 2002; Jacobs et al. 2004; Nardo et al. 2005):

- They summarize complex or multi-dimensional issues,
- They are easier to interpret than many separate indicators,
- They facilitate ranking countries on complex issues.

One of the main arguments against composite indicators is that the selection of sub-indicators and weights is not transparent and contentious. Hudrlikova (2013) summarizes common weighting schemes for the indicators construction. The composite measure depends crucially on the weights attached to each criterion considered. Some details of methodology of composite indicators construction for countries can be found, f.i., in Lafortune et al. (2018) and Sachs et al. (2018)—for the Global SDG Index and in publications of the Joint Research Center of the European Commission. We are aware of difficulties in the process of composite indicator construction for various fields of interest: environment, economy, society or technological development. Individual criteria being considered are often hardly measurable or dependent.

2 Some Methodological Issues

Let us use the following notations:

f_{ik}—the original value of the kth criterion (sub-index) for ith object, $i = 1, \ldots, \mathrm{N}$, $k = 1, \ldots, \mathrm{K}$.
w_k—the weight for the k-th criterion, $\sum_{k=1}^{\mathrm{K}} w_k = 1, \forall k \; w_k \geq 0$.

We assume that the values of the criteria for objects under comparison are measured using interval or ratio scales. Second important assumption is that the number of objects N is not lower than the number of the criteria K, that is: N ≥ K.

2.1 Normalization

Criteria values are not usually measured in the same units. The way data are normalized plays influential role on statistical characteristics of the criteria values and final multi-criteria ranking of objects. We use rescaling equations Eq. 1 or Eq. 2 according to the original direction of the criterion; all rescaled values of criteria are then expressed as ascending variables (i.e., higher values denote better performance) from the range [0, 1].[1]

[1]This transformation in Nardo et al. (2008, p. 28) is called Min-Max.

If for the best performance, minimal value of the original criterion is expected, and then, normalization is done through Eq. 1.

$$a_{ik} = \frac{\max_i f_{ik} - f_{ik}}{\max_i f_{ik} - \min_i f_{ik}}. \tag{1}$$

If for the best performance, maximal value of the original criterion is expected, and then, normalization is done through Eq. 2.

$$a_{ik} = \frac{f_{ik} - \min_i f_{ik}}{\max_i f_{ik} - \min_i f_{ik}}. \tag{2}$$

The values of the criteria should be non-negative. If original raw values are negative, before normalization, they can be transformed, f.e. using the transformation defined by Eq. 3.

$$a_{ik} := f_{ik} + \left|\min_i f_{ik}\right|. \tag{3}$$

It is important to notice that to avoid the sensitivity of final results on extreme outlying values of variables (criteria), it is recommended to censor the data at the level of 2.5 and 97.5 percentile.[2] In our research, we did not apply this approach. The way of normalization chosen in this research was found to have no influence on absolute correlation coefficients among variables. It effects the dispersion and relative dispersion characteristics.

Another well-known proposition is standardization (z-score)—subtracting the mean value of the criterion across objects and dividing by standard deviation. The effect we obtain is the variable with zero mean and standard deviation equal to one. The disadvantage of such transformation is that we obtain both positive and negative values of the criteria. To apply simple average weighting (SAW) method, introduced by Churchman and Ackoff (1954), in such a case, additional transformations of data will be necessary—the method assumes non-negative criteria values.

2.2 *Weighting Using Informative Capacity of the Criteria and PCA Approach*

Constructing the composite index, we aggregate the data. Composite index is constructed as a weighted mean of normalized criteria values as in Eq. 4. The maximal

[2]It means that any observed value lower than 2.5 percentile is raised to the level of 2.5 percentile and any observed value greater than 97.5 percentile is lowered to the level of 97.5 percentile (Nardo et al. 2008).

value of index corresponds to the best object.

$$Q_i = \sum_{k=1}^{K} w_k a_{ik}. \quad i = 1, \ldots, N \tag{4}$$

The weights are essentially value judgments about the relative importance of different performance sub-indicators and have the property to make explicit the objectives underlying the construction of a composite (see Jacobs et al. 2004). There is no agreed methodology in what way we should weight individual indicators. The weights attached to different performance criteria, as they influence the value of the composite index, can change the final ranking of a particular object. Often equal weights are applied, for simplicity. The approach may be valid if there are not enough theoretical, statistical or empirical grounds for diversification of the weights.

Multivariate statistical methods recommend to choose weights for criteria (sub-indicators) according to their *discriminatory strength* measured, f.i. by relative variation. The correlation structure among sub-indicators is closely related to the issue of weights. Paruolo et al. (2013), Becker et al. (2017) suggest Pearson's correlation ratio as a measure of *criterion importance*, answering the question about the relative reduction in variance of the composite indicator to be expected by fixing a criterion value. Another advice is *informative capacity* of the individual indicator. We propose to measure informative capacity using variance inflation factors (VIF) coefficients, which are defined as diagonals of the inverse to the matrix of correlation coefficients among sub-indicators—Eq. 5. VIFs are positively related with the squared multiple correlation coefficient of individual sub-indicator with the set of other sub-indicators taken into account.

$$VIF_k \equiv r^{kk} = \frac{1}{1 - \delta_k^2}, \tag{5}$$

where r^{kk} is the kth diagonal element of the inverse to correlation matrix of sub-indicators and δ_k is the multiple correlation coefficient of the kth sub-indicator with the others. Greater value of VIF corresponds to lower informative capacity of the sub-indicator. $VIF_k = 1$ means that kth individual indicator is independent of the set of other sub-indicators, $VIF_k \to \infty$ indicates exact linear relationship of the k-th individual indicator with the others. The proposition of relative normalized weights for sub-indicators, proportional to their informative capacity, is formulated in Eq. 6:

$$w_k^{ic} = \frac{\min_k VIF_k}{VIF_k} / \sum_{k=1}^{K} \frac{\min_k VIF_k}{VIF_k}. \tag{6}$$

In this work, we leave the discriminatory strength of sub-indicators aside as in economic or social research criteria are often correlated. We are interested in the comparison of the multi-criteria results of the two methods: SAW and proposed in this work—multi-criteria principal components (MCPC), when using equal weights, w_k^{ic}

weights and weights based on principal component analysis (PCA) of the covariance matrix among sub-indicators.

Let us introduce some main formulations related to PCA (see f.e. Jolliffe (2002). $\mathbf{A} = [a_{ij}] i = 1, \ldots, N\ j = 1, \ldots, K$ is a NxK matrix of normalized values of K criteria/sub-indices for N objects (through transformation described in Eq. 1 and Eq. 2, respectively). Centering the columns of the $\mathbf{A}$ matrix, i.e., subtracting the mean value $\overline{a}_j$, $j = 1, \ldots, K$, for each criterion from the corresponding values of observations for objects, we obtain the matrix $\mathbf{A_c} = \mathbf{A} - [\overline{a}_1 \cdot \mathbf{1}_N \ldots \overline{a}_K \cdot \mathbf{1}_N]$, where $\mathbf{1}_N$ is $N \times 1$ vector of ones. Covariance matrix, representing the interdependence of the criteria, can be expressed as in Eq. 7.

$$\mathbf{\Sigma} = 1/N\,\mathbf{A}_c^T \mathbf{A}_c \tag{7}$$

Eigenvalue decomposition formulated by Eq. 8 let us express covariance matrix using its eigenvalues and eigenvectors.

$$\mathbf{\Sigma} = \mathbf{V}\mathbf{\Lambda}\mathbf{V}^T, \tag{8}$$

where $\mathbf{\Lambda} = diag(\lambda_1, \ldots, \lambda_K)$ is diagonal matrix of eigenvalues of $\mathbf{\Sigma}$ ordered in non-increasing sequence on the main diagonal and the matrix $\mathbf{V} = [\mathbf{v}_1, \ldots, \mathbf{v}_K]$ consists of K orthonormalized eigenvectors associated with eigenvalues of $\mathbf{\Sigma}$. Variances of criteria values being the elements on the main diagonal of $\mathbf{\Sigma}$ can be thus expressed in the following way (see Eq. 9):

$$\sigma_k^2 = \sum_{l=1}^{K} v_{kl}^2 \lambda_l \quad k = 1, \ldots, K. \tag{9}$$

The goal of PCA is to pick out linear combinations of the criteria, orthogonal to each other, which decompose the total variance of the criteria set. The first principal component is the one with the biggest variance. The total variance of the set of criteria is equal to the sum of eigenvalues of $\mathbf{\Sigma}$. The proportion of the eigenvalue in the sum is used as a measure of the level on which the associated principal component explains the total variance. If the considered criteria/sub-indices are linearly independent, then all eigenvalues of $\mathbf{\Sigma}$ are equal.

Substituting the definition of covariance matrix formulated in Eqs. 7–8 and solving for $\mathbf{\Lambda}$, we obtain the expression for the $N \times K$ matrix $\mathbf{Y}$ of principal components in Eq. 10:

$$\mathbf{Y}^T\mathbf{Y} \equiv \mathbf{\Lambda} = \frac{1}{N}\mathbf{V}^T\mathbf{A}_c^T\mathbf{A}_c\mathbf{V} \Rightarrow \mathbf{Y} = \frac{1}{\sqrt{N}}\mathbf{A}_c\mathbf{V} \tag{10}$$

Applying PCA, we reduce the number of principal components finally considered to $K^* < K$, f.e. using the criterion of cumulative percentage of explained variance being at least 75%, or scree plot. Ultimately, the NxK* matrix $\mathbf{Y}^*$ defined in Eq. 11

is the NxK* matrix of principal components taken into account:

$$\mathbf{Y}^* = \frac{1}{\sqrt{\mathrm{N}}}\mathbf{A}_c\mathbf{V}^*, \tag{11}$$

where the "*" beside the matrix symbol means that its number of columns is equal to K*. Similarly, the variance of each of the criteria, explained using K* principal components, can be expressed as

$$\sigma_k^{*2} = \sum_{l=1}^{\mathrm{K}^*} v_{kl}^2\lambda_l \quad k = 1, \ldots, \mathrm{K}. \tag{12}$$

Weights for the criteria using PCA are often calculated as in Eq. 13 (see Chao and Wu 2017), i.e., a portion of the total variance explained by K* principal components corresponding to *k*th criterion:

$$w_k^{PC} = \frac{\sigma_k^{*2}}{\sum_{k=1}^{\mathrm{K}^*} \sigma_k^{*2}}. \tag{13}$$

Nardo et al. (2005) pointed out some disadvantages of weighting the criteria using PCA. Among them, they mention

- the method can be used only with correlated sub-indicators,
- the weights are sensitive to modifications of basic data,
- the method is sensitive to the presence of outliers.

We present results of multi-criteria rankings using PCA weighting to be compared with equal weighting, informative capacity weighting and proposed MCPC method being the modification of SAW based also on PCA application.

2.3 Multi-criteria Principal Components (MCPC) Method

Let us denote by $\boldsymbol{q} = [Q_1, \ldots, Q_{\mathrm{N}}]^T$—a vector of composite index values for objects, and by **W**—KxK diagonal matrix with weights w_j, $j = 1, \ldots, \mathrm{K}$, on the main diagonal. Constructing the composite index for K criteria using Eq. 4, we find that it holds

$$\begin{aligned}\boldsymbol{q} = \mathbf{AW} \cdot \mathbf{1}_{\mathrm{K}} &= \left(\sqrt{\mathrm{N}}\mathbf{Y}^*\mathbf{V}^{*T} + [\overline{a}_1 \cdot \mathbf{1}_{\mathrm{N}} \ldots \overline{a}_{\mathrm{K}} \cdot \mathbf{1}_{\mathrm{N}}]\right)\mathbf{W} \cdot \mathbf{1}_{\mathrm{K}} \\ &= \sqrt{\mathrm{N}}\mathbf{Y}^*\mathbf{V}^{*T}\mathbf{W} \cdot \mathbf{1}_{\mathrm{K}} + \mathbf{c}\end{aligned} \tag{14}$$

where

$\mathbf{1}_r$—a vector of ones of size r × 1,

$\mathbf{c} = c \cdot \mathbf{1}_{\mathrm{N}}$—a N × 1 vector with all elements equal to $c = \sum_{j=1}^{K} \overline{a}_j w_j$.

As the calibrating coefficient $\sqrt{\mathrm{N}}$ in the Eq. 13 as well as the same constant added to each element of the vector will not change the final ranking, we propose to calculate the synthesizing variable $\mathbf{S} = [S_i], i = 1, \ldots, \mathrm{N}$ in the way as in Eq. 14:

$$\mathbf{S} = \mathbf{Y}^{*}\left(\mathbf{V}^{*T}\mathbf{W}\right) \cdot \mathbf{1}_{\mathrm{K}^{*}} = \mathbf{Y}^{*}\mathbf{w}^{*} \tag{15}$$

where $\mathbf{w}^{*} = \left(\mathbf{V}^{*T}\mathbf{W}\right) \cdot \mathbf{1}_{\mathrm{K}^{*}}$ is a final vector of principle components combination. We can then multiply the elements of synthesizing variable by $\sqrt{\mathrm{N}}$ and add c to scale its elements similarly as other weighting methods do.

The steps of the proposed procedure MCPC are the following:

1. Calculate the elements of the covariance matrix between criteria after normalization.
2. Calculate the eigenvalues and eigenvectors of the covariance matrix $\mathbf{\Sigma}$.
3. Determine the number $\mathrm{K}^{*} < \mathrm{K}$ of principal components significant in explaining total variance of the criteria set.
4. Calculate the elements of the matrix $\mathbf{Y}^{*}$ (Eq. 11); the columns of this matrix are principal components vectors.
5. Compute the synthesizing variable $\mathbf{S} = [S_i]i = 1, \ldots, \mathrm{N}$ elements using the formula in Eq. 15.
6. Determine the final ranking of objects in descending order of S_i values.

3 Results of EU Countries Comparisons with Respect to "Healthy Lives and Well-Being"

In September 2015, the United Nations General Assembly adopted the 2030 Agenda for Sustainable Development[3] that includes 17 Sustainable Development Goals (SDGs).[4] In this work, we are interested in Goal 3: "Ensure healthy lives and promote well-being for all at all ages." United Nations formulated precisely thirteen important targets to attain by 2030 year:

T1. Reduce the global maternal mortality ratio to less than 70 per 100,000 live births.

T2. End preventable deaths of newborns and children under 5 years of age, with all countries aiming to reduce neonatal mortality to at least as low as 12 per 1000 live births and under-5 mortality to at least as low as 25 per 1000 live births.

---

[3]"*Transforming our world* ..." Resolution of UN General Assembly, September 2015.

[4]Source: https://www.un.org/sustainabledevelopment/sustainable-development-goals/.

T3. End the epidemics of AIDS, tuberculosis, malaria and neglected tropical diseases and combat hepatitis, water-borne diseases and other communicable diseases.
T4. Reduce by one third premature mortality from non-communicable diseases through prevention and treatment and promote mental health and well-being.
T5. Strengthen the prevention and treatment of substance abuse, including narcotic drug abuse and harmful use of alcohol.
T6. Halve the number of global deaths and injuries from road traffic accidents
T7. Ensure universal access to sexual and reproductive health-care services, including for family planning, information and education and the integration of reproductive health into national strategies and programmes.
T8. Achieve universal health coverage, including financial risk protection, access to quality essential health-care services and access to safe, effective, quality and affordable essential medicines and vaccines for all.
T9. Substantially reduce the number of deaths and illnesses from hazardous chemicals and air, water and soil pollution and contamination.
T10. Strengthen the implementation of the World Health Organization Framework Convention on Tobacco Control in all countries, as appropriate.
T11. Support the research and development of vaccines and medicines for the communicable and noncommunicable diseases that primarily affect developing countries, provide access to affordable essential medicines and vaccines
T12. Substantially increase health financing and the recruitment, development, training and retention of the health workforce in developing countries, especially in least developed countries and small island developing States.
T13. Strengthen the capacity of all countries, in particular developing countries, for early warning, risk reduction and management of national and global health risks.

To measure the level of "healthy" and "well-being," we have used eleven indicators developed by EUROSTAT.[5] We are aware of the fact that the following list of indicators is not completely comprehensive:

[SDG_03_10] *Life expectancy at birth.* Life expectancy at birth is defined as the mean number of years that a new-born child can expect to live if subjected throughout his life to the current age specific probabilities of dying.
[SDG_03_20] *Share of people with good or very good perceived health.* The indicator is a ***subjective*** measure; the share of the population aged 16 or over perceiving itself to be in "good" or "very good" health.
[SDG_03_30] *Smoking prevalence.* The indicator measures the share of the population aged 15 years and over who report that they currently smoke boxed cigarettes, cigars, cigarillos or a pipe.
[SDG_03_40] *Death rate due to chronic diseases*. The rate is calculated by dividing the number of people under 65 dying due to a chronic disease by the

[5]Precise definitions of indicators and remarks on the original sources of data can be found on https://ec.europa.eu/eurostat/web/sdi/main-tables.

total population under 65. Unit of measurement: number per 100,000 persons aged less than 65.

[SDG_03_41] *Death rate due to tuberculosis, HIV and hepatitis*. The indicator measures the standardized death rate of tuberculosis, HIV and hepatitis. Unit of measurement: number per 100,000 persons.

[SDG_03_60] *Self-reported unmet need for medical examination and care*. The indicator measures the share (%) of the population aged 16 and over reporting unmet needs for medical care due to one of the following reasons: "Financial reasons," "Waiting list" and "Too far to travel" (all three categories are cumulated).

[SDG_02_10] *Obesity rate by body mass index (BMI)*. The indicator measures the share of obese adults based on their body mass index (BMI $\geq$ 30).

[SDG_08_60] *People killed in accidents at work*. The indicator measures the number of fatal accidents per 100 000 persons in employment.

[SDG_11_20] *Population living in households considering that they suffer from noise*. The indicator measures the proportion of the population who declare that they are affected either by noise from neighbors or from the street.

[SDG_11_40] *People killed in road accidents*. The indicator measures the number of fatalities caused by road accidents per 100,000 persons.

[SDG_11_50a and SDG_11_50b] *Exposure to air pollution by particulate matter*. The two formulated indicators measure the population weighted annual mean concentration, in µg/m3, of particulate matter at urban background stations in agglomerations—PM10 and PM2.5.

Let us summarize shortly the main trends observed in indicator values within last decade or so. Life expectancy was observed as continuously increasing from 77.7 in 2002 to 81 years in 2016. In 2017, it decreased a little bit to 80.9 years (Fig. 1). Interesting indicator SDG 3.20 monitoring subjective perception of good health shows that starting from the year 2015 EU people each year feel better (Fig. 2). Other indicators show that exposure to unhealthy lifestyles in EU is in general decreasing. Air pollution by PM2.5 as we can observe on Fig. 3 during the period 2011–2016 had

Fig. 1 Life expectancy at birth in years SDG 3.10, total—the mean values for 28 European countries. *Source* Own calculations based on data from https://ec.europa.eu/eurostat/web/sdi/main-tables

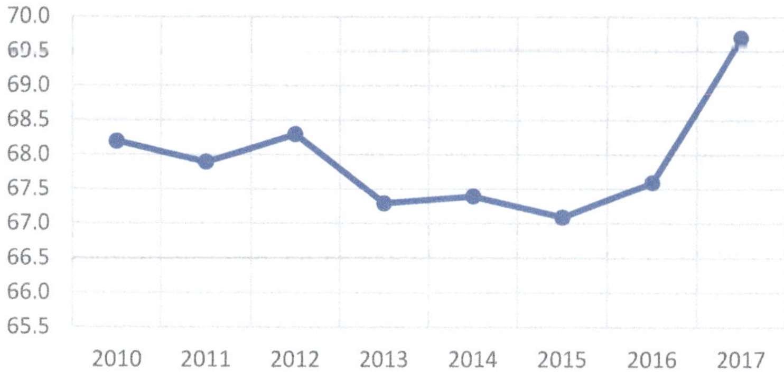

Fig. 2 Share of people with good or very good perceived health SDG 3.20, total—the mean values for 28 European countries. *Source* Own calculations based on data from https://ec.europa.eu/eurostat/web/sdi/main-tables

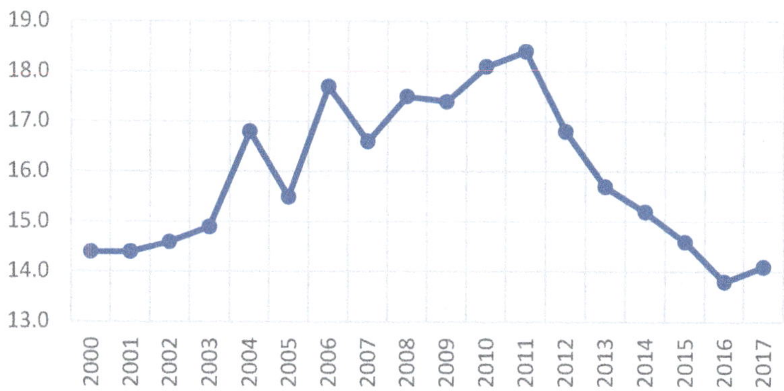

Fig. 3 Exposure to air pollution PM2.5 (diameters less than 2.5 μm) in $\mu g/m^3$—the mean values for 28 European countries. *Source* Own calculations based on data from https://ec.europa.eu/eurostat/web/sdi/main-tables

decreasing tendency. In 2017, it increased regrettably from 13.8 $\mu g/m^3$ observed in 2016 to 14.1 $\mu g/m^3$—the WHO goal to reduce this value below 10 $\mu g/m^3$ seems to be not available.

Table 1 presents chosen statistical characteristics of the Goal 3 indices. Looking at dispersion measured by relative quartile deviation, the EU countries are much diversified with respect of subjectively evaluated unmet needs for medical care. The other factors with large relative deviation are the death rate due to communicable diseases and road accidents.

We have prepared rankings of 28 EU countries using individual sub-indices and data available for the year 2017. The results are presented in Table 2.

We would like to comment on the results for the two following cases:

Table 1 Statistical characteristics of the Goal 3 indices in 2017 for EU countries

Index description	European Union—28 countries mean	Unit	min	Q1	Median	Q3	max	Relative quartile deviation (%)
sdg3.10 Life expectancy	80.9	Years	74.8	78.0	81.4	82.1	83.4	2.6
sdg3.20 Good health	69.7	%	43.9	61.7	70.3	74.4	83.3	9.1
sdg3.30 Smoking	26	%	7.0	20.8	26.5	29.0	37.0	15.6
sdg3.40 Chronic diseases	119	Per 100 thou.	78.7	98.8	112.7	159.3	243.7	26.9
sdg3.41 Tuberculosis	2.6	Per 100 thou.	0.7	1.2	1.8	3.1	10.5	54.2
sdg3.60 Unmet needs	1.7	%	0.1	0.9	1.7	3.3	11.8	71.3
sdg2.10 BMI	52	%	44.9	50.0	55.3	57.0	62.9	6.4
sdg8.60 Accidents at work	1.68	Per 100 thou.	0.5	0.9	1.9	2.6	4.5	42.2
sdg11.20 Noise	17.5	%	8.2	12.5	15.3	18.9	26.1	21.0
sdg11.40 Road accidents	4.9	Per 100 thou.	2.5	3.9	5.2	6.5	10.0	25.5
sdg11.50a PM10	21.6	$\mu g/m^3$	10.0	17.3	20.4	26.1	37.3	21.6
sdg11.50b PM2.5	14.1	$\mu g/m^3$	4.9	11.2	12.9	19.0	23.8	30.2

Source Own calculations; data from https://ec.europa.eu/eurostat/web/sdi/main-tables

1. Rankings of 28 EU countries excluding the goal SDG 11.50 (ten criteria) because complete data were not available for Greece, Lithuania and Malta,
2. Rankings of 25 EU countries and 12 criteria, excluding Greece, Lithuania and Malta.

Multi-criteria rankings were performed in six variants:

- using classical SAW method with equal weights, with weights calculated using informative capacity measure based on VIFs, weights established using PCA,
- using MCPC method when choosing three or four principal components with weights for criteria established as in SAW method.

Table 2 Rankings of the 28 EU countries in 2017 for the Goal 3 individual criteria

Country	sdg 3.10	sdg3.20	sdg3.30	sdg3.40	sdg3.41	sdg3.60	sdg2.10	sdg8.60	sdg11.20	sdg11.40	sdg11.50a	sdg11.50b
Austria	10	14	17	12	20	3	8	21	20	12	11	15
Belgium	12	8	3	8	6	16	5	11	15	16	14	13
Bulgaria	28	18	26	24	16	16	23	27	4	27	25	24
Croatia	21	22	25	22	22	14	24	22	2	26	25	19
Cyprus	6	2	17	2	13	12	12	2	17	20	22	16
Czechia	19	21	21	19	4	7	27	12	12	16	17	18
Denmark	17	11	3	13	8	8	4	7	21	3	5	5
Estonia	20	25	9	21	24	28	16	9	1	6	2	2
Finland	10	15	7	9	1	24	25	8	7	11	1	1
France	3	16	26	11	15	8	2	23	16	15	10	9
Germany	17	19	11	14	13	5	10	5	28	7	8	12
Greece	14	10	28	18	12	27	20	10	23	22	–	–
Hungary	24	23	15	28	18	8	19	18	5	21	20	23
Ireland	6	1	3	6	8	20	22	14	3	5	3	4
Italy	2	3	11	3	23	15	1	19	7	18	22	20
Latvia	27	27	24	25	28	26	21	20	13	24	7	14
Lithuania	25	28	21	26	27	12	18	25	10	22	16	–
Luxembourg	8	12	8	4	6	5	6	23	24	10	13	7
Malta	5	6	10	10	17	3	26	1	26	9	–	–
Netherlands	9	5	3	7	1	1	3	3	27	4	11	8
Poland	22	24	23	20	18	21	15	16	6	25	24	24

(continued)

Table 2 (continued)

Country	sdg 3.10	sdg3.20	sdg3.30	sdg3.40	sdg3.41	sdg3.60	sdg2.10	sdg8.60	sdg11.20	sdg11.40	sdg11.50a	sdg11.50b
Portugal	12	26	13	17	25	18	13	26	25	19	9	9
Romania	26	13	17	27	26	25	28	28	22	28	21	22
Slovakia	23	17	13	23	8	19	14	16	9	14	18	17
Slovenia	16	20	17	16	3	23	11	13	11	13	19	21
Spain	1	9	15	4	21	1	9	15	14	8	15	11
Sweden	4	4	1	1	5	11	7	6	17	1	4	3
UK	15	7	2	15	8	21	16	4	19	2	6	6

Source Own calculations; data from https://ec.europa.eu/eurostat/web/sdi/main-tables

Matrices of Pearson correlation coefficients between indices are presented in Tables 3 and 4. In the first case, with ten criteria and 28 countries, the condition index of the correlation matrix being the ratio of maximal to minimal eigenvalue was equal to 174.46. Maximal correlation observed was between SDG 3.10 *Life expectancy* and SDG3.30 *Chronic diseases*, equal to 0.9612. Correlation coefficients between eight pairs of indices were greater than 0.6 in absolute value. Negative, but not so strong, correlations with other indices were observed for the SDG11.20—subjective *Suffering from noise*. In the second research, with 12 sub-indices and 25 countries, condition index for the correlation matrix was equal to 216.22. Thirteen coefficients exceeded 0.6, and two of them were greater than 0.9.

Weights obtained using three methods for the two rankings are presented in Table 5.

Comparing weights established using informative capacity (IC) with PCA weights, we can notice that in both cases, IC rigidly limits the influence of sub-indices which are heavily dependent—SDG3.10 and SDG3.40. The highest weights using IC were assigned for the index SDG11.20, negatively correlated with other indices. In case of IC, relatively high values are assigned to the SDG3.60 which is variable and weakly correlated with other indices. PCA weights differ much. When the number of indices increases, the distribution of PCA weights becomes more uniform.

In the first research, we have identified four principal components of the covariance matrix, explaining 81.9% of the total variance. Figure 4a presents the composite indices constructed according to the three discussed weightings for SAW method and Fig. 4b composite indices for three corresponding variants of weighting for MCPC method. The EU countries are sorted according to decreasing values of SAW with equal weights and MCPC with equal weights.

See Table 6 for Pearson correlation coefficients between composite indices obtained in six analyzed variants. The most convergent results were for SAW and MCPC with PCA weights. The lowest correlation was observed between SAW with IC weights and MCPC with equal weights. Final rankings obtained are presented in Table 7.

PCA weighting gave almost the same positions in rankings as equal weighting for SAW and similarly for MCPC, although the results of the two methods vary. Weighting according to informative capacity gave results which differ much from other weighting propositions for SAW as well as for MCPC method.

We compared the rankings results using Kendall τ coefficients (see Table 8). Conclusions are approved—lowest values correspond with ranks obtained by equal weighting and the results of weighting according to informative capacity. We have found that the most sensitive positions in rankings were for Netherlands, Malta, Greece, Slovenia and Hungary—the difference in positions exceeded seven for various SAW and MCPC rankings. The least sensitive were rankings for Romania, Latvia (27th or 28th position) and Cyprus (7th or 8th). On the first place in almost all rankings was Sweden, then Ireland and Italy. In the first research, we did not take into account the indices for air pollution.

Table 3 Pearson correlation coefficients between sub-indices in 2017—case of 28 EU countries and ten sub-indices

	sdg3.10	sdg3.20	sdg3.30	sdg3.40	sdg3.41	sdg3.60	sdg2.10	sdg8.60	sdg11.20	sdg11.40
sdg3.10	1	0.6412	0.4287	0.9612	0.5364	0.2748	0.5486	0.5179	−0.4183	0.7005
sdg3.20	0.64	1	0.3967	0.6502	0.7044	0.2689	0.2584	0.3571	−0.2172	0.3662
sdg3.30	0.43	0.40	1	0.4463	0.3160	0.1853	0.2888	0.4880	−0.2026	0.7332
sdg3.40	0.96	0.65	0.45	1	0.5827	0.2501	0.5266	0.5495	−0.3789	0.6750
sdg3.41	0.54	0.70	0.32	0.58	1	0.3450	0.1952	0.4793	−0.1389	0.4405
sdg3.60	0.27	0.27	0.19	0.25	0.35	1	0.2890	−0.0105	−0.2948	0.1786
sdg2.10	0.55	0.26	0.29	0.53	0.20	0.29	1	0.1770	−0.2768	0.4559
sdg8.60	0.52	0.36	0.49	0.55	0.48	−0.01	0.18	1	−0.2426	0.7067
sdg11.20	−0.42	−0.22	−0.20	−0.38	−0.14	−0.29	−0.28	−0.24	1	−0.2827
sdg11.40	0.70	0.37	0.73	0.68	0.44	0.18	0.46	0.71	−0.28	1

Source Own calculations

Table 4 Pearson correlation coefficients between sub-indices in 2017—case of 25 EU countries and 12 sub-indices

	sdg3.10	sdg3.20	sdg3.30	sdg3.40	sdg3.41	sdg3.60	sdg2.10	sdg8.60	sdg11.20	sdg11.40	sdg11.50a	sdg11.50b
sdg3.10	1	0.5768	0.4809	0.9591	0.4703	0.4036	0.6588	0.4663	−0.3708	0.7161	0.4295	0.5590
sdg3.20	0.58	1	0.47	0.58	0.64	0.46	0.34	0.27	−0.13	0.36	0.08	0.21
sdg3.30	0.48	0.47	1	0.49	0.36	0.02	0.30	0.58	−0.27	0.73	0.67	0.68
sdg3.40	0.96	0.58	0.49	1	0.51	0.36	0.62	0.52	−0.35	0.69	0.39	0.55
sdg3.41	0.47	0.64	0.36	0.51	1	0.55	0.21	0.46	−0.09	0.44	0.04	0.12
sdg3.60	0.40	0.46	0.02	0.36	0.55	1	0.37	0.03	−0.43	0.11	−0.26	−0.18
sdg2.10	0.66	0.34	0.30	0.62	0.21	0.37	1	0.30	−0.45	0.51	0.19	0.26
sdg8.60	0.47	0.27	0.58	0.52	0.46	0.03	0.30	1	−0.12	0.74	0.46	0.53
sdg11.20	−0.37	−0.13	−0.27	−0.35	−0.09	−0.43	−0.45	−0.12	1	−0.28	−0.24	−0.23
sdg11.40	0.72	0.36	0.73	0.69	0.44	0.11	0.51	0.74	−0.28	1	0.77	0.78
sdg11.50a	0.43	0.08	0.67	0.39	0.04	−0.26	0.19	0.46	−0.24	0.77	1	0.92
sdg11.50b	0.56	0.21	0.68	0.55	0.12	−0.18	0.26	0.53	−0.23	0.78	0.92	1

Source Own calculations

Table 5 Weights for sub-indices measuring Goal 3: "Healthy Lives and Well-being" in 2017

Index description	First ranking			Second ranking		
	Equal weights (%)	IC weights (%)	PCA weights (4 PC) (%)	Equal weights (%)	IC weights (%)	PCA weights (3 PC) (%)
sdg3.10 Life expectancy	10	2.11	15.41	9.09	1.78	12.99
sdg3.20 Good health	10	8.07	9.10	9.09	11.98	5.63
sdg3.30 Smoking	10	6.61	4.39	9.09	10.52	5.32
sdg3.40 Chronic diseases	10	2.20	13.47	9.09	2.09	10.83
sdg3.41 Tuberculosis	10	8.95	8.94	9.09	10.26	7.11
sdg3.60 Unmet needs	10	15.83	5.32	9.09	10.88	4.95
sdg2.10 BMI	10	19.77	12.70	9.09	13.80	9.08
sdg8.60 Accidents at work	10	8.50	7.49	9.09	11.38	4.69
sdg11.20 Noise	10	23.82	13.39	9.09	17.85	10.30
sdg11.40 Road accidents	10	4.13	9.78	9.09	2.92	8.68
sdg11.50a PM10	–	–	–	4.55	2.80	9.50
sdg11.50b PM2.5	–	–	–	4.55	3.73	10.90

Source Own calculations

The second research, carried out also for 2017 year, was complemented by two air pollution indices. In this research, we have identified three principal components of the covariance matrix explaining 78.9% of the total variance. We compared 25 EU countries—without Greece, Lithuania and Malta. The resulting composite indices are presented on graphs on Fig. 5a, b. See Table 9 for Pearson correlation between composite indices and Table 10 for Kendall τ rank correlation coefficients. The most similar results were for SAW and MCPC methods with equal weights, the least similar—versions compared with SAW and IC weights (Table 11).

The least sensitive was position of Sweden (always the first), Ireland (always the second), Romania (24th or 25th) and Poland (19th or 20th). Results in this research were more stable, in general. Rankings obtained by MCPC method comparing with SAW are less sensitive on changes in weights.

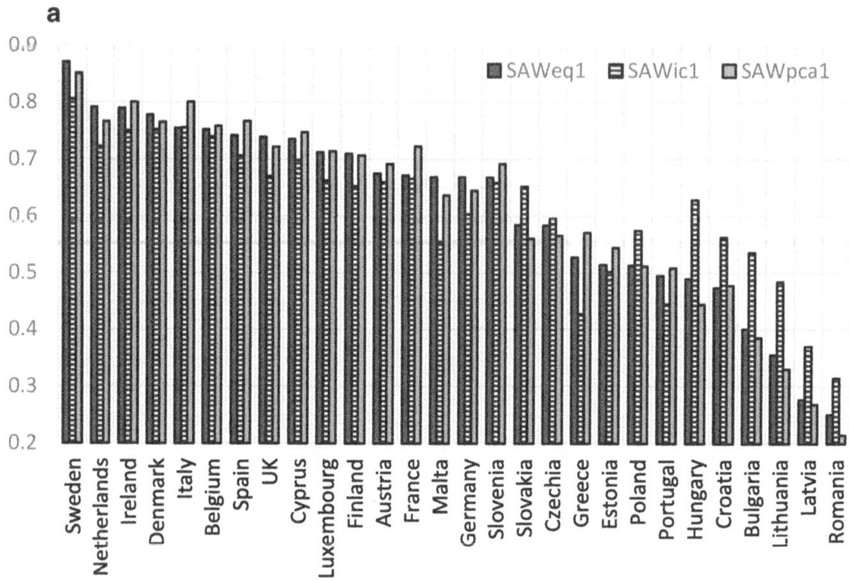

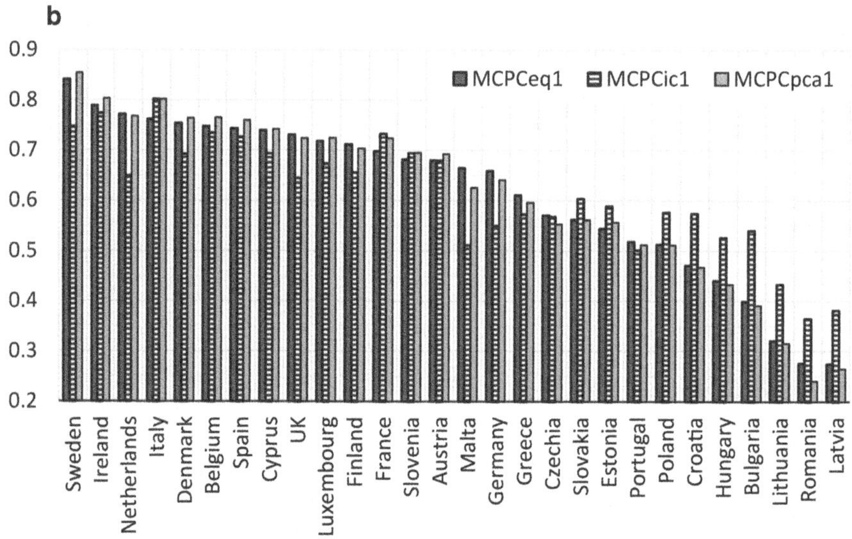

Fig. 4 **a** Composite indices for SAW method with different weightings—Goal 3—1st research for 2017. *Source* Own calculations. **b** Composite indices for MCPC method with different weightings—Goal 3, 1st research for 2017. *Source* Own calculations

Table 6 Pearson correlation coefficients between composite indices, 1st research for 2017

	SAWeq	SAWic	SAWpca	MCPCeq	MCPCic	MCPCpca
SAWeq	1	0.9107	0.9892	0.9884	0.886729	0.984758
SAWic	0.9107	1	0.889997	0.85806	0.899761	0.875427
SAWpca	0.9892	0.8900	1	0.994636	0.925529	0.998356
MCPCeq	0.9884	0.8581	0.9946	1	0.8971	0.996275
MCPCic	0.8867	0.8998	0.9255	0.8971	1	0.927053
MCPCpca	0.9848	0.8754	0.9984	0.9963	0.9271	1

Source Own calculations

4 Conclusions

The first research made possible the classification of EU countries from the point of view criteria of health and well-being in 2017 year. Only air-pollution indices were not analyzed. The most diversifying were composite indices obtained using SAW and MCPC with weights selected according to PCA—relative dispersion was about 27.5%. We have chosen three classes, using thresholds 0.5 and 0.75 (values close to the first and the third quartiles) for the value of SAW index. The resulting classification allows us to conclude that:

- Sweden, Italy, Ireland, Spain, Netherlands, Denmark and Belgium were in 2017 in the top class of EU countries considering criteria of the Goal 3 of sustainable development.
- Romania, Latvia, Lithuania, Bulgaria, Hungary and Croatia are included in the class with the lowest values of the developed composite indices.

Saisana et al. (2005) proposed a measure constructed as the average of the absolute differences in countries' ranks with respect to a reference ranking over chosen group of countries. Calculating the range of country ranks obtained for different weights, we have identified in our research countries which ranks were the most sensitive on weights (differences in ranks equal 7 or more):

- In the first research (without air-pollution indices): Netherlands, Malta, Slovenia, Greece, France, Hungary and Estonia
- In the second research: Italy and Cyprus.

Table 7 Rankings obtained using SAW and MCPC methods—1st research for 2017

Country	SAW			MCPC		
	Equal weighting	IC weighting	PCA weighting (4 PC)	Equal weighting	IC weighting	PCA weighting (4 PC)
Austria	12	12	13	14	10	14
Belgium	6	5	7	6	4	5
Bulgaria	25	22	25	25	22	25
Croatia	24	20	23	23	18	23
Cyprus	9	8	8	8	8	8
Czechia	18	18	18	18	20	20
Denmark	4	3	6	5	9	6
Estonia	20	23	20	20	16	19
Finland	11	14	12	11	12	12
France	13	10	9	12	5	10
Germany	15	17	15	16	21	15
Greece	19	26	17	17	19	17
Hungary	23	16	24	24	23	24
Ireland	3	4	3	2	2	2
Italy	5	2	2	4	1	3
Latvia	27	27	27	28	27	27
Lithuania	26	24	26	26	26	26
Luxembourg	10	11	11	10	11	9
Malta	14	21	16	15	24	16
Netherlands	2	6	5	3	13	4
Poland	21	19	21	22	17	22
Portugal	22	25	22	21	25	21
Romania	28	28	28	27	28	28
Slovakia	17	15	19	19	15	18
Slovenia	16	13	14	13	7	13
Spain	7	7	4	7	6	7
Sweden	1	1	1	1	3	1
UK	8	9	10	9	14	11

Source Own calculations

Table 8 Kendall τ coefficients between ranks—1st research performed for 2017

	SAWeq	SAWic	SAWpca	MCPCeq	MCPCic	MCPCpca
SAWeq	1					
SAWic	0.7884	1				

(continued)

Table 8 (continued)

	SAWeq	SAWic	SAWpca	MCPCeq	MCPCic	MCPCpca
SAWpca	0.8995	0.8042	1			
MCPCeq	0.9312	0.7831	0.9365	1		
MCPCic	0.6508	0.7249	0.7196	0.6772	1	
MCPCpca	0.9048	0.7884	0.9418	0.9524	0.7249	1

Source Own calculations

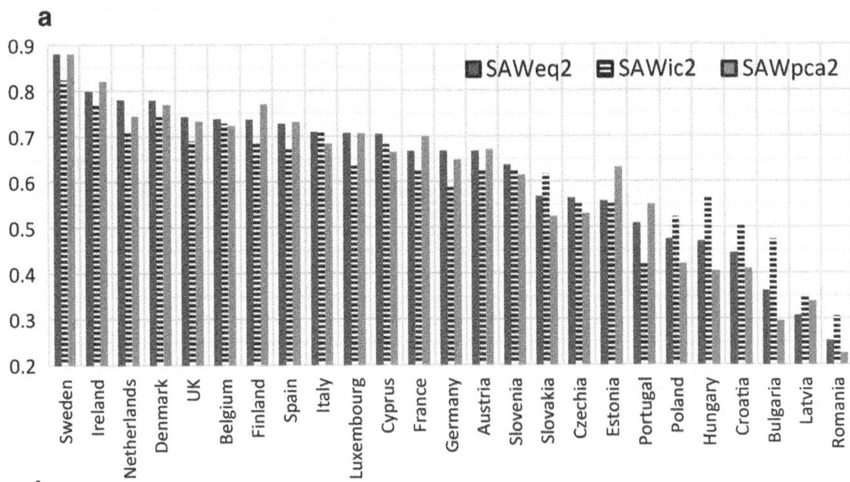

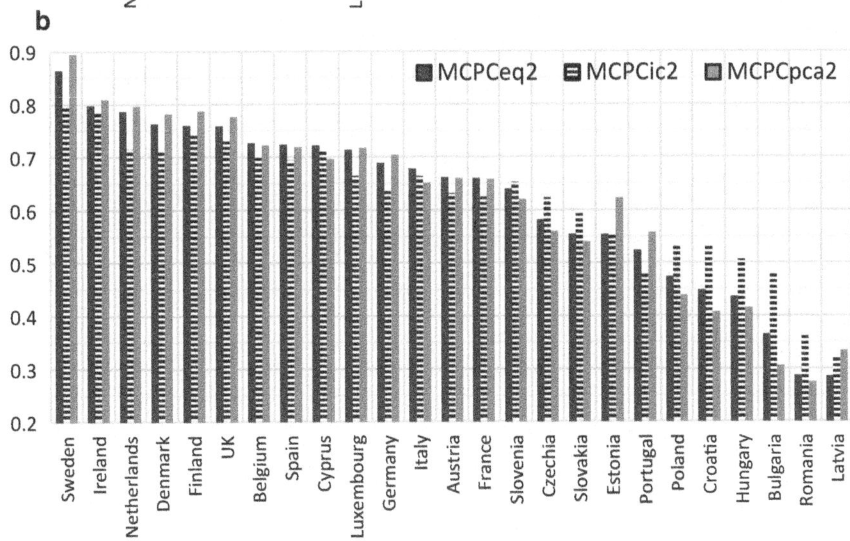

Fig. 5 **a** Composite indices for SAW method with different weightings—Goal 3—2nd research for 2017. *Source* Own calculations. **b** Composite indices for MCPC method with different weightings—Goal 3—2nd research for 2017. *Source* Own calculations

Table 9 Pearson correlation coefficients between composite indices, 2nd research for 2017

	SAWeq	SAWic	SAWpca	MCPCeq	MCPCic	MCPCpca
SAWeq	1	0.955485	0.980119	0.994569	0.965863	0.980843
SAWic	0.9555	1	0.903123	0.933654	0.961403	0.895014
SAWpca	0.9801	0.9031	1	0.976249	0.921555	0.989911
MCPCeq	0.9946	0.9337	0.9762	1	0.971137	0.986199
MCPCic	0.9659	0.9614	0.9216	0.9711	1	0.930947
MCPCpca	0.9808	0.8950	0.9899	0.9862	0.9309	1

Source Own calculations

Table 10 Rankings obtained using SAW and MCPC methods—2nd research for 2017

Country	SAW			MCPC		
	Equal weighting	IC weighting	PCA weighting (3 PC)	Equal weighting	IC weighting	PCA weighting (3 PC)
Austria	14	14	12	13	14	12
Belgium	6	4	8	7	8	7
Bulgaria	23	22	24	23	23	24
Croatia	22	21	21	21	20	22
Cyprus	11	9	13	9	6	11
Czechia	17	18	18	16	16	17
Denmark	4	3	4	4	7	5
Estonia	18	19	15	18	18	15
Finland	7	8	3	5	3	4
France	12	12	10	14	15	13
Germany	13	16	14	11	13	10
Hungary	21	17	22	22	21	21
Ireland	2	2	2	2	2	2
Italy	9	6	11	12	11	14
Latvia	24	24	23	25	25	23
Luxembourg	10	11	9	10	10	9
Netherlands	3	5	5	3	5	3
Poland	20	20	20	20	19	20
Portugal	19	23	17	19	22	18

(continued)

Table 10 (continued)

Country	SAW			MCPC		
	Equal weighting	IC weighting	PCA weighting (3 PC)	Equal weighting	IC weighting	PCA weighting (3 PC)
Romania	25	25	25	24	24	25
Slovakia	16	15	19	17	17	19
Slovenia	15	13	16	15	12	16
Spain	8	10	7	8	9	8
Sweden	1	1	1	1	1	1
UK	5	7	6	6	4	6

Source Own calculations

Table 11 Kendall τ coefficients between ranks—2nd research

	SAWeq	SAWic	SAWpca	MCPCeq	MCPCic	MCPCpca
SAWeq	1					
SAWic	0.8733	1				
SAWpca	0.8733	0.7733	1			
MCPCeq	0.9267	0.8133	0.8667	1		
MCPCic	0.8467	0.8133	0.8133	0.9067	1	
MCPCpca	0.8867	0.7600	0.9067	0.9200	0.8400	1

Source Own calculations

References

Becker W et al (2017) Weights and importance in composite indicators: closing the gap. Ecol Ind 80:12–22. https://doi.org/10.1016/j.ecolind.2017.03.056

Chao Y-S, Wu C-J (2017) Principal component-based weighted indices and a framework to evaluate indices: results from the medical expenditure panel survey 1996 to 2011. PLoS ONE 12(9):e0183997. https://doi.org/10.1371/journal.pone.0183997

Churchman CW, Ackoff RL (1954) An approximate measure of value. J Oper Res Soc Am 2(2):172–187. https://doi.org/10.1287/opre.2.2.172

Hudrlikova L (2013) Composite indicators as a useful tool for international comparison: the Europe 2020 example. Prague Econ Pap 4:459–473

Jacobs R et al (2004) Measuring performance: an examination of composite performance indicators. Centre for Health Economics, CHE technical paper series 29, https://www.york.ac.uk/che/publications/in-house/archive/2000s/

Jolliffe IT (2002) Principal component analysis, 2nd edn. Springer, New York

Lafortune G, Fuller G, Moreno J, Schmidt-Traub G, Kroll C (2018) SDG index and dashboards. Detailed methodological paper. https://github.com/sdsna/2018GlobalIndex/raw/master/2018GlobalIndexMethodology.pdf

Nardo M et al (2005) Tools for Composite Indicators Building, http://publications.jrc.ec.europa.eu/repository/handle/JRC31473

Nardo M et al (2008) Handbook on Constructing Composite Indicators: Methodology and User Guide, OECD Publishing. http://publications.jrc.ec.europa.eu/repository/handle/JRC47008
Paruolo P et al (2013) Ratings and rankings: voodoo or science? J R Stat Soc A 1763(Part 3):609–634
Sachs J et al (2018) SDG index and dashboards report 2018. Bertelsmann, New York
Saisana M, Tarantola S (2002) State-of-the-art report on current methodologies and practices for composite indicator development. Report EUR 20408 EN. European Commission–Joint Research Centre, Ispra
Saisana M, Saltelli A, Tarantola S (2005) Uncertainty and sensitivity analysis techniques as tools for the quality assessment of composite indicators. J R Stat Soc A 168(Part 2):307–323
Transforming our world: the 2030 Agenda for Sustainable Development, Resolution adopted by the General Assembly of United Nations on 25 Sept 2015. https://www.un.org/ga/search/view_doc.asp?symbol=A/RES/70/1&Lang=E

The Comparison of Income Distributions for Women and Men in the European Union Countries

Joanna Landmesser

Abstract The purpose of this study was to compare personal income distributions in countries of the European Union, taking into account gender differences. Using data from the EU-SILC project, the gender income gap for 28 European countries was examined. First, we examined the income inequalities of men and women in each country using the Oaxaca–Blinder decomposition procedure. The unexplained part of the gender pay gap gave us information about the wage discrimination. Second, we extended the decomposition procedure to different quantile points along the whole income distribution. To construct the counterfactual distribution, we used the recentered influence function—regression approach. We found that there exists an important diversity in the size of the gender pay gap across members of the European Union. The results obtained for these countries allowed us to group them into four clusters using the agglomerative clustering algorithm. The results of decomposition were analyzed and compared across the formulated groups of countries.

Keywords Income inequalities · Gender pay gap · Classification

1 Introduction

The persistence of gender differences in personal incomes is one of the best documented facts in labor economics. For 2017, the gender gap in hourly earnings is estimated to be 16% for the EU-28 (Eurostat 2019). There are considerable differences at country level, with the gender pay gap ranging from just over 3.5% in Romania, Luxembourg and Italy, to 25.6% in Estonia, followed by the Czech Republic and Germany.

The gender pay gap is a subject of interest in numerous studies. Despite the diversity of research, many aspects of the gap are still not sufficiently explored. By exploiting data from the EU Statistics on Income and Living Conditions (EU-SILC) survey, a detailed picture of income inequality in Europe can be obtained.

J. Landmesser (✉)
Warsaw University of Life Sciences—SGGW, Warsaw, Poland
e-mail: joanna_landmesser@sggw.pl

K. Jajuga et al. (eds.), *Classification and Data Analysis*, Studies in Classification, Data Analysis, and Knowledge Organization,
https://doi.org/10.1007/978-3-030-52348-0_18

The purpose of our study is to compare personal income distributions for men and women in countries of the EU and to discuss whether there exists a significant diversity in this respect across the countries.

Over the years, many decomposition techniques are developed that attribute fractions of the gap to gender differences in observed employee characteristics. In this way, adjusted gaps can be achieved that represent the unexplained part of gender pay gap. The classic Oaxaca–Blinder decomposition focuses on the gap in average hourly earnings between male and female workers (Oaxaca 1973; Blinder 1973). Numerous studies concentrate on the decomposition of the average values for incomes, for example, Jurajda (2003) for the Czech Republic and Slovakia, Pena-Boquete et al. (2010) for Italy and Spain, Chatterji et al. (2011) for the United Kingdom, Śliwicki and Ryczkowski (2014) for Poland. A number of papers adopt a cross-country perspective. Using EU-SILC data Oczki (2016) and Hedija (2017) showed that the gender pay gap varies among the EU countries. In the studies conducted by Boll and Lagemann (2019) or Leythienne and Ronkowski (2018), the gap is analyzed based on the Structure of Earnings Survey (EU-SES).

Other approaches undertake gender comparisons at different quantiles of the wage distribution (e.g., Albrecht et al. (2003) for Sweden, Landmesser (2016) for Poland). The results of studies show evidence of the glass ceiling and the sticky floor effects. Arulampalam et al. (2006) examined the gender wage gap in 11 European countries using the European Community Household Panel Survey (ECHPS). The gap widened toward the top of the wage distribution in most of the countries and, in a few cases, it also widened at the bottom of the distribution. Nicodemo (2009) analyzed the gap in France, Greece, Italy, Portugal and Spain, using the EU-SILC and the ECHPS datasets. She found a positive wage gap in all countries, the greater part of which cannot be explained by observed characteristics. The gender gap was larger at the bottom and smaller at the top of the distribution in most countries. Also, Christofides et al. (2013) used EU-SILC data and estimated the adjusted gap for 26 European countries. Despite the many differences between the individual studies, they all conclude that the gender pay gap exhibits a remarkable diversity across European countries.

In this paper, we examine the differences at various quantile points along the income distribution. As a decomposition method, we apply the recentered influence function—regression method (Firpo et al. 2009). We enrich the existing literature by undertaking comparisons of the gap components in 28 EU countries, using data from the EU-SILC project in 2014.

The outline of the remainder of the study is as follows. Section 2 describes the econometric methods used in the analysis. The empirical data set and the obtained results are presented and discussed in Sect. 3, and Sect. 4 concludes.

2 Research Methodology

Our hypothesis is that there exists an important diversity in size and shape of the gender pay gap across members of the European Union. The observed differences should be analyzed in an adequate manner. Starting with the works of Oaxaca (1973) and Blinder (1973), decomposition methods are becoming more popular. Their idea is to split the observed gap into parts that have a meaningful economic interpretation.

In the Oaxaca–Blinder approach, a decomposition of the gap in a part explained by differences in individual characteristics of male and female employees, and a remaining unexplained part is performed. Let income Y_g be a linear function of people's characteristics X_g ($g = M$ in men's group and $g = W$ in women's group): $y_g = X_g\beta_g + v_g$. The Oaxaca–Blinder decomposition for the average income inequality between two groups at the aggregate level can be expressed as

$$\hat{\Delta}^{\mu} = \bar{Y}_M - \bar{Y}_W = \bar{X}_M\hat{\beta}_M - \bar{X}_W\hat{\beta}_W = \underbrace{\bar{X}_M(\hat{\beta}_M - \hat{\beta}_W)}_{\text{unexplained effect}} + \underbrace{(\bar{X}_M - \bar{X}_W)\hat{\beta}_W}_{\text{explained effect}}. \quad (1)$$

The first term on the right hand side of the equation represents the amount of discrimination and is the result of differences in the estimated parameters. It is called the unexplained effect and exists only because the market evaluates differently the identical bundle of traits if possessed by different demographic groups. The second term on the right hand side expresses the difference of the potentials of both groups. A drawback of this approach is that it focuses only on average effects which may lead to a misleading assessment if the effects of covariates vary across the income distribution. Also, the standard assumption used in this decomposition that the outcome variable Y is linearly related to the covariates X is questionable and is often not fulfilled in practice.

The mean decomposition analysis may be extended to the case of differences between the two distributions. Some modifications of the Oaxaca–Blinder method have developed Juhn et al. (1993), DiNardo et al. (1996), Donald et al. (2000), Machado and Mata (2005) or Fortin et al. (2010). They define the distribution function $F_{Y_g}(y)$ as the conditional distribution of Y and the joint distribution of all elements of X: $F_{Y_g}(y) = \int F_{Y_g|X_g}(y|X) \cdot F_{X_g}(X)\,dx$. Then, they construct the counterfactual distribution that is the distribution of incomes that would prevail for women if they had the distribution of men's characteristics: $F_{Y_W^C}(y) = \int F_{Y_W|X_W}(y|X) \cdot dF_{X_M}(X)$. The difference in the observed income distributions between men and women can be decomposed as

$$F_{Y_M}(y) - F_{Y_W}(y) = \underbrace{[F_{Y_M}(y) - F_{Y_W^C}(y)]}_{\text{unexplained effect}} + \underbrace{[F_{Y_W^C}(y) - F_{Y_W}(y)]}_{\text{explained effect}}. \quad (2)$$

Such a decomposition can be performed by the use of the recentered influence function–regression method (Firpo et al. 2009). This method is similar to a standard regression, except that the dependent variable Y is replaced by the recentered influence function of the statistic of interest:

$$RIF(y, Q_\tau) = Q_\tau + IF(y, Q_\tau) = Q_\tau + \frac{\tau - \mathrm{I}\{y \leq Q_\iota\}}{f_Y(Q_\tau)}, \tag{3}$$

where $IF(y, Q_\tau)$ is the so-called influence function. The *RIF* for a quantile is simply an indicator variable I{.} for whether the outcome variable y is smaller or equal to the quantile Q_τ. The conditional expectation of RIF can be modeled as a linear function of the explanatory variables. First, we have to compute the sample quantile $\hat{Q}_\tau$ and to estimate the density $\hat{f}_Y(\hat{Q}_\tau)$ using kernel methods. Then, we calculate the *RIF* of each observation and run regressions of the *RIF* on the vector X. The estimates of models for proportions are locally inverted back into the space of quantiles. This provides a way of decomposing quantiles using regression models for proportions (Fortin et al. 2010). The aggregated and detailed decomposition for any unconditional quantile is then:

$$\hat{\Delta}^\tau = \bar{X}_M(\hat{\beta}_{M,\tau} - \hat{\beta}_{W,\tau}) + (\bar{X}_M - \bar{X}_W)\hat{\beta}_{W,\tau} =$$
$$= \sum_{j=1}^{k} (\bar{X}_{jM}(\hat{\beta}_{jM,\tau} - \hat{\beta}_{jW,\tau}) + (\bar{X}_{jM} - \bar{X}_{jW})\hat{\beta}_{jW,\tau}). \tag{4}$$

In this study, our strategy can be summarized as follows: First, for each EU country, we compute the gender gap in average incomes, and the Oaxaca–Blinder decomposition of these gaps into explained and unexplained parts is executed. This is useful for indicative purposes. Then, we examine the differences at various quantile points along the income distribution applying the RIF-regression method. In this process, people characteristics included in our dataset are used as explanatory factors. Finally, after assessing of the gender pay gap for all EU countries, an attempt will be made to group them using agglomerative hierarchical clustering algorithm. The squared Euclidean distance will be used to measure the dissimilarity between each pair of observations, and dissimilarity between clusters of observations will be assessed using Ward's minimum variance method. The results of decomposition will be analyzed and compared across the formulated groups of countries.

3 Empirical Data and Results

Our analysis relies on the 2014 EU-SILC cross-sectional data (research proposal 234/2016-EU-SILC). The EU-SILC is an instrument aiming at collecting comparable cross-sectional and longitudinal multidimensional microdata on income, poverty, social exclusion and living conditions. For this reason, it is a powerful instrument for a comparative analysis of income distribution in all 28 European Union countries. The total number of observations is 174,378 (88,398 men and 85,980 women).

The annual gross employee (cash or near cash) incomes (in Euro) of men were compared with those obtained by women. The gross employee income corresponds to wages and salaries paid for the time worked, remuneration for the time not worked, enhanced rates of pay for overtime, payments for fostering children, supplementary payments (e.g., thirteenth month payment). It includes any social contributions and income taxes payable to social insurance schemes or tax authorities. In our empirical analysis, a logarithm of the annual income constitutes the outcome variable. The sample size, average income and Gini coefficient values for the cross-country sample can be found in Table 1.

In Table 2, the explanatory variables are described. As individual worker characteristics, age, education level and marital status were included. As job-related characteristics, contract type, working time, position in the firm and size of the enterprise were taken into account.

Figure 1 presents the unadjusted gender pay gap (raw differential) and the results of the aggregate Oaxaca–Blinder decomposition of inequalities between men's and women's log incomes for 28 EU countries.

We have found that there is a positive difference between the mean values of log incomes for men and women for all 28 countries. The mean log income differential is the largest in Germany (0.625) and the smallest in Slovenia (0.112). The country heterogeneity is not limited to the size of the gap but also concerns its composition. The difference between the mean log income values was decomposed into two components: the first one explaining the contribution of the different values of models coefficients (the unexplained part), and the second one explaining the contribution of the attributes differences (the explained part). The unexplained effect is huge (and positive) for the states with the low raw differential and is small for the states with the high raw differential. Its share ranges from 30.8% in Luxembourg to 170.1% in Lithuania. This part of the gender pay gap gives us information about the discrimination. The explained gap is negative in ten countries (among others in Latvia, Lithuania, Slovenia, Estonia, Portugal and Poland). The negative value of this component means that the difference of the average log incomes between men and women is reduced by the women's characteristics. In 18 countries, the explained part is positive, that is, it increases the overall gap, with the largest share in Luxembourg (69.2%). Only in five countries, the explained part exceeds the unexplained part of the overall gap. However, the unexplained part is nowhere identified to be negative.

Since the Oaxaca–Blinder technique focuses only on average effects, we carried out the decomposition of inequalities along the distribution of log incomes for men and women using the RIF-regression method. The total differences between the values of log incomes were computed, and the results are shown in Table 3. They are expressed in terms of percentiles (the symbols *p10*, ..., *p90* stand for 10th, ..., 90th percentile).

Table 1 Sample size, average annual income (in Euro) and Gini coefficient values for each country

Country	n	n men	n women	Average Y_M	Average Y_W	Gini men	Gini women
DK	5604	2775	2829	60,725.83	47,143.99	0.280	0.280
LU	3932	2110	1822	56,353.29	43,849.64	0.330	0.357
NL	4912	2433	2479	49,364.79	30,024.32	0.302	0.286
FI	8923	4187	4736	46,093.75	33,300.53	0.302	0.302
SE	5477	2662	2815	45,007.91	34,044.25	0.321	0.321
IE	3759	1773	1986	44,876.80	31,165.20	0.338	0.323
AT	4798	2547	2251	44,253.14	28,841.96	0.265	0.379
BE	4677	2334	2343	43,554.01	32,171.61	0.393	0.384
DE	10,128	4999	5129	42,368.01	24,741.80	0.324	0.385
UK	8179	3965	4214	39,785.26	25,168.27	0.369	0.379
FR	9251	4589	4662	33,494.49	24,012.36	0.303	0.309
IT	12,715	6741	5974	30,007.71	22,655.93	0.309	0.319
CY	3869	1876	1993	26,396.44	18,920.69	0.392	0.392
ES	8493	4378	4115	25,225.28	18,640.89	0.368	0.403
MT	4033	2381	1652	21,985.54	15,955.17	0.321	0.321
SI	9344	4882	4462	20,512.60	18,207.41	0.315	0.315
EL	3687	2059	1628	17,658.02	13,934.03	0.328	0.297
PT	5208	2511	2697	16,341.15	12,697.77	0.385	0.385
EE	5506	2663	2843	13,064.08	8899.18	0.402	0.365
CZ	6501	3443	3058	12,413.62	8865.36	0.283	0.304
HR	3601	1983	1618	10,350.23	8871.80	0.301	0.281
SK	5755	2847	2908	10,292.61	7980.63	0.272	0.275
PL	9908	5180	4728	9619.24	7947.84	0.341	0.325
LV	4968	2270	2698	9491.75	7584.34	0.379	0.379
LT	4196	1998	2198	7919.51	6617.24	0.347	0.347
HU	8054	4061	3993	7565.97	6208.86	0.347	0.319
BG	4058	2018	2040	4646.59	3747.83	0.345	0.249
RO	4842	2733	2109	4306.87	3768.30	0.239	0.315
Total	174,378	88,398	85,980	–	–	–	–

Source Own elaboration

For each country, there are positive differences between the values of log incomes for men and women along the whole log income distribution. Then, the calculated differences were decomposed into the sum of the unexplained and explained components (the results are presented in Figs. 2, 3, 4 and 5).

We also examined and compared the sources of the explained and the unexplained gap, thus providing additional insights into the sources of the pay differential along the income distribution. The detailed decomposition made it possible to isolate the

Table 2 Definitions of variables

Variable	Description
age	Age in years
educlevel	Education level, 1—primary, …, 5—tertiary
married	Marital status, 1—married, 0—unmarried
permanent	Type of contract, 1—permanent job/work contract of unlimited duration, 0—temporary contract of limited duration
parttime	1—person working part-time, 0—person working full-time
manager	managerial position, 1—supervisory, 0—non-supervisory
big	Number of persons working at the local unit, 1—more than 10 persons, 0—less than 11 persons

Source Own elaboration

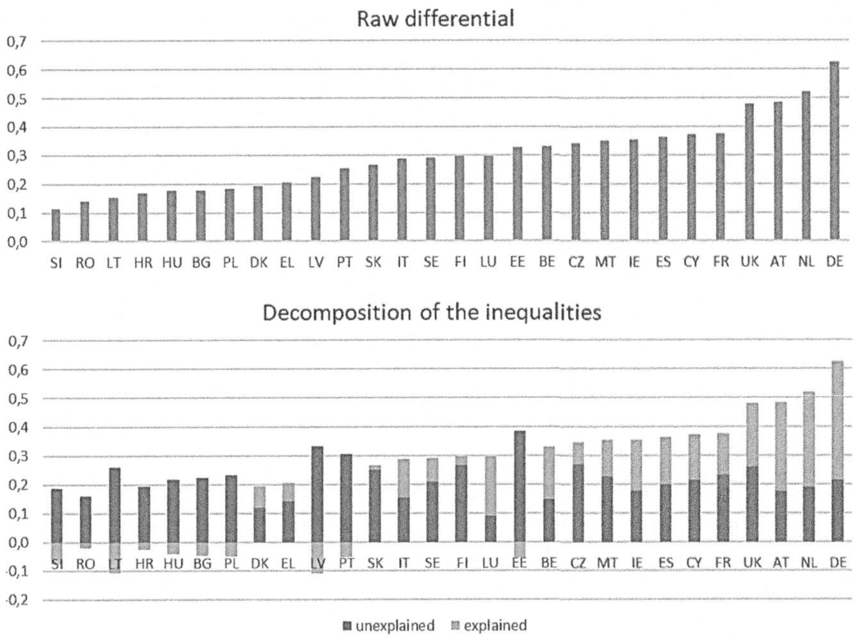

Fig. 1 Unadjusted gender pay gap (raw differential) and the results of the Oaxaca–Blinder decomposition of inequalities between men's and women's log incomes. *Source* Own elaboration

factors explaining the inequality observed to a different extent. Because of lack of space in this paper, we present the results of the detailed decomposition only for two countries—Poland and Germany—and only for the 20th, 50th and 80th percentiles (see Table 4).

The strong effect of different education levels of men and women can be noticed. The negative values of explained components for Poland mean that the differences

Table 3 Total differences between the log incomes along the distribution

Country	*p10*	*p20*	*p30*	*p40*	*p50*	*p60*	*p70*	*p80*	*p90*
AT	0.61	0.68	0.57	0.51	0.46	0.41	0.37	0.34	0.37
BE	0.56	0.41	0.33	0.28	0.25	0.25	0.25	0.26	0.30
BG	0.16	0.12	0.14	0.16	0.17	0.17	0.19	0.22	0.24
CY	0.47	0.65	0.42	0.39	0.35	0.37	0.32	0.20	0.16
CZ	0.42	0.37	0.37	0.29	0.26	0.26	0.26	0.28	0.29
DE	1.01	0.88	0.73	0.62	0.56	0.49	0.43	0.40	0.42
DK	0.08	0.15	0.16	0.17	0.20	0.21	0.24	0.28	0.32
EE	0.26	0.22	0.32	0.32	0.35	0.35	0.35	0.38	0.44
EL	0.25	0.27	0.22	0.16	0.14	0.14	0.14	0.19	0.23
ES	0.54	0.55	0.48	0.38	0.32	0.29	0.26	0.21	0.20
FI	0.30	0.31	0.26	0.26	0.27	0.30	0.33	0.34	0.35
FR	0.72	0.46	0.28	0.24	0.25	0.25	0.26	0.28	0.32
HR	0.19	0.17	0.16	0.10	0.15	0.16	0.11	0.10	0.09
HU	0.24	0.10	0.11	0.12	0.15	0.15	0.15	0.20	0.23
IE	0.39	0.44	0.39	0.38	0.33	0.31	0.31	0.30	0.34
IT	0.38	0.43	0.38	0.28	0.23	0.22	0.21	0.22	0.25
LT	0.11	0.07	0.15	0.17	0.20	0.19	0.14	0.14	0.16
LU	0.55	0.31	0.30	0.29	0.28	0.23	0.20	0.19	0.20
LV	0.14	0.18	0.20	0.23	0.22	0.22	0.20	0.20	0.23
MT	0.56	0.35	0.31	0.26	0.23	0.21	0.18	0.19	0.30
NL	0.69	0.66	0.59	0.52	0.48	0.43	0.42	0.42	0.45
PL	0.30	0.09	0.15	0.15	0.14	0.14	0.14	0.16	0.20
PT	0.44	0.15	0.17	0.25	0.26	0.28	0.26	0.18	0.20
RO	0.10	0.14	0.12	0.18	0.13	0.17	0.21	0.12	0.13
SE	0.44	0.37	0.27	0.21	0.20	0.22	0.24	0.27	0.27
SI	0.27	0.09	0.12	0.14	0.11	0.09	0.05	0.04	0.12
SK	0.26	0.29	0.25	0.25	0.27	0.21	0.23	0.31	0.27
UK	0.61	0.61	0.52	0.49	0.45	0.45	0.41	0.37	0.39

Source Own elaboration

of the average log incomes between men and women are reduced by the women's different education levels (higher than for men). It seems that differences in education levels mitigate the gap in Poland. The opposite is true for Germany. On the other hand, the values of *parttime*, *manager* and *big* attributes possessed by men and women increase the income inequality along the whole distribution in both countries (see the positive explained component values), but this effect is much stronger for Germany. In Poland and Germany, women are discriminated against men because of

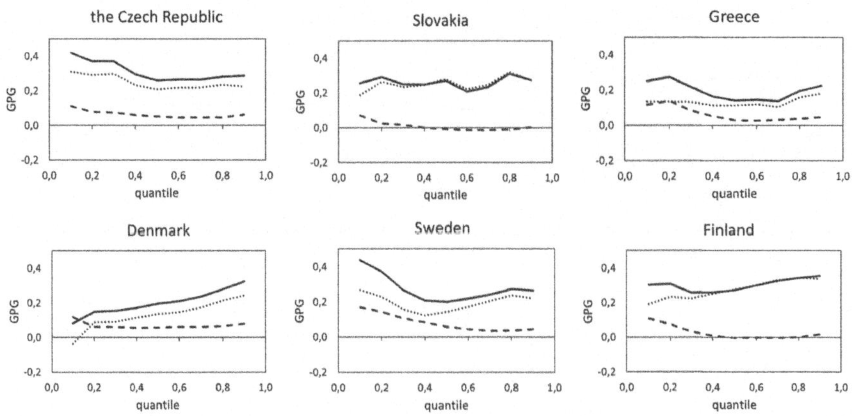

Fig. 2 Log income gaps for men and women versus quantile rank in the group 1 (solid lines—the total gap, dashed lines—the explained effect, dotted lines—the unexplained effect). *Source* Own elaboration

their marital status (the positive unexplained components values for variable *married*) but not because of age or education levels.

After assessing the gender pay gap (the raw, the explained and the unexplained gap) for all 28 countries, an attempt was made to group them using agglomerative hierarchical clustering algorithm. The use of the Ward's minimum variance method with the squared Euclidean distance allowed the grouping of countries into clusters. Four groups were identified:

- Group 1: the Czech Republic, Slovakia, Greece, Denmark, Sweden, Finland,
- Group 2: Poland, Hungary, Bulgaria, Slovenia, Romania, Croatia, Lithuania, Latvia, Estonia, Portugal,
- Group 3: Ireland, Luxembourg, Belgium, France, Italy, Spain, Malta, Cyprus,
- Group 4: the United Kingdom, the Netherlands, Austria, Germany.

The shapes of income gap are examined in Figs. 2, 3, 4 and 5, where solid lines represent the total income gap, dashed lines denote the explained component, and dotted lines indicate the unexplained effect.

Group 1 consists mainly of the countries from the north and east of Europe (welfare states and post-socialist states). It is characterized by the low total gender pay gap of irregular shape along the income distribution. There is the bigger unexplained effect than the explained one. The effect of coefficients is positive, and its share is high in the whole range of the income distribution. This is the result of differences in the 'market prices' of individual characteristics of men and women, interpreted as the labor market discrimination. The explained effect is positive, although very low.

Group 2, the largest group, consists mainly of the former socialist states of Eastern Europe. For most countries in this group, the total effect is low, but it widens at the bottom and/or at the top of the income distribution, suggesting sticky floor (e.g., in Poland, Slovenia, Portugal) and/or glass ceiling effects (especially in Estonia,

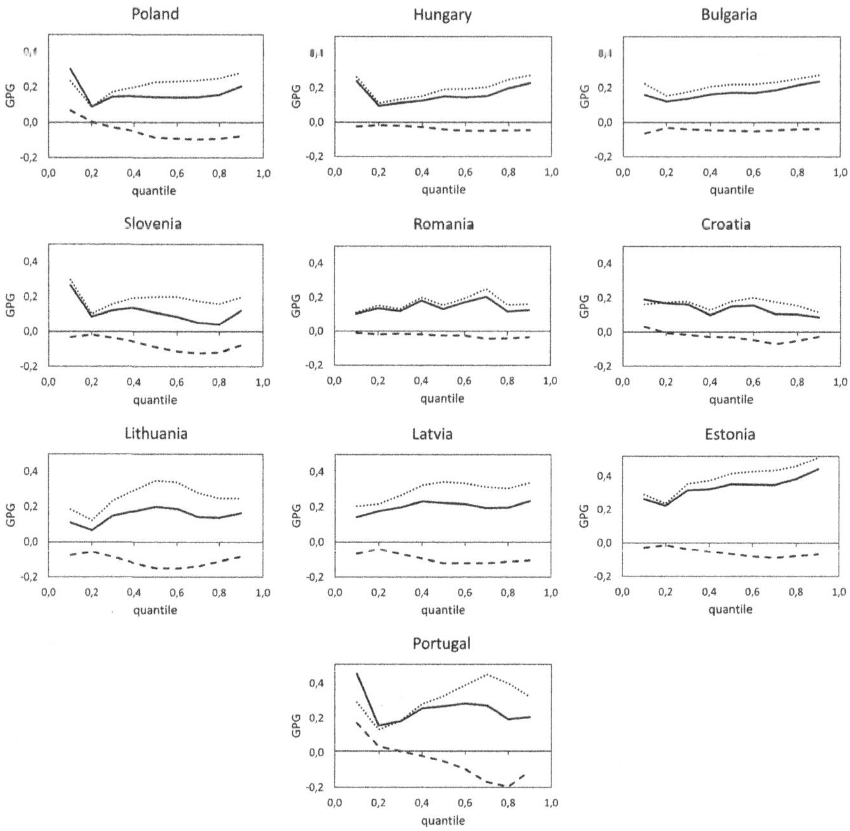

Fig. 3 Log income gaps for men and women versus quantile rank in the group 2 (solid lines—the total gap, dashed lines—the explained effect, dotted lines—the unexplained effect). *Source* Own elaboration

characterized by an increase of the income inequalities as we move toward the top of the income distribution). The unexplained effect is bigger than the explained one. The share of the unexplained part is very high. The explained differential (the effect of characteristics) is negative, which means that the properties possessed by both people's groups decrease the inequalities.

Group 3 consists mainly of the highly developed countries of Western Europe with high GDP per capita. In most from this countries the total gender gap is higher than before, decreasing along the distribution (larger at the bottom of the distribution and smaller at its top). The gender differences in characteristics are positive, which means that the different values of characteristics of men and women increase the income inequalities. The explained effect is bigger than the unexplained effect at the bottom of the log income distribution (except Cyprus). For the higher income ranges, the unexplained effect often prevails. Both effects, the explained and the unexplained, are always positive, increasing the income discrepancies.

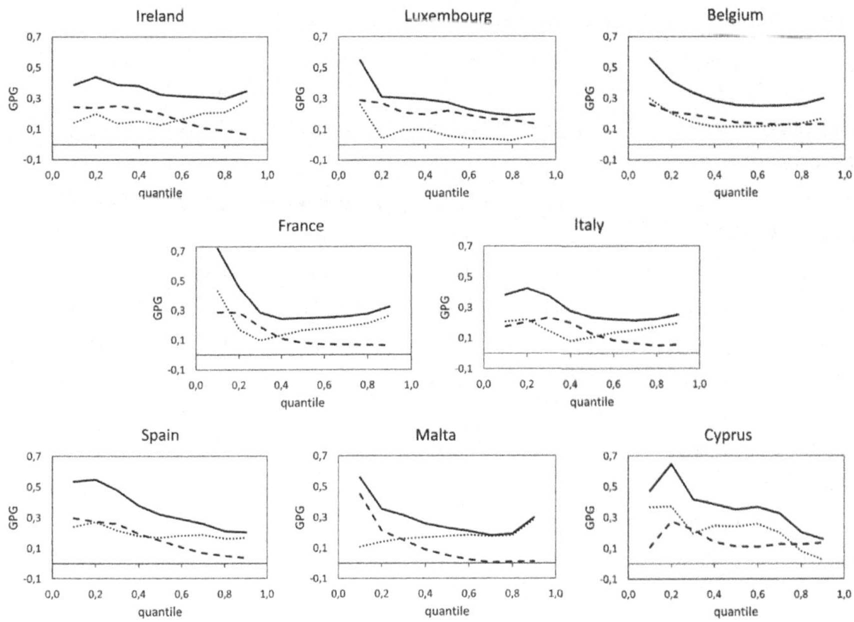

Fig. 4 Log income gaps for men and women versus quantile rank in the group 3 (solid lines—the total gap, dashed lines—the explained effect, dotted lines—the unexplained effect). *Source* Own elaboration

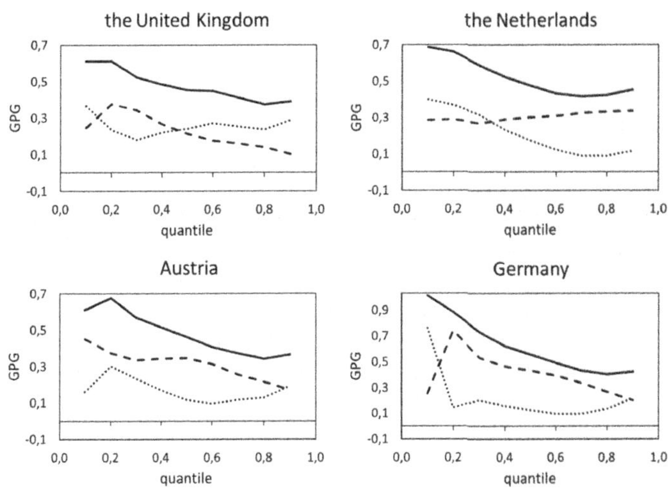

Fig. 5 Log income gaps for men and women versus quantile rank in the group 4 (solid lines—the total gap, dashed lines—the explained effect, dotted lines—the unexplained effect).
Source Own elaboration

Table 4 Results of the detailed RIF-regression decomposition

	Poland						Germany					
	p20		p50		p80		p20		p50		p80	
Difference	0.090	***	0.143	***	0.158	***	0.884	***	0.555	***	0.403	***
Explained												
age	−0.002		−0.003		−0.003		−0.009	**	−0.005	**	−0.006	**
educlevel	−0.034	***	−0.108	***	−0.110	***	0.039	***	0.026	***	0.020	***
married	0.000		0.001		0.000		−0.009	***	−0.004	***	−0.002	**
permanent	−0.003		−0.004		−0.002		0.031	***	0.013	***	0.004	***
parttime	0.036	***	0.017	***	0.010	***	0.519	***	0.319	***	0.201	***
manager	0.002	***	0.006	***	0.011	***	0.042	***	0.030	***	0.029	***
big	0.003	***	0.005	***	0.003	***	0.127	***	0.055	***	0.023	***
Total	0.003		−0.085	***	−0.092	***	0.741	***	0.433	***	0.269	***
Unexplained												
age	−0.089	*	−0.320	***	−0.364	***	−0.428	***	−0.188	***	−0.291	***
educlevel	0.005		−0.310	***	−0.107	**	−0.657	***	−0.114	**	0.104	**
married	0.085	***	0.096	***	0.116	***	0.321	***	0.130	***	0.089	***
permanent	−0.006		−0.037	*	−0.022		−0.243	***	−0.326	***	−0.189	***
parttime	−0.003		−0.001		0.001		0.044	***	0.047	***	0.026	***
manager	0.010	*	0.014	**	0.033	***	−0.018		0.003		0.013	
big	0.016		0.015		0.014		−0.565	***	−0.144	***	−0.032	
cons	0.068		0.771	***	0.581	***	1.688	***	0.716	***	0.415	***
Total	0.087	***	0.228	***	0.250	***	0.142	***	0.123	***	0.134	***

Source Own elaboration
***,**,* - significance at 1%, 5%, 10% level respectively

The last group, the group 4, is made up of the United Kingdom, the Netherlands, Austria and Germany. These are countries with very high GDP per capita and highly segregated labor markets in which a significant proportion of women works part-time. In this case, the large total gap and the large explained effect have a decreasing shape and are rapidly falling as we move toward the top of the income distribution. The unexplained part is positive and at a moderate level, presenting the existing effect of discrimination on the labor market—higher among the poorest, then lower among the richest.

4 Conclusions

Gender income discrepancies are persistent all over Europe. However, a considerable country heterogeneity emerges. We started our analysis using the Oaxaca–Blinder method for the decomposition of the average values for log incomes. Similar like in Boll and Lagemann (2019) or Oczki (2016), we found that in all 28 countries is a positive difference between the mean income values for men and women. Most Eastern European states are exhibiting gaps clearly below average, the West European countries exhibit huge gaps, and moderate gaps are found for Scandinavian countries. The total log income gap on average amounted to 0.298 (on the country level, the unadjusted gap ranged from 0.112 in Slovenia to 0.625 in Germany). The gap that is attributable to different (observable) characteristics of women and men (explained gap) was on average 0.083. The adjusted gap that compares men and women with similar characteristics amounted to 0.215. Thus, a greater portion of the overall gap was unexplained. The unexplained effect was huge for the states with the low raw differential and was small for the states with the high raw differential. On the other hand, what is statistically 'explained' is not necessarily free from discrimination. Genders might face unequal access to wage-attractive jobs (e.g., managerial positions, full-time jobs). The explained gap was negative in the countries with the lowest income discrepancies (e.g., Slovenia, Lithuania, Bulgaria, Poland). The negative value of this component means that the difference of the average log incomes between men and women is reduced by the women's characteristics.

Then, we extended the decomposition procedure to different quantile points along the income distribution by the use of the RIF-regression approach. After assessing the raw, the explained and the unexplained gaps for all 28 countries, we grouped them into four clusters using the agglomerative clustering algorithm.

Group 1 consisted mainly of the countries from the north and east of Europe. It was characterized by the low total gender pay gap. The conducted decomposition showed that the unexplained component quantitatively dominated in the whole range of the income distribution. The gap was poorly explained by gender differences in observable characteristics. Group 2 consisted of the former socialist states of Eastern Europe. The total effect was low, but it widened at the bottom and/or at the top of the income distribution, suggesting sticky floor and/or glass ceiling effects [compare Arulampalam et al. (2006)]. The explained differential was negative, which means that female characteristics are superior to the male ones [similar results in this field have been obtained in Christofides et al. (2013)]. In group 3, consisting of the highly developed countries of Western Europe, the total gender gap was larger at the bottom of the log income distribution and smaller at the top [similar to the results obtained by Nicodemo (2009)]. For the lower income ranges, the explained effect was bigger than the unexplained. Both effects, the explained and the unexplained, are always positive, increasing the income discrepancies. Group 4 was made up of the United Kingdom, the Netherlands, Austria and Germany—countries with very high GDP per capita and highly segregated labor markets. The large total gap and the large explained effect were rapidly falling as we move toward the top of the income distribution. The unexplained part presented the moderate effect of discrimination.

One should be aware that the results obtained depend on the selection of explanatory variables to the estimated models. Concerning the contributions of characteristics, a gendered sorting into atypical employment widens the pay gap. Women work more often in part-time and temporary jobs than men, which is associated with lower earnings. On the other hand, differences in education levels mitigate the gap in most countries (in all countries but Germany, women are on average more highly educated than men). A rich body of literature confirms severe earnings losses of women due to family related breaks (e.g., Boll et al. 2017).

Summarizing, there exists an important diversity in the size and composition of the gender income gap across members of the EU. The findings confirm the trade-off between low gender pay gaps and high female employment rates. Countries with flexible working conditions enable women to enter the market which results in a high female employment rate, but this comes at the cost of severe wage deductions. By contrast, in countries with a low compatibility of family and career (e.g., due to a poor childcare infrastructure), only women with a high earnings potential access the labor market. This results in low gender pay gaps as for Eastern Europe countries.

The gender discrimination may lead to loss in productivity and wealth, so the resulting inequalities pose a serious challenge for the society. Reducing the gender pay gap is one of the key priorities of EU gender policies. However, without a good understanding of the causes of the gap, policy-makers are unable to design the right policy mix for addressing the issue. The analysis of large amounts of individual data on job and worker characteristics is a necessary step on this way.

References

Albrecht J, Bjorklund A, Vroman S (2003) Is there a glass ceiling in Sweden? J Labor Econ 21(1):145–177. https://doi.org/10.1086/344126

Arulampalam W, Booth AL, Bryan ML (2006) Is there a glass ceiling over Europe? Exploring the gender pay gap across the wage distribution. Ind Labor Relat Rev 60(2):163–186. https://doi.org/10.1177/001979390706000201

Blinder A (1973) Wage discrimination: reduced form and structural estimates. J Hum Resour 8(4):436–455. https://doi.org/10.2307/144855

Boll C, Lagemann A (2019) The gender pay gap in EU countries—new evidence based on EU-SES 2014 Data. Intereconomics 54(2):101–105. https://doi.org/10.1007/s10272-019-0802-7

Boll C, Jahn M, Lagemann A (2017) The gender lifetime earnings gap—exploring gendered pay from the life course perspective. J Income Distrib 25(1):1–53

Chatterji M, Mumford K, Smith PN (2011) The public-private sector gender wage differential in Britain: evidence from matched employee-workplace data. Appl Econ 43(26):3819–3833. https://doi.org/10.1080/00036841003724452

Christofides L, Polycarpou A, Vrachimis K (2013) Gender wage gaps, 'sticky floors' and 'glass ceilings' in Europe. Labour Econ 21:86–102. https://doi.org/10.1016/j.labeco.2013.01.003

DiNardo J, Fortin NM, Lemieux T (1996) Labor market institutions and the distribution of wages, 1973–1992: a semiparametric approach. Econometrica 64(5):1001–1044. https://doi.org/10.2307/2171954

Donald SG, Green DA, Paarsch HJ (2000) Differences in wage distributions between Canada and the United States: an application of a flexible estimator of distribution functions in the presence of covariates. Rev Econ Stud 67:609–633. https://doi.org/10.1111/1467-937X.00147

Eurostat (2019) The unadjusted gender pay gap, 2017. https://ec.europa.eu/eurostat/statistics-explained/index.php/Gender_pay_gap_statistics. Accessed 29 Sept 2019

Firpo S, Fortin NM, Lemieux T (2009) Unconditional quantile regressions. Econometrica 77(3):953–973. https://doi.org/10.3982/ECTA6822

Fortin N, Lemieux T, Firpo S (2010) Decomposition methods in economics. NBER WP 16045, Cambridge. https://economics.ubc.ca/files/2013/05/pdf_paper_nicole-fortin-decomposition-methods.pdf. Accessed 29 Sept 2019

Hedija V (2017) Sector-specific gender pay gap: evidence from the European Union countries. Econ Res Ekon Istraz 30(1):1804–1819. https://doi.org/10.1080/1331677X.2017.1392886

Juhn C, Murphy KM, Pierce B (1993) Wage inequality and the rise in returns to skill. J Polit Econ 101:410–442. https://doi.org/10.1086/261881

Jurajda S (2003) Gender wage gap and segregation in enterprises and the public sector in late transition countries. J Comp Econ 31(2):199–222. https://doi.org/10.1016/S0147-5967(03)00040-4

Landmesser JM (2016) Decomposition of differences in income distributions using quantile regression. SiTns 17(2):331–348. https://doi.org/10.21307/stattrans-2016-023

Leythienne D, Ronkowski P (2018) A decomposition of the unadjusted gender pay gap using Structure of Earnings Survey data. Eurostat, Statistical Working Papers. https://ec.europa.eu/eurostat/documents/3888793/8979317/KS-TC-18-003-EN-N.pdf/3a6c9295-5e66-4b79-b009-ea1604770676. Accessed 17 Dec 2019. https://doi.org/10.2785/796328

Machado JF, Mata J (2005) Counterfactual decomposition of changes in wage distributions using quantile regression. J Appl Econom 20:445–465. https://doi.org/10.1002/jae.788

Nicodemo C (2009) Gender pay gap and quantile regression in European families. IZA DP 3978, Bonn. http://ftp.iza.org/dp3978.pdf. Accessed 29 Sept 2019

Oaxaca R (1973) Male-female wage differentials in urban labor markets. Int Econ Rev 14(3):693–709. https://doi.org/10.2307/2525981

Oczki J (2016) Gender pay gap in Poland. Ekonomia międzynarodowa 14:106–113. https://doi.org/10.18778/2082-4440.14.03

Pena-Boquete Y, De Stefanis S, Fernandez-Grela M (2010) The distribution of gender wage discrimination in Italy and Spain: a comparison using the ECHP. Int J Manpower 31(2):109–137. https://doi.org/10.1108/01437721011042232

Śliwicki D, Ryczkowski M (2014) Gender pay gap in the micro level—case of Poland. MIBE (QME) XV(1):159–173

Common Stochastic Mortality Trends for Multiple European Populations

Justyna Majewska and Grażyna Trzpiot

Abstract The main hypothesis for multi-population mortality models is that mortality rate differences for any two populations having similar socioeconomic status and close connections with each other do not diverge indefinitely over time. Recent evaluation studies have demonstrated that multi-population mortality models are superior to individual mortality forecasting models. However, the key point is to understand, extract and model the common trends driving the mortality patterns for a group of countries to improve the national long-term mortality forecasts. The aim of the paper is twofold. Firstly, the discussion on different approaches to identify the existence of the common mortality trends is provided. Secondly, the mortality time-varying indicator derived from the Lee–Carter model is used to obtain the similarities of different countries via a semi-parametric comparison approach. Two and multi-countries cases are provided.

Keywords Mortality · Modeling · Multi-population

1 Introduction

The importance of mortality modeling and forecasting has over the last years grown and became advanced because of rapid population aging. Despite the global decreasing mortality trend, each country still behaves differently from others to some extent. However, due to globalization, populations across the world become more and more linked to each other. It is crucial to extract and understand the common trends driving the mortality patterns for a group of countries that can improve the national long-term mortality forecasts.

J. Majewska (✉) · G. Trzpiot
University of Economics in Katowice, Katowice, Poland
e-mail: justyna.majewska@ue.katowice.pl

G. Trzpiot
e-mail: grazyna.trzpiot@ue.katowice.pl

K. Jajuga et al. (eds.), *Classification and Data Analysis*, Studies in Classification, Data Analysis, and Knowledge Organization,
https://doi.org/10.1007/978-3-030-52348-0_19

The idea of using common mortality trends for some populations was described for the first time by Li and Lee (2005). They pointed out that the convergence in mortality levels for closely related populations can lead to improper mortality forecasts if the forecasts for individual populations are obtained in isolation from one another. The historical similarities in long-run mortality patterns are likely to continue in the future, if the determinants of mortality, such as the socioeconomic, environmental or behavioral factors, will have similar trends. Although the short-term country-specific deviations are expected.

The common mortality trend can be modeled in different ways. The most popular is the application in modeling and forecasting two-sex mortality within the same country or region. The second application is to forecast mortality coherently for the selected low-mortality countries. In the third one, the experience of the low-mortality group of populations can be used to model and forecast mortality in a higher-mortality group of countries, and then, the group of low-mortality countries is referred then as a reference group.

Identifying a common trend is important when looking at, for example, Poland. National statistical office forecasts mortality based on the assumption that the decline in mortality in Poland will continue in a similar way to that observed in Western Europe in the past. Thus, when developing the mortality forecast for Poland, the directions of changes are observed in developed European countries, such as Belgium, Denmark, Germany, Ireland, Greece, Spain, France, Italy, Netherlands, Austria, Portugal, Finland, Sweden, UK, Norway, and Switzerland (GUS 2014).

This paper aims to highlight the long-term trends of general mortality levels in the selected European countries and to investigate the common stochastic trends driving the mortality levels. Our main interest lies in identifying the common trends via analyzing time-varying indicator curves. A semi-parametric approach of Fang et al. (2016) in order to compare nonlinear curves and model common trend is incorporated. The death data sets are collected from the Human Mortality Database for selected European populations by gender.

2 Mortality Modeling

To respond to the challenges associated with population aging, numerous models for mortality modeling and forecasting have been developed in recent decades.

Lee–Carter model For many years, the LC model (Lee and Carter 1992) has been regarded as a benchmark mortality model (Booth and Tickle 2008; Shang et al. 2011; Stoeldraijer et al. 2013). Model assumes that the mortality rate m_{xt} at age x and in year t can be decomposed as

$$\log m_{xt} = a_x + b_x k_t + \varepsilon_{xt}. \tag{1}$$

where a_x is the derived age pattern averaged acruss years, b_x stands for the sensitivity of the mortality rates to the change of k_t, reflecting how fast the mortality rate changes over ages, k_t represents the only time-varying index of mortality level, and ε_{xt} is the residual term at age x in year t with $E(\varepsilon_{xt}) = 0$ and $Var(\varepsilon_{xt}) = \sigma_\varepsilon^2$.

Unobserved parameters a_x, b_x and k_t in (1) mean that the LC model is over-parameterized, and therefore, two normalization constraints are imposed:

$$\sum_t k_t = 0 \quad \sum_x b_x = 1. \tag{2}$$

Parameters k_t and b_x are obtained via singular value decomposition. The evolution of k_t can be fitted by ARIMA techniques, e.g., the Box–Jenkins procedure. Time-varying index is described well by random walk with drift:

$$k_t = k_{t-1} + d + e_t, \tag{3}$$

where d is the drift parameter reflecting the average annual change and e_t is an uncorrelated error.

Given an h-step ahead forecasting k_{t+h} forecast, the mortality rates in future period $t + h$ are derived via the following formula:

$$m_{x,t+h} = \exp(a_x + b_x k_{t+h}). \tag{4}$$

The LC model has become widely used, and there have been various extensions and modifications proposed to capture the main features of the dynamics of the mortality intensity.

The concept of multi-population mortality modeling was developed on the basis of the LC method. The time-varying parameter k_t is the key parameter in this approach. Lee and Li (2005) stated that in order to avoid long-run divergence in mean mortality forecasts for a group using this method, it is a necessary and sufficient condition that all populations in the group have the same b_x and the same drift term for k_t.

3 Identification of Common Stochastic Mortality Trend

3.1 Spatial Clustering Method

The method of creation of homogenous spatial clusters of countries according to relevant socioeconomic predictor variables can be used to obtain long-term mortality projections for all countries from a cluster, in line with the common patterns and interdependencies detected in the long-run mortality experiences of the countries. This approach was implemented by Lazar et al. (2016) and Majewska (2017). In both cases, the SKATER (Spatial 'K'luster Analysis by Tree Edge Removal, Assunção et al. 2006) algorithm was used. In this case, the spatial clustering algorithm should involve some variables, representing the main indicators for similarities in the historic mortality rates of countries, or predictors for these similarities, like socioeconomic status, including employment, income, education and economic well-being, the quality of the health system and the ability of people to access it, health behaviors, social factors, genetic factors and environmental factors including overcrowded housing, lack of clean drinking water and adequate sanitation.

Lazar et al. (2016) and Majewska (2017) identified some groups of homogeneous contiguous clusters—countries from Europe—according to selected socioeconomic variables. Their results vary slightly because of a different group of socioeconomic variables, and the changes concern the individual countries, mainly those with high life expectancy. In general, countries from Northern Europe create one group. Countries from Central Europe (e.g., Germany, Austria, Switzerland, Czech Republic) have a similar trend of the elderly mortality evolution. The next cluster groups developed countries located in Western and Southern Europe. These countries have recorded comparable levels for the mortality rates and experienced a similar evolution of mortality in the long run. Spatial cluster consisted of countries with lower life expectancy and higher mortality (i.e., Bulgaria, Latvia, Lithuania, Poland, Slovakia) is quite evident. The analyses were conducted for data starting in the 1960s. As stated by Lazar et al. (2016) by using a spatial clustering algorithm, we avoid an ad hoc grouping of the countries expected to record similar improvements in mortality. Such a method of identification of common mortality trends seems to be very useful in further steps of multi-population modeling.

3.2 Semi-parametric Method

Fang et al. (2016) presented a semi-parametric approach in order to estimate and forecast mortality rates. Approach is useful when the differences among similar mortality curves mainly rely on shifted time axis and vertical re-scaling, and comparison of similar curves could be simplified by quantifying differences through parameters describing horizontal and vertical shifts. Denoting the underlying curves by f_1 and f_2, the semi-parametric comparison of the nonlinear curves is written then as:

$$f_2(x) = \theta_1 f_1\left(\frac{t - \theta_2}{\theta_3}\right) + \theta_4. \tag{5}$$

where we assume that f_2 has a similar pattern to f_1 and $\theta = (\theta_1, \theta_2, \theta_3, \theta_4)$ are shape deviation parameters. More detailed discussions on semi-parametric comparison of regression curves can be found in Härdle and Marron (1990).

Thus, when N noisy regression curves Y_i, $i = 1, \ldots, N$ exhibit some similar patterns:

$$Y_i = f_i(t) + \varepsilon_i, \tag{6}$$

where f_i denote unknown smoothing regression functions, ε_{xt}—independent errors with mean 0 and variance σ_i^2, a relationship among these curves is described as:

$$f_i(x) = \theta_{i1} g\left(\frac{t - \theta_{i2}}{\theta_{i3}}\right) + \theta_{i4}. \tag{7}$$

In Eq. (7) $\theta = (\theta_1, \theta_2, \theta_3, \theta_4)$ are unknown parameters describing shape deviations, and g is a unknown function specifying the common shape of these curves, which can be interpreted as a reference curve.

Model for two-country case As the time-varying parameter k_t is derived from LC model in Eq. (1), then first country's mortality trend via the second trend is described by:

$$k_c(t) = \theta_1 k_j\left(\frac{t - \theta_2}{\theta_3}\right) + \theta_4 \tag{8}$$

where $k_c(t)$ is the time-varying indicator for the first country, $k_j(t)$ is the time-varying indicator for the second one, and θ are shape deviation parameters, specifically:

θ_1 is the general trend adjustment, θ_2 is the time-delay parameter, θ_3 is the time acceleration parameter, θ_4 is the vertical shift parameter.

To find the optimal solution for shape deviation parameters, the following loss function is minimized:

$$\min_\theta \int_{t_c} \left\{\hat{k}_c(t) - \theta_1 \hat{k}_j\left(\frac{u - \theta_2}{\theta_3}\right) + \theta_4\right\}^2 w(u)du. \tag{9}$$

where $\hat{k}_c(t)$ and $\hat{k}_j(t)$ are the nonparametric estimates of the original time-varying indicators, and t_c is the time interval of the first country's mortality data.

The comparison region needs to satisfy the following condition, in order to make sure the parameter estimation is compared only in the common region defined by $w(u)$:

$$w(u) = \prod_{t_j} 1_{[a,b]}\left\{\left(\frac{u - \theta_2}{\theta_3}\right)\right\}. \tag{10}$$

where is t_j the time interval of the second mortality data and

$$a \geq \inf(t_j) \quad b \leq \sup(t_j). \tag{11}$$

In order to estimate the parameters by the nonlinear least squares estimation criterion in Eq. (9), we first obtain the estimates of kc and kt by nonparametric local linear smoothing, denoted by $\hat{k}c$ and $\hat{k}t$, respectively. Then, we set up the initial estimates $\theta^0 = (\theta_1^0, \theta_2^0, \theta_3^0, \theta_4^0)$ and solve the nonlinear least squares estimation problem by iteratively updating the estimates until convergence.

Model for multi-country case The two-case approach is extended to multi-country case. We assume that the curves share some common trend and can be represented in the form:

$$k_i(t) = \theta_{i1} k_0\left(\frac{t - \theta_{i2}}{\theta_{i3}}\right) + \theta_{i4}, \tag{12}$$

where $k_0(t)$ is a reference curve, understood as common trend and $\theta_i = (\theta_{i1}, \theta_{i2}, \theta_{i3}, \theta_{i4})$ are shape deviation parameters. In order to be able to interpret the reference curve k_0 as a mean trend, we can use the normalizing constraints on the parameter θ_i as

$$N^{-1} \sum_{i=1}^{N} \theta_{i1} = N^{-1} \sum_{i=1}^{N} \theta_{i3} = 1. \tag{13}$$

$$N^{-1} \sum_{i=1}^{N} \theta_{i2} = N^{-1} \sum_{i=1}^{N} \theta_{i4} = 0. \tag{14}$$

In order to estimate θ_i for each country, i can be determined by minimizing the least squares criterion as

$$\int \left\{ k_i(t) - \theta_n k_0\left(\frac{t - \theta_{i2}}{\theta_{i3}}\right) - \theta_{i4} \right\}^2 w_i(t) dt \tag{15}$$

where w_i is chosen to ensure that two functions are evaluated over the common domain as in the two-case approach.

The common trend for multi-country case is estimated as follow. For given parameters θ_i, $i = 1, \ldots, n$, the functional relationship in (12) implies that

$$k_i(\theta_{i3} t + \theta_{i2}) = \theta_{i1} k_0(t) + \theta_{i4} \tag{16}$$

Thanks to the normalizing conditions on θ_{i1} and θ_{i4}, this implies that

$$k_0(t) = N^{-1} \sum_{i=1}^{N} k_i(\theta_{i3} t + \theta_{i2}) \tag{17}$$

where w_i is chosen to ensure that the two functions are evaluated over the common domain as in the case of two countries. That is, if k_i is appropriately transformed with respect to the individual parameters θ_i, then k_i is simply the average. In practice, k_i can have measurement errors and also are available at a different number of time points.

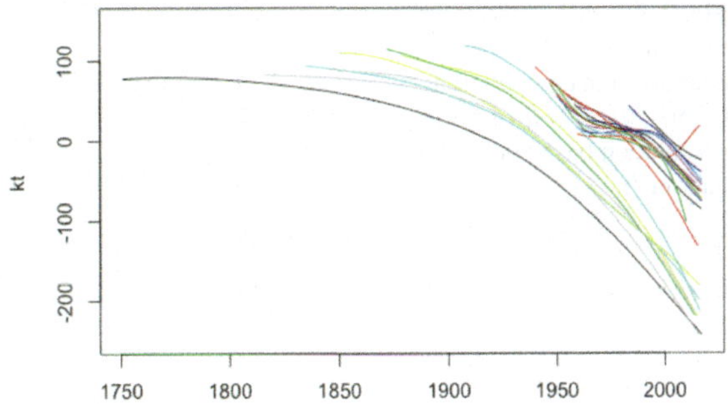

Fig. 1 Smoothed mortality movements for European countries—female population (1750–2016). *Source* Own preparation based on HMD dataset

Then, the functional mean can be estimated more efficiently with nonparametric smoothing, which essentially gives rise to a weighted average estimate.

Following Fang et al. (2016) to initialize k_0, the trimmed mean of the sample estimates can be chosen, based on the middle 50% of the countries in terms of the length of the recording period. The estimation of k_0 and k_i is done with a local linear kernel smoothing method to account for measurement error.

4 Common Mortality Trends in Europe

4.1 European Mortality Trends

The demographic data sets are collected from the Human Mortality Database (HMD). The mortality rates are gender–age specific starting from 0 to 110+. Sample sizes are different: The sample size ranges from 50 years to 262 years (e.g., for Sweden). Recall that the mortality rate is the number of deaths per 1000 living individuals per the calendar year. We use the log mortality.

Smoothed mortality trends (smoothed k_t estimated from LC model) for selected 22 European countries[1] (separately for male and female) are illustrated in Figs. 1 and 2. The longest curve in both cases represents time-varying mortality in Sweden. Slightly higher but similar in shape time-varying mortality indicators are presented for countries from Western and Northern Europe (e.g., Denmark, France, Finland, Netherlands, Norway) and Southern Europe (Spain, Italy). These are the low-mortality countries. The mortality of other countries (like Poland, Slovakia, Czechia and Hungary) may be expected to join the low-mortality group in the future. The selection of European countries is also based on the results of cluster analysis described shortly in Chap. 2.

[1] Austria, Belarus, Bulgaria, Czechia, Denmark, Estonia, Finland, France, Germany, Hungary, Italy, Latvia, Lithuania, Luxembourg, Netherlands, Norway, Poland, Portugal, Slovakia, Slovenia, Spain, Sweden.

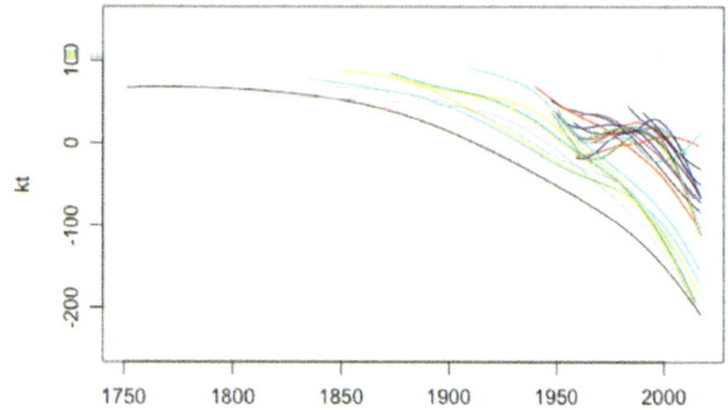

Fig. 2 Smoothed mortality movements for European countries male population (1750–2016). *Source* Own preparation based on HMD dataset

The shortest curves are smoothed individual country-level mortality trends in East Europe. These countries have less available time-series data, higher mortality and more volatile past mortality trends. Similarities are evident. Countries from Eastern Europe experienced parallel economic and social progress. In the post-communist countries, the mortality rates have started to decrease significantly in the decade between 1980 and 1990. However, there are some outliers where mortality rates increase over the last years. Mortality rates in Lithuania and Latvia are the highest in Europe, and they exhibit opposite tendencies.

4.2 *Two-Country Mortality Case*

In the two-population case, Poland and selected European low-mortality countries are analyzed. It is expected that Poland may catch up the low-mortality countries in the future. As a graphic illustration, the case of Poland and Austria is presented in detail.

European countries were selected on the basis of quantifying differences through parameters describing horizontal and vertical shifts between the mortality curves of Poland and these countries as it is plotted in Fig. 3 for Austria and Poland. In

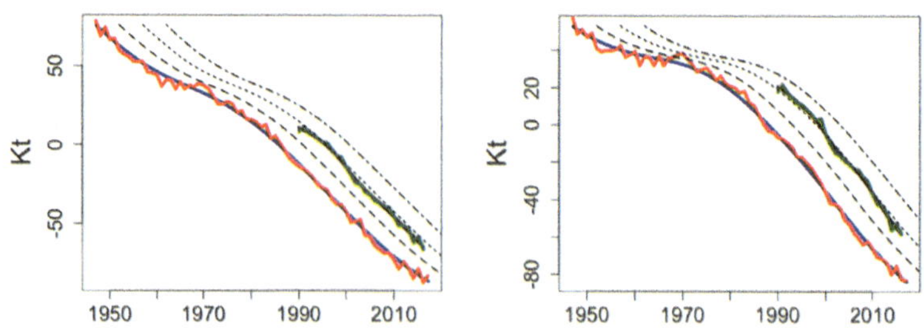

Fig. 3 Austria and Poland mortality trends and smoothed versions: female (left) and male (right). The dotted lines from left to right are shifted Austria's smoothed trends of 5, 10 and 15 years forward, respectively. *Source* Own preparation based on HMD dataset and MuPoMo package

Table 1 Estimated parameters of the model (5)—female population

Poland versus country	Estimates $\theta = (\theta_1, \theta_2, \theta_3, \theta_4)$
Netherlands	(1.26, −45.18, 0.99, 7.50)
Denmark	(1.30, −52.09, 0.99, 5.49)
Sweden	(1.98, −60.10, 0.99, 6.47)
Finland	(1.00, −37.08, 0.99, 6.21)
France	(0.80, −44.94, 0.99, 2.65)
Spain	(0.53, −23.10, 1.00, 4.74)
Portugal	(0.60, −13.17, 1.00, 5.39)
Austria	(0.71, −8.85, 1.00, 0.00)
Italy	(0.62, −38.37, 1.00, 5.96)

Source Own preparation based on HMD dataset and MuPoMo package

the case of Poland, we reduce time period from 1958–2016 to 1990–2016 to avoid high-mortality period related to the communism era.

Poland's female k_t may reach a similar behavioral area as a female in Austria by shifting horizontally by 8.85 years ($\theta_2 = -8.85$) and Poland's male—by 10.0 years ($\theta_2 = -10.0$). Thus, we see that for females, Austria's mortality trend is almost 9 years earlier than Poland, for males, it is 10 years. For these two countries, the analysis focuses only on time delay—the vertical shift is not necessary ($\theta_4 = 0$).

However, there are bigger differences. Estimated parameters for each pair of countries (Poland vs. one of the low-mortality European countries) are presented in Tables 1 and 2.

From Fig. 4, we see that after the curve shifts (based on the optimal value of θ), the *kt* of Poland female fits quite well in the k_t of Austria of years around from 1995 to 2005. The very similar results were obtained for male (Fig. 5).

Table 2 Estimated parameters of the model (5)—male population

Poland versus country	Estimates $\theta = (\theta_1, \theta_2, \theta_3, \theta_4)$
Netherlands	(1.99, −50.56, 0.99, 4.20)
Denmark	(2.05, −50.72, 0.99, 5.87)
Sweden	(2.49, −60.96, 0.98, 6.75)
Finland	(1.54, −36.88, 1.00, 7.54)
France	(1.00, −45.00, 1.00, 7.46)
Spain	(0.43, −26.63, 1.00, 7.23)
Portugal	(0.58, −9.49, 1.00, 4.65)
Austria	(1.00, −10.00, 1.00, 0.00)
Italy	(1.00, −35.00, 1.00, 3.76)

Source Own preparation based on HMD dataset and MuPoMo package

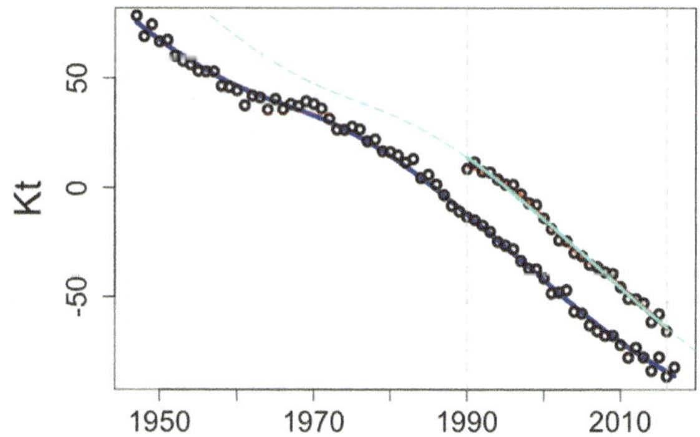

Fig. 4 Fit of Poland's mortality trend via Austria's historical data (female population). Dots represent the original *kt* from Austria and Poland and their smoothed trends. *Source* Own preparation based on HMD dataset and MuPoMo package

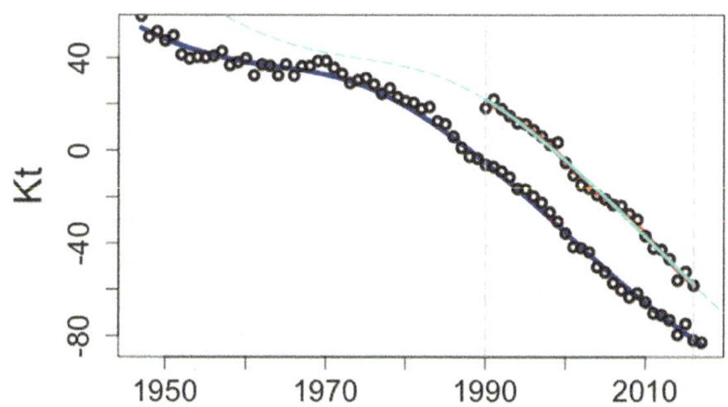

Fig. 5 Fit of Poland's mortality trend via Austria's historical data (male population). Dots represent the original *kt* from Austria and Poland and their smoothed trends. *Source* Own preparation based on HMD dataset and MuPoMo package

A wide range of time delay of Poland's mortality trend in relation to individuals European populations trends leads to the next stage—the multi-population case.

4.3 European Common Low-Mortality Trend

In order to derive common mortality trend, the parameters of the simplified 4-parameter model (12) were estimated. The reference curve is obtained using historical mortality data from the following countries: Austria, Denmark, Finland, France, Italy, Netherlands, Norway, Portugal, Spain and Sweden. Figures 6 and 7 display the smoothed estimates of *kt* from these countries with an initial estimate of the reference curve overlaid.

Since a common mortality trend is available, it could be applied to help improve the estimation of individual case. In Figs. 8 and 9, the newly estimated Poland mortality trend via semi-parametric comparison with the common trend is presented. The (black) solid line is the common trend or updated reference curve, and the (cyan) dashed line is the estimated Poland mortality trend based on the common trend. It confirms that there exists a time delay of Poland's mortality trend around 40 years later than European low-mortality trend. Due to the estimation of parameters, we

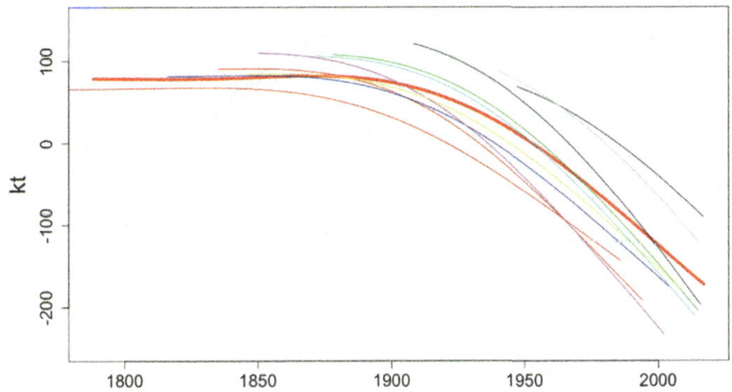

Fig. 6 European common mortality trend (bold solid line) compared with individual mortality trends—the case of male European populations. *Source* Own preparation based on HMD dataset and MuPoMo package

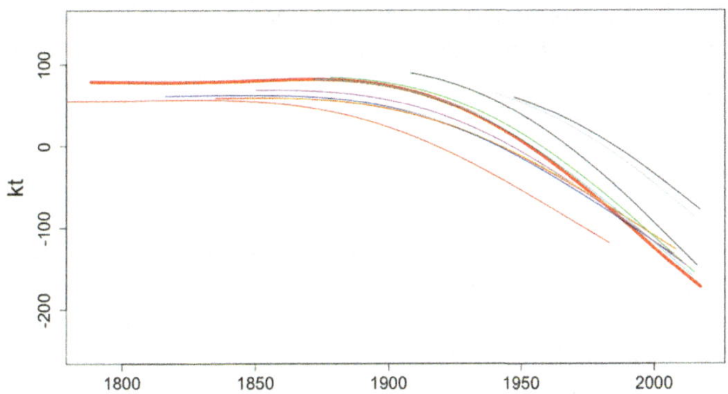

Fig. 7 European common mortality trend (bold solid line) compared with individual mortality trends—the case of female European populations. *Source* Own preparation based on HMD dataset and MuPoMo package

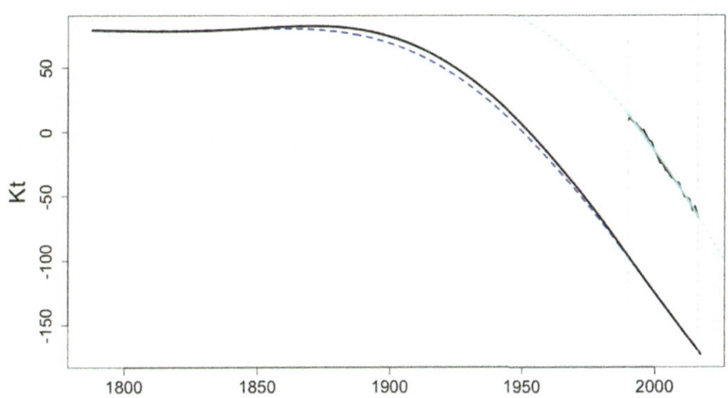

Fig. 8 Common mortality trend and estimated Poland mortality trend based on the common trend—female population. *Source* Own preparation based on HMD dataset and MuPoMo package

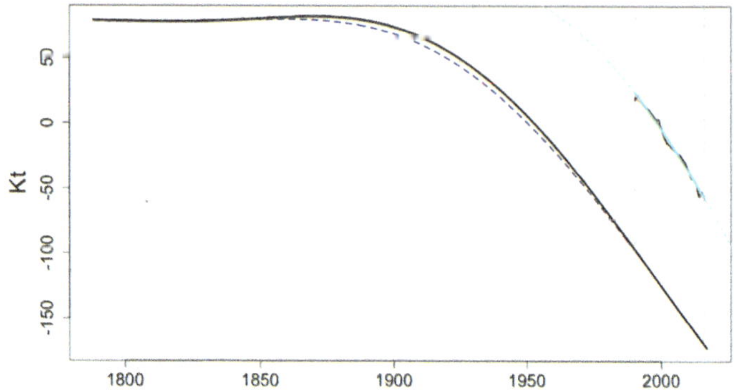

Fig. 9 Common mortality trend and estimated Poland mortality trend based on common trend—female population. *Source* Own preparation based on HMD dataset and MuPoMo package

could extend forecasting horizon of Poland approximately 40 years through the information from common trend (Figs. 8 and 9, with the (cyan) dashed line).

5 Conclusions

One important advance in mortality modeling and forecasting in the last decades is the further development of multi-population methods. Models have been developed to avoid unrealistic crossovers or divergence in future mortality between countries or sexes, which can result from applying mortality projection models to single populations/countries. The key issue is the identification group of populations with similar mortality trends, and this paper investigates the existence of some common stochastic trends in Europe. The semi-parametric method can be used for forecasting mortality time-varying index for those countries for which it is expected that they may catch up to and join the low-mortality group in the future.

The analysis results in the conclusion that Poland's mortality trend is about 41 years later for females and 45 years for males than common trends in selected countries from West, North and South Europe. However, it should be noted that the method is useless for countries with a quite different shape of mortality time-varying parameters than common trends, e.g., for Latvia and Lithuania, because they exhibit quite different a tendencies in contrast with other countries. The main advantage of the semi-parametric method is the possibility to consider different population sizes of an individual country, and thus, a semi-parametric comparison of common mortality trend with each individual nation-level one will be a promising way with respect to forecasting.

References

Assunção RM, Neves MC, Camara G, Freitas CDC (2006) Efficient regionalization techniques for socio-economic geographical units using minimum spanning trees. Int J Geogr Inf Sci 20:797–811
Booth H, Tickle L (2008) Mortality modelling and forecasting: a review of methods. Ann Actuar Sci 3(1–2):3–43
Fang L, Härdle WK, Park J (2016) A mortality model for multi-populations: a semi-parametric approach. SFB 649 Discussion Paper 2016-023, Humboldt-Universität zu Berlin, pp 1–31
GUS (2014) Prognoza ludności na lata 2014–2050. https://stat.gov.pl/obszary-tematyczne/ludnosc/prognoza-ludnosci/prognoza-ludnosci-na-lata-2014-2050-opracowana-2014-r-,1,5.html. Accessed 3 Aug 2019
Härdle W, Marron J (1990) Semiparametric comparison of regression curves. Ann Stat 18:63–89
Lazar D, Buiga A, Deaconu A (2016) Common stochastic trends in European mortality levels: testing and consequences for modeling longevity risk in insurance. Roman J Econ Forecast 2:152–168
Lee R, Carter L (1992) Modeling and forecasting the time series of US mortality. J Am Stat Assoc 87:659–671
Li N, Lee R (2005) Coherent mortality forecasts for a group of populations: an extension of the Lee-Carter method. Demography 42(3):575–594
Majewska J (2017) An EU cross-country comparison study of life expectancy projection models. Selected papers from the 2016 Conference of European Statistics Stakeholders. Special issue, pp 83–93
Shang HL, Booth H, Hyndman R (2011) Point and interval forecasts of mortality rates and life expectancy: a comparison of ten principal component methods. Demogr Res 25(5):173–214
Stoeldraijer L, van Duin C, van Wissen L, Janssen F (2013) Impact of different mortality forecasting methods and explicit assumptions on projected future life expectancy: the case of the Netherlands. Demogr Res 29(13):323–354

Impact of the Selected Factors on the Men and Women Wages in Poland in 2014. The Conjoint Analysis Application

Aleksandra Matuszewska-Janica

Abstract Numerous studies on the labour market relate to the wages. They indicate a significant diversity of factors affecting the remuneration in groups of employees with different characteristics. The aim of the study is to assess the relative importance of selected variables (attributes of employees) on the level of wages. The analysis is conducted for the employees working in Poland in 2014 in the enterprises with at least ten workers. We also take into account the sub-samples included only men or only women. Additionally, an assessment of the impact of outliers on changes in relative importance of analysed attributes was carried out. The analysis is applied the relative importance measure practised in conjoint analysis. The data employed in the study are from the Eurostat's *Structure of Earnings Survey*. The results indicated primarily two variables that have greater importance for the diversity of the wages. They are economic activity (industry, according to the NACE rev. 2) and occupation (according to the ISCO-08). Their joint relative importance is greater than 50%. It is also noticeable that along with the elimination of the outliers the importance of the enterprises' economic activity is increasing and importance of the level of occupation is decreasing. In the samples encompassed men are noted higher relative importance of variables as follows: economic activity of enterprise and size of the enterprise. In turn, in the samples encompassed women we observe higher relative importance for occupation, educational level, age group and contractual working time (full- or part-time).

Keywords Labour market · Wages · Conjoint analysis

1 Introduction

Wages are an important element of the economy functioning. In the economic sense, wages are the price of the labour. In this context, in the absence of control, they

A. Matuszewska-Janica (✉)
Department of Econometrics and Statistics, Warsaw University of Life Sciences—SGGW, Warsaw, Poland
e-mail: aleksandra_matuszewska@sggw.pl

K. Jajuga et al. (eds.), *Classification and Data Analysis*, Studies in Classification, Data Analysis, and Knowledge Organization,
https://doi.org/10.1007/978-3-030-52348-0_20

are determined by supply and demand (see Hicks 1963, p. 1). As defined by the National Labour Inspectorate (Polish body supervising compliance with labour law), remuneration (wage) is a periodic monetary compensation paid in exchange for work by an employer to an employee, and its amount is determined by the type, quantity and quality of work, respectively. A number of conditions related to the performance of work means that the amount of remuneration is influenced by a huge number of factors. The subject of remuneration determinants is discussed in many research papers; however, the important dimension of a given determinant among other determinants is usually omitted.

The main purpose of this study is to indicate the importance of factors (attributes) affecting the level of remuneration in Poland in 2014. Two additional elements were included in the analysis. The first is the division of employees into male and female groups. It results from the fact that the average wages calculated for these groups differ significantly. During previous analyses, the question arose how large are the differences between determinants of the male and female wages. The presented analysis helps to assess the impacts of selected attributes on the men and women remuneration and compare them. The second is to check whether there are differences in the assessment of the importance of factors affecting wages, if outliers are eliminated from the sample.

The presented analysis based on the data collected by Eurostat within the Structure of Earnings Survey (SES). In the study are employed econometric models and relative importance of the attributes measure, which is applied in the *conjoint analysis* (see, i.e., Bak 2004).

2 Background for the Wages Analysis

Remuneration (wage) has several functions (see Borkowska 2012): (1) income and (2) social which are perceived by employees, (3) cost and (4) incentive (motivational), which are recognised by employers. Usually, an individual's wage is income that covers the cost of living and satisfies his and his family's material and immaterial needs (the income function). From the employer point of view, the remuneration and its derivatives are a component of costs (cost function). Social function includes the following elements (see Borkowska 2006, pp. 357–358):

- shaping (structuring) a good working climate, preventing wage conflicts (very important function also behind employer);
- preventing a decrease in the real value of remuneration;
- paying all full-time employees a minimum wage (as required by law);
- preventing deep pay inequalities.

Mosley et al. (2015) pointed out that well-rewarded employees are prone to a large extend to perform a better job that those that do not get paid well. Borkowska (2006, p. 358) lists four sub-function within the motivational function. Firstly, wages should motivate people to work and attract to the enterprise. Secondly, wages should also provide incentives to be loyal to an employer. Thirdly, wages ought to stimulate employees to the high work results (higher productivity of the employees). And finally, they need to stimulate employees to develop their competencies, which will result in the development of the enterprise.

The economic function of remuneration indicates that wage (remuneration) is equivalent to work. So the level of remuneration and the differences in wages among workers depend on the employee's workload and effect. Wages should be also related to an appropriate level according to the economic development in the country. In turn, employers should apply appropriate compensation systems so that wages are clear and acceptable to employees (see *Compensation*—Encyclopaedia of Management).

There are many theories that are the background to the remuneration analysis. We can list neoclassical theory and Keynesian theory, labour market segmentation theory, efficiency wage theory, rent-sharing and rent-extraction theory among others (see Krynska and Kopycinska 2015, see also Zieliński 2012). Concerning the determinants of the individuals' remuneration and differences among individuals' remuneration, three theories are most often taken into consideration. They are:

- human capital theory (see Schultz 1961, 1971; Becker 1962, 1964; Mincer 1974, see also Kunasz 2004);
- discrimination theory (see Becker 1957, Arrow 1972a, b, 1973);
- preference theory (see Hakim 1998, 2000, 2002, 2004, 2006; Charles and Grusky 2004; Jacobs and Gerson 2004; Kan 2007).

As point Nafukho et al. (2004), based on the articles (Lucas 1988, 1990) "the fundamental principle underpinning human capital theory is the belief that peoples' learning capacities are of comparable value to other resources involved in the production of goods and services". Human capital theory indicated that education and schooling are seen as investments. Such investments prepare the workforce and contribute to increase the productivity of employed individuals and organisations (see Nafukho et al. 2004). In reference to this theory and the functions of remuneration, we can maintain that factors increasing the employee's human capital also contribute to the increase in his remuneration. Two leading factors are education and experience.

We can understand discrimination at the labour market if the situation exists when some characteristics which are unrelated to the individuals' talent, skills, and drive, restrict access to the same opportunities like amount of wage, access to a job position among others (see D'Amico 1987). As Becker (1957) explains, there is exists on the labour market some group of employees or employers who perceive members of some groups (characterised by some attributes like race or sex among others) as a worse workers. Then their work or skills are not measured and assessed the same. This is a very difficult problem to analyse in quantitative terms because we usually do not have enough information about the characteristics that differentiate employees and their job positions.

Preference theory explains that diversity of lifestyle preferences and choices in modern societies have a significant impact on the individuals' position on the labour market. Individuals represent different preferences regarding work and family and such preferences have influenced on the labour market activity, employment choices, education choices among others (see, i.e., Hakim 2004). In such situation, Hakim (2006) describes as follows: “There are no sex differences in cognitive ability but enduring sex differences in competitiveness, life goals, the relative emphasis on agency versus connection”. This stays largely in opposite to the many feminist theories, so the preference theory is perceived by some as a controversial theory.

Factors that have an impact on the remuneration level we can divide into three groups.

- Individual characteristics of employee (e.g. age, sex, job seniority, the type and level of education, occupation, full- or part-time job, type of job contract, family social and economic status, preferences).
- Enterprise characteristics (type of industry, public or private sector, size of the enterprise, activity of the trade unions among others).
- Characteristics of the environment (e.g. economic situation in the region or in the country, structure of the labour market, family policy).

Knowledge of factors affecting the level of remuneration has not only a cognitive aspect but also helps in formulating strategies related to the labour market policy and supports management in enterprises. As is mentioned earlier, it is not possible to include in the quantitative analyses many factors (attributes) affecting the level of remuneration. This is mainly due to limited data availability or difficult measurement of some characteristics. Therefore, analysts usually limit themselves to a certain group of variables (factors).

3 Data and Applied Methods

The study was conducted based on the EU *Structure of Earnings Survey* (SES) individual data (microdata) of 2014.[1] The characteristics included in the SES refer to the enterprises with at least 10 employees. It is necessary to mention that information collected in the SES databases is pulled from the enterprises' registers.[2] The

[1] *Structure of Earnings Survey* (SES) is the 4-yearly Eurostat's periodic survey. It has been conducted regularly in the EU Member States, candidate countries and countries of the European Free Trade Association (EFTA) since 2002. Currently, the last available microdata are from 2014. This survey is conducted in Poland by the Central Statistical Office (CSO) every two years. The surveys conducted by CSO are more extensive than those reported by Eurostat. However, Eurostat's reports have one advantage, namely they provide comparability of the collected data among different countries. In turn, the presented analysis is a part of the larger research that involves the determinants of men and women wages across European countries. This is a premise to use of the Eurostat data and not CSO.

[2] Data regarding wages that are comparable among the EU countries can also be obtained from other large surveys, such as *Labour Force Survey* (LFS) or *EU Statistics on Income and Living Conditions* (EU-SILC). The advantage of these surveys is the opportunity to take into account a larger number of variables, e.g. relating to the family status or to the previous status in the labour market among others. By contrast, the SES database offers only the limited number of variables. This survey does

presented analysis is divided into two stages. In the first step, parameters of linear econometric models are estimated using the generalised least squares method with heteroskedasticity correction. In the second stage are calculated relative importance of selected attributes (variables).

As is mentioned, in the first stage estimated parameters of the regression are as follows:

$$lnW_{si} = \beta_{s0} + \sum_{j}^{k} \sum_{l=1}^{n_j} \beta_{sjl} X_{jlsi} + \varepsilon_{si} \tag{1a}$$

where W_{si}—gross hourly wages of ith individual from s-th sample ($s = p, f, m$); X_{jlsi}—jth dummy explanatory variable for ith individual from sth sample; p—whole sample; m—sample with male individuals; f—sample with female individuals; $f + m = p$; k—number of variables (attributes); n_j—number of analysed variants of the jth variable.

We can present the regression (1a) with estimated parameters as follows:

$$ln\hat{W}_{si} = b_{s0} + \sum_{j}^{k} \sum_{l=1}^{n_j} b_{sjl} X_{jlsi} \tag{1b}$$

We are able to obtain 11 explanatory variables (attributes) from the SES database (see Table 1). All of them are binary variables, but some of the groups of variables have more than 2 variants.

The variable that counts the most number of variants is the economic activity of the enterprise. The variants represent the selected major groups taken from the Statistical Classification of Economic Activities in the European Community, commonly referred to as NACE rev. 2 (*nomenclature statistique des activités économiques dans la Communauté européenne*). The variants included in the analysis are as follows:

(B)—mining and quarrying; (C)—manufacturing (reference variant); (D)—electricity, gas, steam and air conditioning supply; (E)—water supply, sewerage, waste

not encompassed variables related, i.e. to the family status or total job seniority. Such variables have a significant impact on the level of wages as it is pointed in the articles (Witkowska 2014) or (Kompa and Witkowska 2018) among others. But they (LFS and EU-SILC) have also some limitation. Firstly, the respondents are households, while in the SES they are enterprises (the informations are pulled from the enterprises' registers). So the information about individuals wages obtained from the SES are more reliable. Secondly, wages in the Eurostat's LFS datasets are defined as deciles of monthly wages. This makes it impossible to convert the monthly wages into the hourly wages and also limits the application of wide range of methods. In the EU-SILC survey, we have access only to yearly income of respondents. In turn, the SES data are reported as values of the hourly wages. This allows to avoid some problems with determinates analyses, such as unreasonable (excessive) influence of some variables, like part-time work. In addition, in the LFS and EU-SILC datasets we have information not only about employed, but also about unemployed or labour market inactive individuals. Thus, the selection of the appropriate sample to the wage determinants analyses is more complex and multistage than in the SES case. The mentioned premises indicated the SES database as the background of the presented analysis.

Table 1 List of the explanatory variables

No.	Variables (group of variables)	Variants	Reference variant
1	Sex[a]	2	Female
2	Region	6	PL1—central region
3	Occupation (major groups ISCO-08)	9	Group 9 (elementary occupations)
4	Size of the enterprise	3	10–49 employees
5	Collective pay agreement	3	No. collective agreement exists
6	Age group	5	Aged 20–29
7	Highest educational level attained	4	Basic education (G1)
8	Contractual working time (full-time or part-time)	2	Full-time employees
9	Type of employment contract (indefinite duration/temporary or fixed duration)	2	Indefinite duration
10	Length of service in the enterprise (in years)	9	Less than one year
11	Economic activity of enterprise (NACE rev. 2)	18	C—manufacturing

Source Own calculation
[a]The variable used in the model estimated using whole sample (*p*)

management and remediation activities; (F)—construction; (G)—wholesale and retail trade, repair of motor vehicles and motorcycles; (H)—transportation and storage; (I)—accommodation and food service activities; (J)—information and communication; (K)—financial and insurance activities; (L)—real estate activities; (M)—professional, scientific and technical activities; (N)—administrative and support service activities; (O)—public administration and defence, compulsory social security; (P)—education; (Q)—human health and social work activities; (R)—arts, entertainment and recreation; (S)—other service activities.

The SES datasets are omitted tree economic activities: (A)—agriculture, forestry and fishing; (T)—activities of households as employers; undifferentiated goods and services-producing activities of households for own use; (U)—activities of extraterritorial organisations and bodies.

The next variable is occupation (job position). Its variants are taken from the current version of the International Standard Classification of Occupations (ISCO). The variants represent the major groups of this classification:

(1)—managers; (2)—professional; (3)—technicians and associate professionals; (4)—clerical support workers; (5)—service and sales workers; (6)—skilled agricultural, forestry and fishery workers; (7)—craft and related trades workers; (8)—plant and machine operators, and assemblers; (9)—elementary occupations (reference variant).

The SES does not collect the information for the one major group (10)—armed forces occupations.

The variable length of service in the enterprise represents the number of completed years of service in the enterprise participating in the survey. This variable has been transformed into nine binary variables (variants) for the purposes of the presented analysis.

(1)—less than 1 year (reference variants); (2)—one year; (3)—2–4 years; (4)—5-9 years; (5)—10–14 years; (6)—15–19 years; (7)—20–29 years; (8)—30–39 years; (9)—40 years and over.

Age is represented by variable with 5 variants:

(1)—Y0_29, less than 30 years (reference variants); (2)—Y30_39, between 30 and 39 years; (3)—Y40_49, between 40 and 49 years; (4)—Y50_59, between 50 and 59 years; (5)—Y_GE60, 60 years and over.

The variable region refers to the NUTS 1 level (NUTS is Classification of Territorial Units for Statistics—*nomenclature d'unités territoriales statistiques*). The first-level NUTS in Poland is represented by the following regions:

PL1—Central Region (Łódź, Mazovia)—reference variants; PL2—South Region (Lesser Poland, Silesia); PL3—East Region (Lublin, Subcarpatioan, Świętokrzyskie, Podlaskie); PL4—Northwest Region (Greater Poland, West Pomerania, Lubusz); PL5—Southwest Region (Lower Silesia, Opole Voivodeship); PL6—North Region (Kuyavian-Pomeranian, Varmia-Masuria, Pomerania).

The variable education concerns the highest successfully completed level of education (ISCED). ISCED is the International Standard Classification of Education maintained by the United Nations Educational, Scientific and Cultural Organization (UNESCO). Educational levels included in the presented analysis correspond with the version of the classification from 2011. This variable involves 4 variants:

G1—(group 1) basic education (reference variant): 0—less than primary education, 1—primary education; 2—lower secondary education; G2—(group 2) secondary education, 3—upper secondary education, 4—post-secondary (non-tertiary) education; G3—(group 3) tertiary education (up to 4 years): 5—short-cycle tertiary education; 6—ba'chelor or equivalent education; G4—(group 4) tertiary education (more than 4 years): 7—master or equivalent education, 8—doctoral or equivalent education.

The other two variables with more than 2 variants are: size of the enterprise and collective pay agreement. The size of the enterprise is related to the number of individuals employed by the enterprise: E10_49—10–49 employees (reference variant); E50_249—50–249 employees; GT250-250 or more employees.

The collective pay agreement identifies the type of pay agreement covering at least 50% of the employees in the enterprise. SES dataset for Poland includes the following options: A—national level or interconfederal agreement; D—enterprise or single employer agreement; N—no collective agreement exists (reference variant).

In order that to consider the impact of outliers on the relative importance of attributes, it is necessary to identify them. The most popular tool applied for identifying outliers is box and whisker plot (called also box plot with fences). For this purpose are used quartiles and interquartile range (see Aczel and Sounderpandian 2008). Interquartile range (*IQR*) is defined as the difference between the lower (Q_L) quartile and upper (Q_U) quartile:

$$IQR = Q_U - Q_L \tag{2}$$

IQR is used to indicate points on the box plot called fences that are needed for identifying outliers. The values of these fences are calculated as follows:

- $LOF = Q_L - 3 \cdot IQR$, lower outer fence and $LIF = Q_L - 1.5 \cdot IQR$, lower inner fence;
- $UIF = Q_U + 1.5 \cdot IQR$, upper inner fence and $UOF = Q_U + 3 \cdot IQR$, upper outer fence.

Observation is considered as a mild outlier if it is in the range *LOF* to *LIF* or *UIF* to *UOF*. In turn, the value of observation lower than *LOF* or greater than *UOF* designated that we could consider it as an extreme outlier. In the presented study, we have considered upper outliers, greater than *UIF* or *UOF*, respectively. Values of individuals' wages equal or greater than *UIF* ($W_i \geq Q_U + 1.5 \cdot IQR$) we call suspected outliers. In turn, values of individual's wages equal or greater than *UOF* ($W_i \geq Q_U + 3 \cdot IQR$) we call outliers. Table 2 presents numbers of observations in selected samples. It is worth to mention that percentage of outliers is close to 2% and the share of suspected outliers exceeds a little bit 5% of the whole sample. In addition, we can observe that in the groups of top earners men are overrepresented. The share of women equals 39.6% in the set of suspected outliers and 28.1% in the set of outliers. While in the whole samples women represent more than 50% of the individuals.

The assessment of the relative importance of the individual attributes (variables, presented in Table 1) is based on the *conjoint analysis* (see Bak 2004, p. 193; Walesiak et al. 1999; Hair et al. 2014, p. 350):

Table 2 Selected characteristics of the analysed samples

Type of sample	Number of observation	Male	Female	%Female
Whole sample	722,697	359,119	363,578	50.3
Limited sample (1); $W_i < Q_U + 3 \cdot IQR$	710,409	350,281	360,128	50.7
Outliers (1); $W_i \geq Q_U + 3 \cdot IQR$	12,288	8838	3450	28.1
% of outliers (1)	1.7	2.5	0.9	x
Limited sample (2); $W_i < Q_U + 1.5 \cdot IQR$	684,465	336,036	348,429	50.9
Suspected outliers (2); $W_i \geq Q_U + 1.5 \cdot IQR$	38,232	23,083	15,149	39.6
% of suspected outliers (2)	5.3	6.4	4.2	x

Source Own calculation

$$V_j = \frac{\max_{l_j}\{U_{jl_j}\} - \min_{l_j}\{U_{jl_j}\}}{\sum_{j=1}^{k}\left(\max_{l_j}\{U_{jl_j}\} - \min_{l_j}\{U_{jl_j}\}\right)} \cdot 100\% \quad (3)$$

where U_{jl_l} are the estimates of part-worths (utilities) of the lth level of the jth variable.

Individual utilities are identified with regression parameters $U_{jl_j} = b_{jl_j}$, where b_{jl_j} is the estimated parameter by lth level of the jth variable (see formula 1b). In the presented analysis, we adopt that utility for the reference variant (level) of the variable equals zero $U_{jr} = 0$.

4 Results

The relative importance of the analysed variables (attributes) is based on the estimated parameters of the model (1). In the first step, the samples with information for all employees are analysed (general samples—without division into women and men). Obtained results (see Table 3) indicate that two variables (attributes) have the greatest importance for the description of the wages level: economic activity (industry, according to the NACE rev. 2 classification) and the occupation (major groups of the ISCO-08). The relative importance of each mentioned variable is much more than 20%, and their joint relative importance is greater than 50%. These results confirm the outcomes obtained in case of the application of the classification trees. Industry and occupation proved to be the most important explanatory variables in the classification process (see, e.g., Matuszewska-Janica and Witkowska 2013).

Industry has an important impact on the level of the remuneration. The literature points some explanations. Rycx and Tojerow (2007) quote analyses of Dickens and Katz (1987), Krueger and Summers (1987, 1988), Katz and Summers (1989), Ferro-Luzzi (1994) Hartog et al. (1997), Lucifora (1993), Rycx (2002, 2003) and Vainiomäki and Laaksonen (1995), who presented that wages differentials "between workers with the same observable individual characteristics and working conditions but employed in different sectors". So they have proved the existence of sectoral effects on workers' wages (see also Thaler 1989). The inter-industry wage differentials are explained by differences across workers in "unobserved" ability or quality. But they are unobserved to the researcher but not to the worker or firm (see Blackburn et al. 1992; Murphy and Topel 1987 among others). However, the individual features of the employee, the employer as well as job position in the sectors (industries) play an important role. First of all observed wages of differential across industry are caused by the nature of the work offered in these sectors (e.g. which required high or low job qualifications). Other explanation is that the incentive conditions (characteristic of the employer) can vary between sectors. In such situation, two employees with identical productive characteristics and working conditions probably earn different wages (see Rycx and Tojerow 2007).

Table 3 Relative importance of the variables (attributes) in the general samples

Variable (attribute)	Whole sample (p) (%)	Limited sample (1) $W_i < Q_U + 3 \cdot IQR$ (%)	Limited sample (2) $W_i < Q_U + 1.5 \cdot IQR$ (%)
Region	4.6	4.6	4.5
Economic activity of enterprise	23.8	25.1	26.3
Size of the enterprise	6.5	6.7	7.0
Collective pay agreement	1.2	1.5	1.8
Age group	3.6	3.4	3.0
Occupation (major groups ISCO-08)	28.3	25.9	24.4
Highest educational level attained	12.1	12.1	11.5
Length of service in the enterprise (in years)	7.9	8.5	9.1
Contractual working time (full-time or part-time)	1.6	1.5	1.5
Type of employment contract	4.6	4.8	5.1
Sex	5.8	5.8	5.7

Source Own calculation

Job position in the analysis is presented as the major groups from the occupational classification ISCO-08 proposed by the International Labour Organisation. Employees from the high-ranked groups (managers and professionals) usually earn significantly more than others, especially the lower-level employees. Very interesting explanation present De Beyer and Knight (1989). They compare the occupational positions to the factories. They explain that "in some factories, more complex and technologically advanced than others, there is a nexus of relationships among output and the inputs labour, cognitive skill, vocational skill and natural ability, such that these inputs are found in combination and yield high returns". It is strictly connected to the human capital theory, where the education, job experience and natural ability (recognise at the high job-positions) are significant determinants of the wages (so called skill premiums). The level of the wages is also analysed on the background of the productivity (see, e.g., Meager and Speckesser 2011 or Landmann 2004).

It is also noticeable that along with the elimination of the outliers the importance of the enterprises' economic activity increased (from 23.8 to 26.3%) and the importance of the level of occupation decreased (from 28.3 to 24.4%). As was mentioned, the employees at the high positions (primarily managers and professionals) usually earn significantly more than others. Therefore, the removal of the outliers from the sample decreases the meaning of the occupational position on behalf of the other attributes (variables), including economic activity.

Education is another factor that has a fairly significant impact on remuneration. Generally, according to the human capital theory we expect higher wages as the level of education increases. The confirmatory results for the employees in Poland are presented in the article (Roszkowska and Majchrowska 2014) among others. In the presented results, relative importance of the variable level of education equals 12.1%. After elimination of the extreme outliers (in the limited sample 1), these values do not change, but in the limited sample 2 (for $W_i < Q_U + 1.5 \cdot IQR$) the V_j measure equals 11.5% (the decrease by 0.6 p.p.). So we can conclude that the education level has a little bit bigger impact on remuneration in the group of the best-paid employees.

The next two variables for which the relative importance is more than 6% are the size of the enterprise and the job tenure with the current employer (length of service in the enterprise). The statistics show that in the enterprises with the greater number of employees, the wages are higher on average (see Oi and Idson 1999 or Schmidt and Zimmermann 1991 among others). The relative importance of the size of the enterprise increases from 6.5 to 7.0% along with the elimination of the outliers. The variable job tenure is a specific variable, because it is defined as the length of service of the individual in the current enterprise. So we have no information about total job tenure of the individual employees (observations). Generally, longer job tenure is compensated by higher wages according to the human capital theory. In addition, we can this factor to associate with the premium for staying with the same employer. The relative importance of this variable equals 7.9% in the whole sample and 9.1% in the limited sample 2. So, the value of this indicator increases by 1.2 p.p. after removing the extreme and suspected outliers. Such situation can be connected to the fact, that we observe higher mobility in the labour market in the groups of managers and professionals than in other groups of workers. Thus, the elimination from the sample the best-paid employees (usually managers and professionals) can cause by increasing the relative importance of this attribute for others employees.

The next variable included in the models is sex. Its relative importance equals 5.8% in the whole sample and 5.7% in the limited sample (2). So, it stays about at the same level independently to the elimination of the outliers from the sample. The problem of the wage differences related to the sex, called gender wage gap, is widely discussed in the economic literature (see Blau and Kahn 2003; Arulampalam et al. 2007; Newell and Reilly 2001 among others). We can distinguish two components of this gap (see, i.e., Oaxaca 1973; Blinder 1973; Jann 2008). The first is connected to the explained part of it and comes from the differences between characteristics of male and female employees (e.g. women used to work more often in part-time than men and they have shorter job tenure on average). The second is connected to the unexplained part. It is usually connected to the characteristics which cannot be included in the analysis (because of lack of data or because unobservable of the variables most frequently). In the case of variable sex, the result deviates from the assumption. It is expected that the significance of this attribute would decrease along with the elimination of outliers. The analysis presented in the paper (Matuszewska-Janica 2018) indicates that the gender pay gap has been reduced along the elimination of the outliers. So we expect that also the importance of the variable sex is also reduced as the determinant of wages.

The relative importance of the variable region equals 4.6% for the whole sample and 4.5% for limited sample 2 (the difference between these two values is very small). Then, we can conclude that this variable has very similar importance (influence) for the wages level regardless of the outlier's prevalence. The next variable is the type of employment contract. Its relative importance is at the same level as the previous variable (region) and equals 4.6% for the sample with outliers. The relative importance of this variable increases up to 5.1% (by 0.5 p.p.) after excluding the outliers.

Age group is the variable for which the value of the V_j measure equals 3.6% and decreases along with the elimination of the outliers to 3.0%. According to the analyses of the human capital, age is identified with the work experience (employees usually achieve a higher work experience along with increasing age). Therefore, such result is expected that this variable will have a greater impact on the level of remuneration in the samples which included the highest-paid individuals. However, if we compare the relative importance of this variable with the relative importance of the attributes described so far we can see that age has a relatively small impact on the wages.

The lowest values of the V_j measure we obtain for two variables: collective pay agreement and contractual working time (full-time or part-time). The relative importance is smaller than 2% in both cases. So the variability of the gross hourly wages depends least on these variables.

Table 4 presents the relative importance values of the attributes obtained for the male and female sub-samples separately. The highest values of the V_j measure we obtain for the economic activity of enterprise and occupation. Just as for the whole sample, but we can notice sharp differences in the values of this measure in both types of sub-samples (men and women). The relative importance of the economic activity is higher in the male sub-samples (by 5.1 p.p. or more) in comparison with the female sub-samples. We observe, that differences between relative importance of this attribute increase along the elimination of the outliers (from 5.1 to 7.7 p.p.), whereby the value of the V_j measure is much more higher for men (29.7%) than for women (22.0%). So, we conclude that economic activity of the enterprise has a greater impact on the male wages than the female wages.

In turn, the much more higher relative importance of the occupation is observed in the female group (from 32.3% in the sub-sample with outliers to 30.4% in the sub-sample without extreme and suspected outliers). It worth to notice, that only for this attribute (and only in the female sub-samples), we obtain the value of the V_j higher than 30%. The differences between the relative importance calculated for the male and female sub-samples increases along with the elimination of the outliers also for this variable (from 2.8 to 7.6 p.p.), and the dynamic of this increase is much more higher in comparison with the previous variable, whereby we can notice that occupation has a bigger impact on the female wages. On the other hand,

Table 4 Relative importance of the variables (attributes) in the male and female sub-samples

Variable (attribute)	Whole sample (p) (%)		Limited sample (1) $W_i < Q_U + 3 \cdot IQR$		Limited sample (2) $W_i < Q_U + 1.5 \cdot IQR$ (%)	
	Female	Male	Female	Male	Female	Male
Region	5.3	4.9	5.3	5.3	5.2	5.4
Economic activity of enterprise	20.8	25.9	21.4	28.0	22.0	29.7
Size of the enterprise	4.6	10.2	4.8	10.8	5.0	11.3
Collective pay agreement	1.4	1.2	1.7	1.6	1.8	2.1
Age group	7.0	3.4	6.9	3.3	6.6	3.0
Occupation (major groups ISCO-08)	32.3	29.5	30.9	25.7	30.4	22.8
Highest educational level attained	13.9	11.5	14.0	11.2	13.2	10.5
Length of service in the enterprise (in years)	7.9	7.7	8.3	8.3	8.7	9.0
Contractual working time (FT or PT)	2.5	0.3	2.4	0.1	2.5	0.1
Type of employment contract	4.2	5.3	4.3	5.7	4.5	6.1

Source Own calculation

the elimination of the outliers much more reduces the relative importance of this variable in the male sub-samples (decline by 6.7 p.p.).

The next variable is education level. The relative importance of this attribute is greater than 10% regardless of the sample. But we observe the higher values of the V_j in female sub-samples (by about 2.5 p.p. than in male sub-samples). The relative importance of the length of service (job tenure in the present enterprise) stays at similar level in both groups: men and women. The differences between corresponding values are no greater than 0.3 p.p. In both groups, V_j increases along with the elimination of outliers. The greater differences in the relative importance we observe for age variable. The values of V_j measure in women group remain in the range from 7.0 to 6.6%, while in men group it changes from 3.4 to 3.0%. So we observe that this attribute in female sub-samples is twice as important as in the male sub-samples. These three variables (education level, length of service and age) are strictly connected to the human capital theory. If we sum up the values V_j measure for these variables, we can observe that in the females sub-samples it equals circa 29%; when in the male sub-samples, it equals circa 22.5%. The differences between male and female sub-samples fluctuate between 6 and 6.4 p.p. We can conclude that these three attributes have a greater importance in case of female wages than in male wages.

The obtained results indicate a greater relative importance of size of the enterprise in the male sub-samples. The V_j measure equals from 10.2% (in the sample with included outliers) to 11.3% (in the sample without outliers). In case of female sub-samples, it fluctuates from 4.6 to 5%. As it can be seen, the differences in the impact of this variable on the male and female wages are considerable.

Region has a similar relative importance in both men and women group. Its value oscillates around 5%. Taking into consideration type of employment contract, we observe the differences in relative importance in male and female sub-samples greater than 1 p.p. Such difference equals 1.1 p.p. for sub-samples with outliers to 1.6 p.p. in sub-samples without outliers. In turn, the contractual working time (full-time or part-time job) has bigger impact in case of female wages (V_j measure equals 2.5%). As distinct from groups of men, where we observe V_j measure equals 0.3% or less. This is mainly due to the fact that women take part-time work more often than men.

5 Conclusions

Several facts (findings) can be pointed out as a summary of the presented analysis. First of all, economic activity (with respect to the NACE rev.2) and job position (with respect to the ISCO-08) are the main factors (attributes) affecting the level of remuneration in the analysed group of employees. Their total relative importance exceeds 50%. Whereby we notice the value of relative importance V_j measure of job position (occupation) is higher in the female sub-samples. In turn, we observe the greater relative importance of economic activity in the male sub-samples. Secondly, the combined relative importance of the three variables that have a direct reference to the human capital theory (level of education, age and job tenure which in our analysis concerns the length of service in the current enterprise) amounts to circa 29% in the female sub-samples and 22.5% in the male sub-samples. We can interpret such situation that these attributes have more impact on the variability of the women remuneration than in the case of men. Thirdly, we observe that in the samples involving separately men and women, the differences in the relative importance equal at least 2 p.p. for seven variables out of ten analysed. This confirms the observations that many factors influence with varying intensity on wages in these two groups. Finally, the elimination of the outliers from the sample influences on the relative importance of the some variables (attributes), such as industry, job position, size of the enterprise, age or job seniority among others. Hence, it is worth to survey situation in groups of employees across various cross sections.

With regard to the presented analysis, first of all it should be consult that from a wide range of wages determinants the study included only a dozen or so (see Table 1). This is mainly due to the SES database limitations referring to the methodology and Eurostat's guidelines of the survey do not allow to include some types of variables. Thus, the validity of attributes applies only to the selected group of variables. However, the presented results allow to systematise the impacts of individual attributes on the level of wages and comparing such impacts among different groups of labour market participants (in our case male and female employees).

The next stages of the analysis will include comparing the validity of attributes affecting the level of men and women wages among the EU countries. It is also planned to apply other regression models such as multinomial logit models and package conjoint implemented in R program (see Bak and Bartłomowicz 2011; Walesiak and Gatnar 2009 or Pelka and Rybicka 2012).

References

Aczel AD, Sounderpandian J (2008) Complete business statistics, 7th edn. McGraw-Hill/Irwin

Arrow KJ (1972a) Models of job discrimination. In: Pascal AH (ed) Racial discrimination in economic life. D.C. Heath, Lexington, pp 83–102

Arrow KJ (1972b) Some mathematical models of race discrimination in the labor market. In: Pascal AH (ed) Racial discrimination in economic life. D.C. Heath, Lexington, pp 187–204

Arrow KJ (1973) The theory of discrimination. In: Ashenfelter O, Rees A (eds) Discrimination in labor markets. Princeton University Press, Princeton, pp 3–33

Arulampalam W, Booth AL, Bryan ML (2007) Is there a glass ceiling over Europe? Exploring the gender pay gap across the wage distribution. ILR Rev 60(2):163–186. https://doi.org/10.1177/001979390706000201

Bak A (2004) Dekompozycyjne metody pomiaru preferencji w badaniach marketingowych. Wydawnictwo Akademii Ekonomicznej im, Oskara Langego we Wroclawiu, Wroclaw

Bak A, Bartłomowicz T (2011) Implementacja klasycznej metody conjoint analysis w pakiecie conjoint programu R. Prace Naukowe Uniwersytetu Ekonomicznego we Wrocławiu—Taksonomia 176(18):94–104

Becker GS (1957) The economics of discrimination. The University of Chicago Press, Chicago

Becker GS (1962) Investment in human capital: a theoretical analysis. J Polit Econ 70(5, Part 2):9–49. https://doi.org/10.1086/258724

Becker GS (1964) Human capital: a theoretical and empirical analysis, 1993rd edn. The University of Chicago Press, Chicago

Blackburn ML, Neumark D, Blackburn M (1992) Unobserved ability, efficiency wages, and interindustry wage differentials. Q J Econ 107(4):1421–1436. https://doi.org/10.2307/2118394

Blau FD, Kahn LM (2003) Understanding international differences in the gender pay gap. J Labor Econ 21(1):106–144

Blinder A (1973) Wage discrimination: reduced form and structural estimates. J Hum Resour VII(4):436–455. https://doi.org/10.2307/144855

Borkowska S (2006) Wynagradzanie. In: Król H, Ludwiczynski A (eds) Zarzadzanie zasobami ludzkimi. Tworzenie kapitału ludzkiego organizacji, PWN, Warszawa, pp 354–420

Borkowska S (2012) Skuteczne strategie wynagrodzeń-tworzenie i zastosowanie. Wolters Kluwer, Warszawa

Charles M, Grusky DB (2004) Occupational ghettos: the worldwide segregation of women and men. Stanford University Press, Stanford

Compensations. Encyclopaedia of management (in Polish). https://mfiles.pl/pl/index.php/Wynagrodzenie. Online 30 Sept 2019

D'Amico TF (1987) The conceit of labor market discrimination. Am Econ Rev 77(2):310–315

De Beyer J, Knight JB (1989) The role of occupation in the determination of wages. Oxf Econ Pap 41(3):595–618

Dickens WT, Katz LF (1987) Inter-industry wage differences and industry characteristics. In: Lang K, Leonard J (eds) Unemployment and the structure of labour market. Basil Blackwell, Oxford, pp 48–89
Ferro-Luzzi G (1994) Inter-industry wage differentials in Switzerland. Swiss J Econ Stat 130(3):421–443
Hair JF Jr, Black WC, Babin BJ, Andersen RE (2014) Multivariate data analysis, 7th edn. Pearson, Harlow
Hakim C (1998) Developing a sociology for the twenty-first century: preference theory. Br J Sociol 49:137–143. https://doi.org/10.2307/591267
Hakim C (2000) Work-lifestyle choices in the 21st century: preference theory. Oxford University Press, Oxford
Hakim C (2002) Lifestyle preference as determinants of women's differentiated labor market careers. Work Occup 29:428–459. https://doi.org/10.1177/0730888402029004003
Hakim C (2004) Key issues in women's work: female diversity and the polarisation of women's employment. Psychology Press, London
Hakim C (2006) Women, careers, and work-life preferences. Br J Guid Couns 34(3):279–294. https://doi.org/10.1080/03069880600769118
Hartog J, Van Opstal R, Teulings CN (1997) Inter-industry wage differentials and tenure effects in the Netherlands and the U.S. Economist 145(1):91–99
Hicks J (1963) The theory of wages. Palgrave Macmillan, London
Jacobs JA, Gerson K (2004) The time divide: work, family, and gender equality. Harvard University Press, Cambridge
Jann B (2008) The Blinder-Oaxaca decomposition for linear regression models. Stata J 8(4):453–479. https://doi.org/10.1177/1536867X0800800401
Kan MY (2007) Work orientation and wives' employment careers: an evaluation of Hakim's preference theory. Work Occup 34(4):430–462. https://doi.org/10.1177/0730888407307200
Katz LF, Summers LH (1989) Industry rents: evidence and implications. Brookings Pap Econ Activ: Macroecon 1989:209–275
Kompa K, Witkowska D (2018) Factors affecting men's and women's earnings in Poland. Econ Res—Ekonomska istraživanja 31(1):252–269. https://doi.org/10.1080/1331677X.2018.1426480
Krueger AB, Summers LH (1987) Reflection on the inter-industry wage structure. In: Lang K, Leonard J (eds) Unemployment and the structure of labour markets. Basil Blackwell, Oxford
Krueger AB, Summers LH (1988) Efficiency wages and inter-industry wage structure. Econometrica 56(2):259–293
Krynska E, Kopycinska D (2015) Wages in labour market theories. Folia Oecon Stet 15(2):177–190
Kunasz M (2004) Teoria kapitału ludzkiego na tle dorobku myśli ekonomicznej. In: Manikowski A, Psyk A (eds) Unifikacja gospodarek europejskich: szanse i zagrożenia. Uniwersytet Warszawski, Warszawa
Landmann O (2004) Employment, productivity and output growth. ILO working paper, Employment Strategy Paper 2014/17
Lucas RE (1988) On the mechanics of economic development. J Monet Econ 22(1):3–42
Lucas RE (1990) Why doesn't capital flow from rich to poor countries? Am Econ Rev 80(2):92–96
Lucifora C (1993) Inter-industry and occupational wage differentials in Italy. Appl Econ 25(8):1113–1124
Matuszewska-Janica A (2018) Men and women wage differences in Spain and Poland. Monten J Econ 14(1):45–52
Matuszewska-Janica A, Witkowska D (2013) Zróżnicowanie płac ze względu na płeć: zastosowanie drzew klasyfikacyjnych. Prace Naukowe Uniwersytetu Ekonomicznego we Wroclawiu—Taksonomia 279:58–66
Meager N, Speckesser S (2011) Wages, productivity and employment: a review of theory and international data. European Employment Observatory Thematic expert ad-hoc paper, pp 1–73
Mincer JA (1974) The human capital earnings function. In: Schooling, experience, and earnings. NBER, pp 83–96

Mosley DC, MosleyDC Pietri PH (2015) Supervisory management: the art of inspiring, empowering, and developing people. Cengage Learning, Stamford
Murphy KM, Topel RH (1987) Unemployment, risk, and earnings: testing for equalizing wage differences in the labor market. In: Lang K, Leonard JS (eds) Unemployment and the structure of labor market. Basil Blackwell, New York
Nafukho FM, Hairston N, Brooks K (2004) Human capital theory: implications for human resource development. Hum Resour Dev Int 7(4):545–551
Newell A, Reilly B (2001) The gender pay gap in the transition from communism: some empirical evidence. Econ Syst 25(4):287–304. https://doi.org/10.1016/S0939-3625(01)00028-0
Oaxaca R (1973) Male–female wage differentials in urban labor market. Int Econ Rev 14(3):693–709. https://doi.org/10.2307/2525981
Oi WY, Idson TL (1999) Firm size and wages. In: Handbook of labor economics, vol 3(B), pp 2165–2214
Pelka M, Rybicka A (2012) Pomiar i analiza preferencji wyrażonych z wykorzystaniem pakietu conjoint programu R. Przegląd Statystyczny 59(3):302–315
Roszkowska S, Majchrowska A (2014) Premia z wykształcenia i doświadczenia zawodowego według płci w Polsce. Narodowy Bank Polski, Departament Edukacji i Wydawnictw, Warszawa
Rycx F (2002) Inter-industry wage differentials: evidence from Belgium in a cross-national perspective. Economist 150(5):555–568
Rycx F (2003) Industry wage differentials and the bargaining regime in a corporatist country. Int J Manpower 24(4):347–366
Rycx F, Tojerow I (2007) Inter-industry wage differentials: what do we know? Reflets et Perspectives de la vie économique 46(2):13–22. https://doi.org/10.3917/rpve.462.0013
Schmidt CM, Zimmermann KF (1991) Work characteristics, firm size and wages. Rev Econ Stat 3(4):705–710
Schultz TW (1961) Investment in human capital. Am Econ Rev 51(1):1–17
Schultz TW (1971) Investment in human capital. The role of education and of research. The Free Press, New York
Thaler RH (1989) Anomalies: interindustry wage differentials. J Econ Perspect 3(2):181–193
Vainiomäki J, Laaksonen S (1995) Interindustry wage differentials in Finland: evidence from longitudinal census data for 1975–85. Labour Econ 2(2):161–173
Walesiak M, Gatnar E (eds) (2009) Statystyczna analiza danych z wykorzystaniem programu R. PWN, Warszawa
Walesiak M, Dziechciarz JZ, Bak A (1999) An application of conjoint analysis for preference measurement. Argumenta oeconomica 1(7):169–178
Witkowska D (2014) Determinants of wages in Poland. Quant Methods Econ 15(1):192–208
Zieliński M (2012) Rynek pracy w teoriach ekonomicznych. CeDeWu, Warszawa

The manufacturer's authorised representative in the EU is Springer Nature Customer Service Centre GmbH, Europaplatz 3, 69115 Heidelberg, Germany. If you have any concerns regarding our products, please contact ProductSafety@springernature.com

Printed and bound by CPI Group (UK) Ltd, Croydon, CR0 4YY
30/09/2026
02225509-0010